AF556696

ITALIAN PHYSICAL SOCIETY

PROCEEDINGS

OF THE

INTERNATIONAL SCHOOL OF PHYSICS « ENRICO FERMI »

Course XLVII

edited by B. K. Sachs
Director of the Course

VARENNA ON LAKE COMO
VILLA MONASTERO
30th June - 12th July 1969

General Relativity and Cosmology

1971

ACADEMIC PRESS • NEW YORK AND LONDON

SOCIETA' ITALIANA DI FISICA

RENDICONTI

DELLA

SCUOLA INTERNAZIONALE DI FISICA «ENRICO FERMI»

XLVII CORSO

a cura di R. K. SACHS

Direttore del Corso

VARENNA SUL LAGO DI COMO

VILLA MONASTERO

30 Giugno 12 Luglio 1969

Relatività generale e Cosmologia

1971

ACADEMIC PRESS • NEW YORK AND LONDON

ACADEMIC PRESS INC.
111 Fifth Avenue
New York 3, N. Y.

United Kingdom Edition
Published by
ACADEMIC PRESS INC. (London) Ltd.
Berkeley Square House, London W. 1

Library of Congress Catalog Card Number: 79-119468

PRINTED IN ITALY

INDICE

R. GEROCH – Space-time structure from a global viewpoint.

G. F. R. ELLIS – Relativistic cosmology.

D. W. SCIAMA – Astrophysical cosmology.

Introduction.

R. K. SACHS

University of California - Berkeley, Cal.

This volume covers most of the courses and some of the seminars that were given at the 1969 Varenna summer school on relativity and cosmology. They should prove useful in bridging the gap between the standard textbooks and the flood of research articles on these topics.

EHLER's article is an extensive, essentially self-contained introduction both to general relativity and to general-relativistic kinetic theory. It is followed by GEROCH's article on some of the global implications of the theory. The two articles following, by ELLIS and SCIAMA, are complementary. Together they constitute a detailed review of cosmology. The last main lecture series, by THORNE, covers many aspects of the currently hot subject, relativistic stellar structure. The remaining articles cover some of the seminars presented at the summer school.

Varenna is a beautiful place and it is always pleasant to have people from many nations meet in an atmosphere of respect and co-operation. I think many of us were saddened by the constrast between life at Varenna and the chauvinism of the world outside.

1. P. Roe
2. I. Miodek
3. A. Fisher
4. H. Heintzmann
5. B. Bertotti
6. I. Ehlers
7. F. De Felice
8. G. Börner
9. R. Rüdiger
10. R. Bergamini
11. B. Jones
12. G. Sedmak
13. S. A. Bonometto
14. B. H. Voorhees
15. K. Sachs
16. D. Sciama
17. Mrs. Ehlers
18. H. Urbantke
19. G. Steigman
20. J. M. Stewart
21. L. Maraschi
22. L. Padrielli
23. B. Marano
24. R. Göbel
25. D. P. Williams
26. B. R. Sutton
27. G. F. Mitchell
28. T. K. Gujadhua
29. J. Demaret
30. A. Lausberg
31. M. Guyot
32. P. Cazzola
33. R. Klein
34. F. Lucchin
35. P. Fortini
36. J. Shamir
37. M. Kanai
38. K. B. Marathe
39. J. R. Ipser
40. M. A. H. McCallum
41. P. Andrle
42. T. Spruch
43. R. A. Russel-Clark
44. A. Barnes
45. Z. Perjes
46. P. Gräff
47. D. Bramanti
48. R. W. Huff
49. J. Baerentzen
50. J. Rosenberg
51. R. White
52. V. Icke
53. K. S. Thorne
54. C. Anderson
55. M. Von Reinhardt
56. M. Kundt
57. J. R. Pulham
58. J. Kuijpers
59. G. Ellis
60. M. Blagojevic
61. R. McLenaghan
62. J. Heise
63. C. Bartolini
64. E. T. Liang
65. E. Rudolph
66. F. Miglietta
67. J. Silver

SOCIETÀ ITALIANA DI FISICA

SCUOLA INTERNAZIONALE DI FISICA « E. FERMI »

XLVII CORSO - VARENNA SUL LAGO DI COMO - VILLA MONASTERO - 30 Giugno - 12 Luglio 1969

General Relativity and Kinetic Theory (*).

J. EHLERS

University of Texas at Austin - Austin, Tex.

1. – Introduction.

The principal aim of these lectures is to present the general-relativistic kinetic theory of gases in a coherent, rigorous way. This theory is of interest since:

a) it provides a simple model of matter which, in contrast to the conventional fluid description, incorporates the particle structure of matter;

b) it offers a way to complete the Einstein field equation by an additional law, the Liouville or Boltzmann equation, such that a deterministic model for gravitating material systems is obtained;

c) it forms a basis for relativistic thermodynamics of equilibrium and nonequilibrium, particularly, transport processes;

d) it permits a unified treatment of gases consisting of particles with positive mass and those consisting of zero-mass particles (radiations);

e) its assumptions seem reasonably appropriate for a number of real systems, like the system of galaxies, some stellar systems, certain radiations and hot gases and plasmas;

f) it is sufficiently young and incomplete so that further work in it is possible and desirable.

Kinetic theory deals with averages of various kinds; these are represented mathematically as integrals over domains of different dimensions. The appropriate tools for dealing with such integrals and « volume elements » are differential forms. They form so useful instruments also for other purposes, like

(*) Supported in Part by OAR, AF-33(615) 1029.

computing curvature tensors, constructing and analysing exact solutions of relativistic field equations, dealing with symmetry groups, that I use one Section to give what I hope is a reasonably self-contained introduction to them.

In accordance with the overall design of the course, these special topics are preceded by a general summary of Einstein's theory; here I also try to use a modern, geometrical approach.

Component notation for tensors is mostly employed, but conceptual definitions are given in most cases, and sometimes an index-free notation is also used. Throughout the lectures I shall follow the sign conventions recently proposed by WHEELER, MISNER, and THORNE for the curvature tensors.

As for units and dimensions, I put $c = 8\pi G = \hbar = k = 1$, where c = velocity of light, G = Newtonian gravitation constant, $2\pi\hbar$ = Planck constant, k = Boltzmann constant. All physical quantities then are pure numbers.

2. – Summary of Einstein's theory of gravitation and its mathematical apparatus.

2·1. – *General relativity theory* is both a theory of *gravitation* and a theory of the *space-time geometry*. Since the space-time geometry enters into the quantitative description of almost all physical processes, general relativity is, in principle, a fundamental part of physics.

The most characteristic feature of gravitation is that *all neutral, structureless test particles follow the same trajectories*, if started from the same space-time point with the same initial velocity. In this sense, gravitation is the only *universal force* known.

The equality of the ratio (passive gravitational mass/inertial mass) has been established experimentally for Au and Al with an accuracy of 10^{-11} by DICKE, ROLL and KROTKOV [1]. According to this result, generalized to all materials, *it is impossible to separate gravitational and inertial forces by means of local experiments with test particles.* This fact led EINSTEIN to incorporate the gravitational field into the geometry of space-time, or more precisely, to consider the space-time geometry as *defined* through the behaviour of neutral, freely falling test particles, and thereby to *identify the gravitational field with the space-time geometry* (*).

Whereas in Newtonian physics and in special relativity theory the space-time geometry is considered as *given* rigidly once and for all, in general relativity it is treated, in consequence of the identification just considered, as a *physical field interacting with matter.* Thereby an old disparity is removed which troubled, in particular, Ernst MACH: Before general relativity, space-time geometry, manifested in the form of the class of all inertial reference

---

(*) See the second paragraph of Subsection 2·8.

frames, was a universal external field acting on, but not acted upon by, matter; in general relativity it is on equal footing with other fields and matter, « acting » and « suffering ».

If one considers, as examples of physical laws, Newton's laws of mechanics, Maxwell's equations of electromagnetism, Schrödinger's equation or the Born-Jordan canonical commutation relations of quantum mechanics, one recognizes that *some space-time geometry has to be assumed in order that these laws can be formulated* [2]. One needs straight lines, distances, parallelity of vectors at different points, rectangular Cartesian co-ordinates, simultaneity, durations of time intervals, etc.

Analysing such laws from this point of view one finds that *three types of geometrical structure* are needed to formulate physical laws: *Topology*, *connexion*, and *metric*. A *topology* enables one to speak of *nearness* of points or events, of limits, continuity, connectedness and related concepts. A *connexion* on a space (manifold) is a prescription of a *parallel transport* for vectors (and tensors) along curves which enables one to compare the values of vector fields at different points, to form (covariant) derivatives and differential equations. A *metric* assigns *inner products* to vectors, *lengths* to curves, and in relativistic theories determines the *light cones* and *causal relations*.

In Einstein's theory, *connexion and metric of space-time*—and those structures which are implied by them, like curvature—*represent the gravitational field*, and in this way the gravitational field enters inextricably and universally into other laws of physics [2].

Although *gravitation* is by far the weakest of the known four fundamental interactions, it presumably *dominates the behaviour of large masses*, since it has an *infinite range* like electromagnetism and, in contrast to the latter, its *sources are always positive* and attract each other, so that gravitational forces add up to large values in large systems of massive bodies.

2'2. – The *main assumptions* upon which Einstein's theory of gravitation is based, along with interpretation rules to be discussed in later Subsections (particularly, Subsect. 2'7, 2'8 and 2'13), are the following:

A) *Space-time is a four-dimensional, oriented* [3], *differentiable* [4], *connected Hausdorff manifold* X.

B) X *carries a semi-Riemannian metric* g *of signature* $+++-$; X *is time-oriented* [5] *with respect to* g.

C) *The curvature associated with* g *is related to the matter by the Einstein field-equation*

$$G^{ab} + \Lambda g^{ab} = T^{ab} . \tag{1}$$

We discuss these assumptions and some of their implications in turn in the following Subsections [6].

2'3. – As to *assumption A)*, we recall that *a 4-dimensional differentiable manifold X is a set which is locally indistinguishable from the space R^4 of quadruplets of (real) numbers*, though it may have a different type of connectivity globally. Each point p—in the physical interpretation, each *event* like the collision of two particles—is contained in a subset U which can be put into one-to-one correspondence with an open set $\tilde{U}$ in R^4. If $x: U \to \tilde{U}$ is such a map and q is a point of U, then $x(q) = (x^1(q), \dots, x^4(q)) \in \tilde{U} \subset R^4$. The numbers $x^a(q)$, $a = 1, 2, 3, 4$, are the *co-ordinates* of q with respect to the *co-ordinate-system* x, the maps $q \to x^a(q)$ are called *co-ordinate functions*; U is the *domain* and $\tilde{U}$ is the *range* of the co-ordinate-system x. If U and U' are the domains of two admissible co-ordinate systems x, x', then $x(U \cap U')$ and $x'(U \cap U')$ are required to be open subsets of $\tilde{U}$ and $\tilde{U}'$, respectively, and the functions f^a defined by

$$x^{a'}(q) = f^a(x^1(q), \dots, x^4(q)) , \qquad q \in U \cap U', \tag{2}$$

which express the «primed» co-ordinates of q in terms of the unprimed ones, are required to be differentiable [4]. (Notice that (f^a) maps the open set $x(U \cap U')$ of R^4 onto the open set $x'(U \cap U')$ of R^4: the f^a are ordinary, numerical functions.)

A collection of co-ordinate systems $(x, x', \dots)$ such that $X = U \cup U' \cup \dots$ is called an *atlas*. Exhibiting such an atlas, with differentiably related co-ordinate functions as in (2), changes a set X into a differentiable manifold.

X is *topologized* by taking as open sets unions of inverse images of open sets in R^4 under co-ordinate mappings. This means that continuity of point functions and similar properties are to be judged in terms of local co-ordinates.

n-dimensional manifolds are defined analogously.

The physical motivation for assumption A) is that physical events in the space-time regions of experiments or observations can indeed be distinguished by means of four co-ordinates, usually 3 spatial ones and 1 time variable, and that fields have been successfully described by differentiable functions of four such co-ordinates [39].

X is *connected* if any two of its points can be joined by a continuous curve (in X). The space-time manifold is assumed to be connected since disconnected parts of it could not interact, and so we could observe, only the component in which we live.

X is *Hausdorff* if any two distinct events have disjoint neighbourhoods. This property is required for space-time primarily for simplicity; by it one excludes, *e.g.*, the possibility that a geodesic «splits up» into branches [40].

X is called *orientable* if it admits an atlas such that the co-ordinate transformations (2) for all pairs of overlapping domains have positive Jacobians.

If one such atlas is chosen and all those co-ordinate systems which are related to the atlas by transformations with positive Jacobians are called *oriented*, the *manifold is oriented.* (An example of a manifold which is *not* orientable is the *Möbius strip.*)

2'4. – Since a manifold is locally indistinguishable from R^n, one can define *differentiable functions* on it; local co-ordinates x^a are examples of such functions. Similarly, smooth *curves* are defined as mappings of intervals of the real line into the manifold. (Thus a curve is, by definition, parametrized.)

If $k\colon \lambda \to k(\lambda)$ is a curve and f a real function on X, then $\mathrm{d}f(k(\lambda))/\mathrm{d}\lambda$ is the *derivative of f in the direction of k* at the point $k(\lambda)$. Two curves k, k' passing through the point $k(0) = k'(0) = p$ are said to have the *same tangent* at p if

$$\left.\frac{\mathrm{d}f(k'(\lambda))}{\mathrm{d}\lambda}\right|_{\lambda=0} = \left.\frac{\mathrm{d}f(k(\lambda))}{\mathrm{d}\lambda}\right|_{\lambda=0} \tag{3}$$

for all (differentiable) functions f.

An equivalence class of curves all passing through p with the same tangent is called a (tangent) *vector at p.* Since in local co-ordinates x^a a curve is given by functions $x^a(\lambda)$ and since

$$\left.\frac{\mathrm{d}f(k(\lambda))}{\mathrm{d}\lambda}\right|_{\lambda=} = u^a f_{,a}\,,$$

if

$$u^a = \left.\frac{\mathrm{d}x^a}{\mathrm{d}\lambda}\right|_{\lambda=0}, \qquad f_{,a} = \frac{\partial f}{\partial x^a} \quad \text{(evaluated at } p\text{)},$$

we see that *a vector u is characterized by a linear differential operator* $f \to u^a f_{,a}$. The numbers u^a are the *components* of u with respect to the co-ordinate system x. Identifying a vector with the corresponding differential operator we can write

$$u^a = u(x^a)\,. \tag{4}$$

The vectors at p form a vector space, the *tangent space* T_p to X at p. Particular vectors are $\partial/\partial x^a$; $\partial/\partial x^1$ is tangent to the first co-ordinate lines. We have

$$u = u^a \frac{\partial}{\partial x^a} \tag{5}$$

as the representation of the (arbitrary) vector u as a linear combination of the (co-ordinate-) *basis vectors* $\partial/\partial x^a$. The *dimension* of T_p equals that of X.

The linear, homogeneous maps of T_p into R are called covariant vectors or 1-*forms*. They also form a vector space T^*_p, the *dual space* of T^p. We shall use greek letters $\varphi, \psi, \ldots$ to denote forms. The value of φ at u is written $\varphi(u)$.

For each function f, defined in a neighbourhood of p, we define a 1-form $\mathrm{d}f$ by

$$\mathrm{d}f(u) = u(f) = u^a f_{,a} \,, \tag{6}$$

u being any vector at p. $\mathrm{d}f$ is called the *differential*, or *gradient*, of f. In particular, the co-ordinate functions x^a of a co-ordinate system x have gradients $\mathrm{d}x^a$.

From the definition (6) follows

$$\mathrm{d}x^a \left(\frac{\partial}{\partial x^b}\right) = \frac{\partial x^a}{\partial x^b} = \delta^a_b \,. \tag{7}$$

Hence, $(\mathrm{d}x^1, \ldots, \mathrm{d}x^4)$ is the *basis of* T^*_p which is *dual* to the basis $(\partial/\partial x^1, \ldots, \partial/\partial x^n)$ of T_p.

Any 1-form φ can be expanded as

$$\varphi = \varphi_a \, \mathrm{d}x^a \,. \tag{8}$$

The numbers $\varphi_a = \varphi(\partial/\partial x^a)$ are the *components* of φ with respect to x. Then $\varphi(u) = \varphi_a u^a$.

Having obtained the spaces T_p, T^*_p of contravariant and covariant vectors, respectively, one can introduce arbitrary *tensors* at p. A *tensor* R *of type* $\binom{1}{3}$, for example, is defined as a function of one 1-form and three vectors,

$$R\colon (\varphi, u, v, w) \to R(\varphi, u, v, w) \,,$$

which is linear and homogeneous in each argument. R can be expanded with respect to basis elements,

$$R = R^a{}_{bcd} \frac{\partial}{\partial x^a} \otimes \mathrm{d}x^b \otimes \mathrm{d}x^c \otimes \mathrm{d}x^d \,. \tag{9}$$

The numbers $R^a{}_{bcd}$ are the *components of* R.

The sign $\otimes$ indicates *tensor multiplication*. In the familiar component notation we have, *e.g.*,

$$(u \otimes v)^{ab} = u^a v^b \tag{10}$$

for the tensor product of two vectors u, v. In order to describe symmetries of tensors, we shall use the brackets ( ) and [ ] to denote *symmetrization* and *antisymmetrization*, respectively, *e.g.*

(11) $$F_{(ab)} = \tfrac{1}{2}(F_{ab} + F_{ba}) ,$$

(12) $$F_{[ab]} = \tfrac{1}{2}(F_{ab} - F_{ba}) .$$

2'5. – Let us now consider *assumption B*). The metric tensor $g = g_{ab}\,\mathrm{d}x^a \otimes \mathrm{d}x^b$, $g_{ab} = g_{ba}$, defines an *inner product* between vectors,

(13) $$u \cdot v = g_{ab}\, u^a v^b .$$

u is called *timelike, spacelike*, or *lightlike* if u^2 $(= u \cdot u) < 0, > 0, = 0$, respectively. A similar terminology is applied to curves. A co-ordinate system x is *orthonormal at* p if, at p, $g_{ab} = \eta_{ab} = \operatorname{diag}(1, 1, 1, -1)$. The existence of such a co-ordinate system for each point p is guaranteed by the assumption about the *signature* of g. The matrices (g_{ab}) and $(g^{ab}) = (g_{ab})^{-1}$ are used to *lower and raise indices*, respectively.

In each tangent space T_p the *null cone* $\{u: u^2 = 0\}$ consists of two halves separated by the vertex. If it is possible in a space-time X to select, at all points, « *future* » *half-cones* such that the resulting half-cone field is continuous everywhere, then X *is* said to be *time-orientable with respect to* g, and if the future half-cones have been chosen in one of the two possible ways, X is called *time-oriented.* A vector is then called *future-directed*, if it is in the interior of or on the future null cone. (An example of a pair (X, g) which is *not* time-orientable is given in Appendix I.) Time-orientability is necessary for a global distinction between future and past.

2'6. – Assumption B) is made in order that *special relativity*, which is so well supported experimentally, *remains valid locally* in some sense *even in the presence of gravitational fields.* In special relativity most physical laws are formulated as tensor (or spinor) differential equations with respect to inertial frames. This method rests on the mathematical fact that the partial derivatives of the components of a tensor field form again the components of a field, if only inertial co-ordinates are used, since the corresponding co-ordinate transformations are *linear.*

On the basis of assumption A) alone, no preferred class of linearly related co-ordinate systems can be singled out, hence one cannot introduce derivatives of tensors except for a special class of tensors, as will be discussed in Subsect. **3**'2.

With assumption B), however, *one can use the metric tensor to introduce co-ordinate systems which are nearly inertial in the vicinity of some arbitrary event* p. If, at p, $g_{ab,c}=0$, the co-ordinate system is called *geodesic at* p. If a co-ordinate system is orthonormal and geodesic at p, it is said to be *inertial at* p. (An inertial co-ordinate system is one which is inertial at each event.) Assumptions A) and B) imply that *each event* p *admits a co-ordinate system which is inertial at* p. (First) Partial derivatives of tensor field components, taken at p with respect to co-ordinates which are geodesic at p, do form the components of a tensor (at p); this new tensor is called the *covariant derivative* of the original one.

From this intuitively plausible, but formally clumsy definition one can infer an axiomatic characterization of the operation of covariant differentiation (*), and derive expressions for covariant derivatives in arbitrary co-ordinates.

One computes the *Christoffel symbols*

$$\Gamma^a_{bc}=\tfrac{1}{2}g^{ad}(g_{db,c}+g_{dc,b}-g_{bc,d}) \tag{14}$$

and then has the following *formulae for the components of the covariant derivative of a vector field* u *and a* 1-*form field* φ, respectively:

$$u^a_{\ ;b}=u^a_{\ ,b}+\Gamma^a_{bc}u^c\,, \tag{15}$$

$$\varphi_{a;b}=\varphi_{a,b}-\Gamma^c_{ab}\varphi_c\,. \tag{16}$$

Formulae for tensors of higher orders are obtained by considering first tensor products like $u^a\varphi_b$, and then applying sum and product rules.

If a vector field u is defined on a curve $k=(x^a(\lambda))$, one can form its *absolute derivative*

$$\frac{\mathrm{D}u^a}{\mathrm{d}\lambda}=\frac{\mathrm{d}u^a}{\mathrm{d}\lambda}+\Gamma^a_{bc}\frac{\mathrm{d}x^b}{\mathrm{d}\lambda}u^c\,; \tag{17}$$

similarly for other types of tensors.

u, defined on k, is *parallel along* k if

$$\frac{\mathrm{D}u^a}{\mathrm{d}\lambda}=0\,. \tag{18}$$

Since geodesic co-ordinates can alternatively be characterized by $\Gamma^a_{bc}=0$, parallel transport means that components with respect to geodesic co-ordinates (at p) remain unchanged to first order (at p).—Parallel transport preserves inner products.

(*) See ref. [6], Subsection 5·1.

2·7. – Since each event p admits co-ordinates which are inertial at p, special relativity can be expected to hold approximately in a sufficiently small neighbourhood of p, provided physical laws assume their special relativistic forms at p in these co-ordinates. (Einstein's elevator.) This last requirement is sometimes called the *principle of minimal gravitational coupling*; it amounts, in general co-ordinates, to *replacing ordinary partial derivatives in special-relativistic laws by covariant derivatives*, to obtain the general-relativistic form of these laws.

For example, the special-relativistic *Maxwell equations*

$$F_{[ab,c]} = 0\,, \qquad F^{ab}{}_{,b} = J^a \tag{19}$$

are replaced by

$$F_{[ab;c]} = 0\,, \qquad F^{ab}{}_{;b} = J^a\,. \tag{20}$$

Unfortunately, some ambiguity may arise in applying this principle, since higher-order covariant derivatives do, in general, not commute (see Subsect. **2**·9).

We assume provisionally that even in a gravitational field matter can be described by a *stress energy-momentum tensor* T^{ab}. (This will be shown in Subsect. **4**·5 for the special case of gases.) If orthonormal co-ordinates at p are used, with x^4 as the time co-ordinate, then the physical meaning of the components of T^{ab} for an observer at p with 4-velocity $\partial/\partial x^4$ is given by

$$T^{ab} = \left(\begin{array}{c|c} \text{pressure tensor} & \text{momentum density} \\ \hline \text{energy current density} & \text{energy density} \end{array}\right). \tag{21}$$

The special-relativistic conservation law for energy and momentum, $T^{ab}{}_{,b} = 0$, goes over into

$$T^{ab}{}_{;b} = 0\,. \tag{22}$$

(For dilute gases this relation will be derived in Subsect. **4**·11 from 4-momentum conservation during collisions without use of the field-equation (1).) In spite of its motivation, this relation is *not* a conservation law; it could be called the 4-*momentum balance equation* in a gravitational field.

2·8. – Applying the principle of minimal gravitational coupling to the law of inertia for a test particle, we obtain for its worldline $k = (x^a(\lambda))$ with tangent $p^a = \mathrm{d}x^a/\mathrm{d}\lambda$ the condition

(23) $$\frac{\mathrm{D}p^a}{\mathrm{d}\lambda} = 0\,;$$

i.e. the tangent must be parallel along k. Such curves are called *geodesics.* Accordingly, we assume that *freely falling test particles have timelike geodesic worldlines.* This statement may be regarded as a *physical definition of the space-time metric,* since *the set of all timelike geodesics determines a metric of signature* $(+++-)$ *uniquely except for a constant factor* [41].

Applying the WKB approximation to the Maxwell equations (20) with $J^a = 0$, one can derive that *light rays are null geodesics* [43].

The differential equations for geodesics are equivalent to the Euler-Lagrange equations associated with the *Lagrangian*

(24) $$\mathscr{L}(x^a, \dot{x}^a) = \tfrac{1}{2}\, g_{ab}(x^a)\, \dot{x}^a \dot{x}^b\,,$$

that is, they are the extremals of $\int \mathscr{L}\,\mathrm{d}\lambda$. For nonnull geodesics, one can also use

(25) $$\int \mathrm{d}s = \int \sqrt{|2\mathscr{L}|}\,\mathrm{d}\lambda\,.$$

Putting $p_a = \partial \mathscr{L}/\partial \dot{x}^a = g_{ab}\,\dot{x}^b$ we can pass to the *Hamiltonian*

(26) $$\mathscr{H}(x^a, p_a) = \tfrac{1}{2}\, g^{ab}(x^c)\, p_a p_b\,,$$

and use canonical equations. Thus, the computation of geodesics may be regarded as a problem of analytical mechanics.

2·9. – As pointed out in the previous Subsection, *covariant derivatives do,* in general, *not commute.* Rather, we have formulae like

(27) $$u^a{}_{;[bc]} = -\tfrac{1}{2}\, u^d R^a{}_{dbc}\,,$$

where the functions

(28) $$R^a{}_{bcd} = 2(\Gamma^a_{b[d,c]} + \Gamma^a_{e[c}\Gamma^e_{d]b})$$

form the components of a tensor of type $\binom{1}{3}$, the *Riemann curvature tensor.* Its covariant components satisfy

(29) $$R_{abcd} = R_{[ab][cd]}\,, \qquad R_{a[bcd]} = 0\,.$$

Moreover, the curvature tensor satisfies the *Bianchi identity*

$$R^a{}_{b[cd;e]} = 0 . \tag{30}$$

According to the symmetries (29) one can form only one nonvanishing contraction of $R^a{}_{bcd}$, the (symmetric) *Ricci tensor*

$$R_{ab} = R^c{}_{acb} . \tag{31}$$

Its trace

$$R = R^a{}_a , \tag{32}$$

is the *curvature scalar*, and

$$G_{ab} = R_{ab} - \tfrac{1}{2} g_{ab} R \tag{33}$$

is the (symmetric) *Einstein tensor*. From (30) and the identity

$$g_{ab;c} = 0 \tag{34}$$

(which is equivalent to (14)) one gets

$$G^{ab}{}_{;b} = 0 , \tag{35}$$

the *contracted Bianchi identity*.

The formal significance of the curvature tensor for tensor analysis is exhibited by the *Ricci identity* (27). Its fundamental importance for Riemannian geometry emerges from the fact that *all general tensor functions of g_{ab} and its derivatives can be obtained tensor-algebraically from $R^a{}_{bcd}$ and its covariant derivatives* [44].

2·10. – The *geometrical and physical meaning of $R^a{}_{bcd}$* is illustrated by the *geodesic-deviation equation.*

Consider a one-parameter family of geodesics (a strip), and concentrate on two neighbouring members of it. Let η^a be a *connecting vector* between them, joining points with the same value of the affine parameter. For nonnull geodesics we choose η^a perpendicular to the first geodesic.

Then to first order in η^a

$$\frac{D^2\eta^a}{d\lambda^2} = R^a{}_{bcd} u^b u^c \eta^d , \tag{36}$$

where u^a is the tangent vector of the first geodesic [14]. Thus, $R^a{}_{bcd}$ *measures the extent to which the function $\eta^a(\lambda)$ fails to be linear*; this deviation from linearity is called *curvature.* (That this corresponds to the intuitive idea of curvature is perhaps best seen by specializing to the cases of a plane and a sphere.)

Applying (36) to a pair of freely falling test particles we see that their relative acceleration is obtained by transforming their connection vector η^a linearly with $R^a{}_{bcd}u^b u^c$. Hence, $R^a{}_{bcd}$ *measures the differential or tidal gravitational field,* and for an observer with 4-velocity u^a the tensor $R^a{}_{bcd}u^b u^d$ has the same meaning as the second gradient, $\varphi_{,\lambda\mu}$, of the gravitational potential φ in Newtonian theory. In the sense of this correspondence, therefore,

$$(37) \qquad \nabla^2\varphi \simeq R_{ab}u^a u^b ,$$

both sides being traces.

2·11. – We now consider *assumption C*), Einstein's field equation. It establishes a *relation between the gravitational field and its sources* which is analogous to Poisson's law

$$(38) \qquad \nabla^2\varphi = \tfrac{1}{2}\varrho$$

of Newtonian theory (*).

If (38) is accepted as an idealized expression of experiences embraced by Newtonian theory; if the correspondence (37) is used, and if the relation

$$(39) \qquad T_{ab}u^a u^b = \varrho$$

contained in (21) is taken into account, one is led from (38) to

$$(R_{ab} - \tfrac{1}{2}T_{ab})u^a u^b = 0 .$$

If this is to hold for all observers, we must require

$$(40) \qquad R_{ab} = \tfrac{1}{2}T_{ab}$$

as the field equation. Unfortunately, (40), (32), (33) and (35) imply T $(= T^a{}_a) = \text{const}$, which is certainly not true. However, comparison of (22), (35) and (40) suggests to « correct » (40) into [7]

$$(41) \qquad G_{ab} = T_{ab} .$$

This not only makes (22) and (35) compatible, but implies

$$R_{ab} = T_{ab} - \tfrac{1}{2}g_{ab}T ,$$

(*) The factor $\frac{1}{2}$ is due to the conventions stated at the end of the Introduction.

hence

$$R_{ab}u^a u^b = T_{ab}u^a u^b + \tfrac{1}{2}T = \tfrac{1}{2}\,(\text{energy density} + 3\times\text{pressure}) \approx \tfrac{1}{2}\varrho \qquad \text{if } p \ll \varrho,$$

so that the correspondence with (38) is not destroyed as long as $p \ll \varrho$. (In the solar system $p/\varrho \leqslant 10^{-6}$.) Finally, we may even replace (41) by (1), introducing a *cosmological term*, as long as Λ is much smaller than the average density of those systems in which Newtons theory is known to hold.

Apart from the correspondence argument just given the field-equation (1) is supported by the following:

a) If T^{ab} *is derivable from a* generally covariant, scalar *Lagrangian* $L(\psi_A, \psi_{A,a}, \dots; g_{ab})$ according to

$$T^{ab} = \frac{1}{\sqrt{-g}}\,\frac{\delta(\sqrt{-g}\,L)}{\delta g_{ab}}\,, \tag{42}$$

(Here $g = \det(g_{ab})$, and δ indicates variational differentiation) then (1) *is implied by the action principle*

$$\delta\int (R - 2\Lambda + L)\sqrt{-g}\,\mathrm{d}^4x = 0\,, \tag{43}$$

which also yields, on varying the ψ_A's, nongravitational laws (*). For an electromagnetic field, *e.g.*,

$$L = -\tfrac{1}{2}F_{ab}F^{ab} \qquad\qquad (F_{ab} = 2A_{[b,a]})\,.$$

b) (1) *is the most general field equation of the form*

$$D_{ab}[g_{cd}] = T_{ab}\,,$$

in which D_{ab} *is a second-order tensor function of* g_{ab} *satisfying* $D^{ab}{}_{;b} = 0$ *identically* [45].

c) Attempts to construct a gravitation theory along the lines of special relativistic field theory lead, under certain conditions, not only to assumption *B*), but even to Einstein's equation (1) with $\Lambda = 0$. (See, *e.g.*, ref. [9]).

The field-equation (1) requires a *symmetrical* total stress energy-momentum tensor,

$$T^{[ab]} = 0\,. \tag{45}$$

(*) See, *e.g.*, ref. [8].

It is remarkable that a field equation of the form (44) with $D^{ab}{}_{;b}=0$ which recommends itself strongly by simplicity and analogy with Poisson's equation, does not exist for a nonsymmetrical T^{ab}.

2·12. – For specific physical systems the field eq. (1) has to be augmented by an expression for T^{ab} in terms of matter (or field) variables, and further laws for these additional variables have to be laid down, for example Maxwell's equations, or equations of state.

For *deterministic systems* these laws are such that *the evolution of all variables is uniquely determined* (up to co-ordinate and gauge transformations) *by initial data on a spacelike hypersurface* (*). Examples of such systems, which include self-gravitating perfect fluids with an « equation of state » $\mu=f(p)$, will be treated in the sequel (Subsect. **4·7**, **4·11**, **4·18**, **4·20**).

Since the field-equation (1) implies the 4-momentum balance eq. (22), the motion of a body is restricted by the field equation. It is indeed possible to derive, at least within the framework of approximation methods, *equations of motion* from (1) and/or (22), if assumptions are made about the separation of g_{ab} into a background and a self-field and about either T^{ab} or the vacuum field surrounding the body (**). Consequently, *the geodesic hypothesis* introduced in Subsect. **2·8** *must be considered as an implicit definition of a « spherical test body »*, the detailed theories of motion providing conditions under which bodies may be considered as such test bodies.

2·13. – In general relativity theory, *co-ordinates are merely labels to distinguish events*, they do (in general) *not* represent distances or times or otherwise *measurable physical quantities. Measurable quantities are invariants.*

The most important measurable quantities in cosmology and astrophysics are *times, angles, and energies* (of various kinds of radiation). *Times* are expressed as invariant integrals $\int \mathrm{d}s=\int[-g_{ab}\,\dot{x}^a\dot{x}^b]^{\frac{1}{2}}\,\mathrm{d}\lambda$ over sections of world-lines representing clocks; *angles* between directions given by unit vectors $\boldsymbol{e}, \boldsymbol{d}$ contained in an observer's rest space are computed as in Euclidean geometry from $\cos\varphi=\boldsymbol{e}\cdot\boldsymbol{d}=g_{ab}\,e^a d^b$; the *energy* of radiation described by T^{ab} which is *absorbed in a time interval* τ *on a screen* s is given by the invariant integral

$$\int_s u_a T^{ab}\sigma_b\,, \tag{46}$$

where u^a is the 4-velocity of the screen and σ_b denotes the element of the

(*) For the relativistic Cauchy problem see ref. [10], p. 130, and ref. [11].

(**) See, *e.g.*, corresponding reviews in ref. [10] and ref. [12]; see also ref. [13].

timelike strip S (defined in Subsect. **3**.6, eq. (76)) which represents the history of the screen s during the time of observation τ. (See Fig. 1.)

(Spatial) *distance is not a primitive observable in relativity theory,* nor is it a simple, directly measured quantity in astrophysics. Rather, distances are derived quantities which can be defined in different ways in terms of the basic quantities considered above [43].

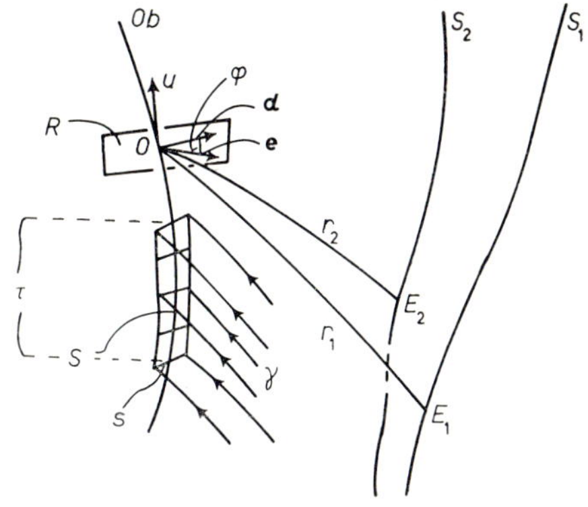

Fig. 1. – Basic observables in general relativity. Two sources S_1, S_2 emit radiation. The observer Ob receives at O the rays r_1, r_2 emitted by S_1, S_2 at E_1, E_2, respectively. He measures the angle φ between r_1, r_2; $\boldsymbol{e}$ and $\boldsymbol{d}$ are the projections of the tangents to r_1, r_2 at O into Ob's instantaneous 3-space R, which is orthogonal to Ob's 4-velocity u at O; $\cos\varphi = \boldsymbol{e}\cdot\boldsymbol{d}$. During a time interval τ of duration $\int_\tau ds$, measured by a clock carried by Ob, Ob also measures the radiation γ which hits a screen initially at s; S is the strip swept out by the screen during τ. The energy received is given by (46); T^{ab} is associated with γ.

The mathematical expressions representing observed (or predicted) numbers contain certain terms which describe *properties of the objects* affecting the measuring device (T^{ab}, *e.g.*), and other terms which describe *properties of the measuring apparatus* (u^a, σ_b, *e.g.*). The physical situation itself is represented as a «figure» in space-time, consisting of tensor fields, world lines, world tubes, hypersurfaces, ...; this «figure» also contains some elements which represent (in a simplified, idealized form, of course) the measuring equipment, and the measured quantities are certain invariants of this «figure» (*).

In particular circumstances one can adapt the co-ordinates to the world-lines which represent object (a star, *e.g.*) and observer, and take x^4 to coincide with proper time on those lines, or choose other co-ordinate conditions to simplify the expressions for the physical quantities of interest. Then co-ordinates may themselves represent observables, exactly or approximately, and things may look as in Newtonian theory. This does not contradict the statement made in the first paragraph of this Subsection, and should not lead to confusion.

(*) Examples which illustrate these remarks particularly clearly are found in ref. [14].

2˙14. – We have discussed how the three types of geometric structure mentioned in the first Subsection are introduced via assumptions A, B and C. The connexion, or parallel transport, appeared in Subsect. **6** as a by-product of the metric. It is nevertheless useful to be aware of the conceptual distinction between a *metric* g_{ab} and a *connexion* Γ^a_{bc}. It is possible to define a connexion, via a system of functions Γ^a_{bc}, on a manifold *independently* of a metric, and to introduce covariant derivatives and geodesics in terms of it. In general, Γ^a_{bc} need not be symmetrical in the lower indices; then $\Gamma^a_{[bc]}$ is a tensor, the *torsion tensor*.

If a metric is given, then there exists precisely one connexion with zero torsion which preserves inner products under parallel displacement [6], namely the *Riemannian connexion* formally given by (14) and physically motivated in Subsect. **6** (*).

It is possible to distinguish even more kinds of geometric structure on a manifold, *e.g. conformal* and *projective* structures, and to analyze which importance they might have for physics. This is mentioned here only in order to emphasize that Einstein's theory of gravitation, based on A, B, C, is *one* particularly simple realization of the fundamental idea of incorporating gravitation into the geometry; or to make the space-time geometry a physical field; *other possibilities exist* and may well be called upon if observations or experiments such as Dicke's measurement of the solar oblateness [15] should disprove Einstein's theory.

3. – Differential forms and integration on manifolds (**).

A classical gas is represented in space-time as a broken complex of world-lines, corners of lines corresponding to (point) collisions. Kinetic theory deals with average properties of such a complex. Such averages will be *integrals*, sometimes taken over *space-time regions*, sometimes over *hypersurfaces in* X (as in (46)), and in other cases over *hypersurfaces in phase space* (to be defined below).

The appropriate mathematical tools for constructing such integrals are *differential forms*, which will be considered now. *Neither a metric nor a connexion is required*; Subsect. **3˙1**-**3˙5** continue the discussion of structures defined on bare n-dimensional manifolds, begun in Subsect. **2˙4**.

(*) Theories with torsion have been considered by several authors; see [49].
(**) For this whole Section compare, *e.g.*, ref. [16].

Our principal aim is to *generalize the fundamental formula*

$$\int_a^b f'(x)\,\mathrm{d}x = f(b) - f(a) \tag{47}$$

of calculus to integrals over suitable subsets of manifolds.

Looking at (47) we recognize that we must consider:

a) what possible *integrands* may be used instead of $f'(x)\,\mathrm{d}x$;

b) which *operation of differentiation* generalizes $f(x) \to f'(x)\,\mathrm{d}x$;

c) how to define suitable *domains of integration* (in (47): $a \leqslant x \leqslant b$), their *boundaries* (in (47): $\{a, b\}$), and the *orientations* of both, separately, and relatively to each other;

d) the *operation of integration*;

e) a *relation between an integral over a domain M and one over M's boundary.*

We treat these points in turn.

3'1. – We recall that a covariant vector is also called a 1-*form*, and written

$$\varphi = \varphi_a\,\mathrm{d}x^a\,. \tag{48}$$

An *m-form* is a completely skew covariant tensor of degree m. A 2-form, *e.g.*, can be expanded as

$$\chi = \chi_{ab}\,\mathrm{d}x^a \otimes \mathrm{d}x^b\,,$$

(compare (9), where $\chi_{ab} = -\chi_{ba}$. Thus

$$\chi = -\chi_{ba}\,\mathrm{d}x^a \otimes \mathrm{d}x^b = -\chi_{ab}\,\mathrm{d}x^b \otimes \mathrm{d}x^a = \tfrac{1}{2}\chi_{ab}(\mathrm{d}x^a \otimes \mathrm{d}x^b - \mathrm{d}x^b \otimes \mathrm{d}x^a)\,.$$

We define the *exterior product*

$$\mathrm{d}x^a \wedge \mathrm{d}x^b = 2\,\mathrm{d}x^{[a} \otimes \mathrm{d}x^{b]} \tag{49}$$

of $\mathrm{d}x^a$, $\mathrm{d}x^b$, so that

$$\chi = \tfrac{1}{2}\chi_{ab}\,\mathrm{d}x^a \wedge \mathrm{d}x^b\,. \tag{50}$$

Similarly one defines

$$\mathrm{d}x^{a_1} \wedge \ldots \wedge \mathrm{d}x^{a_m} = \sum_\sigma (-1)^\sigma\,\mathrm{d}x^{a_{\sigma(1)}} \otimes \ldots \otimes \mathrm{d}x^{a_{\sigma(m)}} = \mathrm{d}x^{a_1 \ldots a_m}\,, \tag{51}$$

where the sum is extended over all permutations σ acting on the indices $a_1, \ldots, a_m$ and $(-1)^\sigma$ is the signature of σ; the last symbol is introduced merely as an abbreviation for the somewhat lengthly exterior product.

Then *an m-form ψ can be expanded as*

$$\psi = \frac{1}{m!}\psi_{a_1\ldots a_m}\,\mathrm{d}x^{a_1\ldots a_m}\,, \tag{52}$$

where $\psi_{a_1\ldots a_m} = \psi_{[a_1\ldots a_m]}$ are the *components* of ψ.

The exterior product (51) is *associative*; *e.g.*

$$(\mathrm{d}x^1\wedge\mathrm{d}x^2)\wedge\mathrm{d}x^3 = \mathrm{d}x^1\wedge(\mathrm{d}x^2\wedge\mathrm{d}x^3)\,,$$

as is easily checked.

The exterior multiplication can be extended uniquely to arbitrary forms by requiring associativity and distributivity. Thus, *e.g.*, the exterior product of the 1-form φ and the 2-form χ given by (48) and (50), respectively, is the 3-form

$$\lambda = \varphi\wedge\chi = \tfrac{1}{2}\varphi_a\chi_{bc}\,\mathrm{d}x^a\wedge\mathrm{d}x^b\wedge\mathrm{d}x^c = \tfrac{1}{2}\varphi_{[a}\chi_{bc]}\,\mathrm{d}x^a\wedge\mathrm{d}x^b\wedge\mathrm{d}x^c\,,$$

whence the components of λ are given by (cf. (52))

$$\lambda_{abc} = 3\varphi_{[a}\chi_{bc]}\,. \tag{53}$$

In multiplying a p-form by a q-form, the numerical factor corresponding to

$$3 = \frac{(1+2)!}{1!2!} \text{ in (53) is } \frac{(p+q)!}{p!q!}\,.$$

The exterior product is not commutative; *if φ, ψ are p and q-forms*, respectively, *then*

$$\varphi\wedge\psi = (-1)^{pq}\psi\wedge\varphi\,. \tag{54}$$

If one admits, purely formally, *sums* of forms of different degrees like $\varphi_a\,\mathrm{d}x^a + \frac{1}{2}\chi_{ab}\,\mathrm{d}x^{ab}$ (*inhomogeneous forms*), one obtains an *associative, noncommutative algebra of dimension 2^n over the reals.* Since we did not use the fact that the $\mathrm{d}x^a$'s are gradients of co-ordinate functions, but only that $(\mathrm{d}x^1, \ldots, \mathrm{d}x^n)$ is a basis of a vector space T^*_p, this algebra can be constructed for any (finite-dimensional) vector space. It is *generated* by $\{1, \mathrm{d}x^1, \ldots, \mathrm{d}x^n\}$, has the defining relations $\mathrm{d}x^a\wedge\mathrm{d}x^a = 0$, and is called the *Graßmann algebra* associated with that vector space.

r 1-forms $\theta^1, \ldots, \theta^r$ are *linearly dependent* if and only if

$$\theta^1 \wedge \theta^2 \wedge \ldots \wedge \theta^r = 0 \ . \tag{55}$$

An m-form ψ may be *contracted* with a vector A, to give an $(m-1)$-form $\varphi = A \cdot \psi$. In terms of components φ is defined by

$$\varphi_{a_2 \ldots a_m} = A^{a_1} \psi_{a_1 a_2 \ldots a_m} \ . \tag{56}$$

We shall need later a component-free definition, equivalent to (56), of this contraction operation. To obtain it, contract both sides of (56) with arbitrary vectors $B_2, \ldots, B_m$:

$$\varphi_{a_2 \ldots a_m} B_2^{a_2} \ldots B_m^{a_m} = \psi_{a_1 a_2 \ldots a_m} A^{a_1} B_2^{a_2} \ldots B_m^{a_m} \ .$$

Recalling the component-free definition of a covariant tensor (Subsect. 2˙4) as a multilinear function of vectors we rewrite the last formula as

$$\varphi(B_2, \ldots, B_m) = \psi(A, B_2, \ldots, B_m) \ ,$$

or

$$(A \cdot \psi)(B_2, \ldots, B_m) = \psi(A, B_2, \ldots, B_m) \ , \tag{57}$$

which is the required definition. In words, $A \cdot \psi$ *is obtained by inserting into the function* ψ *of* m *vector variables the (fixed) vector* A *as the first argument*, leaving the dependence on the remaining $m-1$ variables unchanged.

We leave it to the reader to verify the following:

Lemma. Let $\Omega \neq 0$ be an n-form (at some point of an n-manifold), $L \neq 0$ a vector, and ω an $(n-1)$-form such that $\omega(A_1, \ldots, A_{n-1}) \neq 0$ whenever the set $(L, A_1, \ldots, A_{n-1})$ is linearly independent. Then

$$\omega = aL \cdot \Omega \tag{58}$$

with some real number $a \neq 0$. Conversely, any such ω satisfies the condition stated above.

Corollary. The ω's just characterized have the following property: $\omega(A_1, \ldots, A_{n-1}) = 0$ whenever the set $(L, A_1, \ldots, A_{n-1})$ is linearly dependent.

(Note that *any* $(n-1)$-form ω can be written in the form (58) with $a = 1$ for some unique L, provided $\Omega \neq 0$.)

These assertions become intuitively plausible if Ω and ω are interpreted as (signed) *volume-functions for* n-dimensional and $(n-1)$-dimensional (oriented) *parallelotopes* spanned by $(L, A_1, \ldots, A_{n-1})$ and $(A_1, \ldots, A_{n-1})$, respectively.

The preceding relations may all be considered for forms defined *at one point* p of a manifold (only). We may instead consider *form-fields*, defined *in a region or globally*. The above relations can then be applied *pointwise*; and it is this application to fields which is primarily of interest in general relativity.

We need to consider how forms defined on different manifolds can be related to each other.

Let $g\colon M \to N$ be a *differentiable mapping* of the manifold M into the manifold N, *i.e.* a mapping which, if expressed locally in terms of co-ordinates, is given by differentiable functions. Further, let ψ be an r-form on N. Introducing local co-ordinates (u^λ) around $p \in M$ and (x^a) around $g(p) \in N$, we can represent ψ locally as in (52). By inserting the functions $x^a(u^\lambda)$, which represent g locally, into ψ, we can « *pull back* » this form from N to M, obtaining on M the form

$$g^*\psi = \frac{1}{r!}\,\psi_{a_1 \ldots a_r} \frac{\partial x^{a_1}}{\partial u^{\lambda_1}} \cdots \frac{\partial x^{a_r}}{\partial u^{\lambda_r}}\, du^{\lambda_1} \wedge \ldots \wedge du^{\lambda_r}\,. \tag{59}$$

This pull-back operation is independent of co-ordinate systems and *commutes with the operations* $+$, $\wedge$ for forms.

3'2. – For *functions* f we have defined the gradient

$$\mathrm{d}f = f_{,a}\,\mathrm{d}x^a \tag{60}$$

in Subsect. 2'4. Considering functions as 0-forms we can say that the operator « d » sends 0-forms into 1-forms. This operation can be extended to form fields of arbitrary degree. We define, for an m-form ψ represented as in (52), the *exterior derivative* $\mathrm{d}\psi$ as the $(m+1)$-form

$$\mathrm{d}\psi = \frac{1}{m!}\,\mathrm{d}\psi_{a_1 \ldots a_m} \wedge \mathrm{d}x^{a_1 \ldots a_m}\,; \tag{61}$$

thus with (60) and (54)

$$(\mathrm{d}\psi)_{a_1 \ldots a_{m+1}} = (-1)^m (m+1)\,\psi_{[a_1 \ldots a_m, a_{m+1}]}\,. \tag{62}$$

As indicated by the notation, $\mathrm{d}\psi$ is independent of the co-ordinate system used in (61) to define it (*).

(*) For a definition of d which is explicitly co-ordinate–independent, see [6], p. 89.

Exterior differentiation has the following *properties*:

(63) $$\mathrm{d}(\varphi + \psi) = \mathrm{d}\varphi + \mathrm{d}\psi\,,$$

(64) $$\mathrm{d}(\mathrm{d}\varphi) = 0\,,$$

(65) $$\mathrm{d}(\varphi \wedge \psi) = \mathrm{d}\varphi \wedge \psi + (-1)^p \varphi \wedge \mathrm{d}\psi \qquad (p = \deg \varphi).$$

These properties (which are easily verified by using the definition (61)) *together with the requirement that for functions* d *is given by* (6), *characterize* d *uniquely*. (See, *e.g.*, ref. [6] or [16].)

The algebra of (differentiable) form fields defined on a manifold, equipped with the operations $+$, $\wedge$, d, is called the *Cartan differential algebra* of the manifold.

The pull-back operation for forms defined in (59) *commutes with exterior differentiations*, *i.e.* if $g\colon M \to N$, then

(66) $$g^*(\mathrm{d}\psi) = \mathrm{d}(g^*\psi)\,,$$

as a short computation shows.

The following lemma will be useful later:

Lemma. Let Ω be a n-form field, L a vector field, f a function on an n-manifold. Then

(67) $$\mathrm{d}f \wedge (L \cdot \Omega) = L(f)\,\Omega\,.$$

To prove (67), introduce local co-ordinates and rewrite it in component notation, obtaining

$$n f_{,[a_1} L^b \Omega_{|b|a_2 \dots a_n]} = L^b f_{,b}\, \Omega_{a_1 \dots a_n}\,.$$

(The index b is to be excluded from antisymmetrization, as indicated by vertical bars.) This equation needs to be verified only for $(a_1, \dots, a_n) = (1, \dots, n)$, and that can easily be done.

3'3. – In order to obtain reasonable *domains of integration*, consider first, as an example, a smooth, finite 2-surface M embedded in Euclidean 3-space, bounded by one or several smooth curves (see Fig. 2). Certainly we wish to include such an M in the class of permissible domains of integration.

M is *not* a 2-manifold in the sense specified in Subsect. **2'3**, since a point like q does not have a neighborhood (within M!) which is homeomorphic

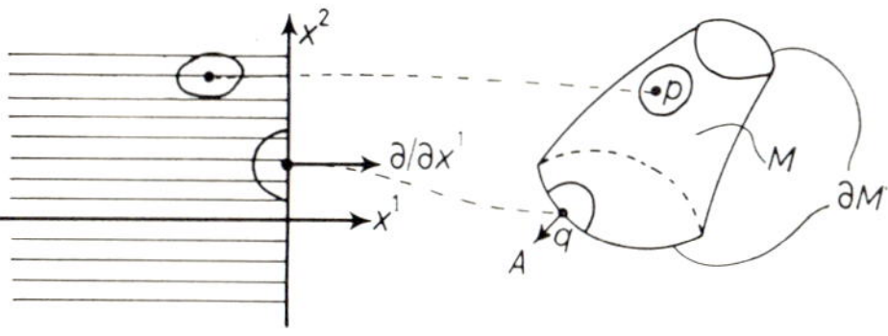

Fig. 2. – A 2-manifold with boundary and the corresponding co-ordinate half-plane.

to an open set in R^2. However, the following assertion is true: Each point of M has a neighborhood which is homeomorphic to an open set of the *half-plane* $\{(x^1, x^2)\colon x^1 \leqslant 0\}$, if open sets in the half-plane are defined as intersections of open sets of the whole plane with that half-plane (*relative topology*).

The example suggests to define an *n-manifold with boundary, M,* by *replacing* in the definition of a manifold given in Subsect. 2·3 *the co-ordinate space R^n by $H^n = \{(u^1, \ldots, u^n\colon u^1 \leqslant 0\}$*, using the relative topology of H^n.

The *boundary* ∂M of M is defined as the $(n-1)$-manifold whose underlying point set consists of those points of M which are mapped into points of the boundary $\partial H^n = \{(0, u^2, \ldots, u^n)\}$ by permissible co-ordinate mappings x. (This set is, in fact, a $(n-1)$-manifold whose co-ordinate mappings are obtained by restricting those of M to the boundary set.)

A manifold is also a manifold with (empty) boundary; « manifold with boundary » is the more general concept.

A closed rectangle A in R^2 is not a manifold with boundary if its differential structure is to be such that the restrictions of the natural co-ordinate functions of R^2 to A are differentiable. The reason is that A's boundary is not a manifold. This example suggests to introduce the notion of a « manifold with boundary and corners ». We shall not proceed beyond this remark, however.

If a manifold with boundary, M, is orientable, then so is ∂M. Given an orientation of M, the *induced orientation* of ∂M is defined as follows. If x is an oriented co-ordinate system of M whose domain N intersects ∂M, then $x|_{N\cap\partial M}$, the restriction of x to the boundary, is defined to be an oriented co-ordinate system of ∂M. (This is a prescription familiar from Gauss's and Stokes's theorems. $\partial/\partial x^1$ at the boundary is an outer normal of H^n.)

We have so far considered manifolds (with or without boundary) *intrinsically*, without regard to *imbeddings* in higher-dimensional manifolds. From the introductory remarks of this Section it is clear, however, that we need to consider submanifolds (of space-time or phase space).

An m-manifold M is called a *submanifold* of an n-manifold N (both with or without boundary) if M is a subset of N and if for each point $p \in M$ there exist co-ordinate systems x of N with domain U and u of M with domain V such that $p \in V \subset U \cap M$, and such that the restrictions of the co-ordinate functions x^a to V, expressed as functions of (u^λ), are differentiable, with

$$\text{rank of } \left(\frac{\partial x^a}{\partial u^\lambda}\right)_{\substack{a=1,\ldots,n\\ \lambda=1,\ldots,m}} = m\,.$$

The functions $x^a(u^\lambda)$ give a *parameter representation* of the part V of M.

Examples of 2-submanifolds of R^3 are spheres and tori.

3·4. – Let M be an *oriented, compact m-manifold with boundary* and let ψ be a (continuous) m-form field on M.

We wish to define $\int_M \psi$, the *integral of ψ over M*.

For a part D of M which is contained in the domain of some oriented co-ordinate system x and for which $\tilde{D} = x(D)$, the co-ordinate range corresponding to D, is Riemann-measurable, we represent ψ as in (52) and define

$$\int_D \psi = \int_{\tilde{D}} \psi_{12\ldots m}\, \mathrm{d}x^1 \wedge \mathrm{d}x^2 \wedge \ldots \wedge \mathrm{d}x^m \,, \tag{68}$$

where the right-hand side is taken to be the Riemann integral of $\psi_{12\ldots m}$ over $\tilde{D}$. The co-ordinate-independence of this integral follows from the standard transformation formula for multiple integrals,

$$\int_{x(D)} \psi_{1\ldots m}\, \mathrm{d}x^1 \wedge \ldots \wedge \mathrm{d}x^m = \int_{x'(D)} \psi_{1\ldots m} \frac{\partial(x^1, \ldots, x^m)}{\partial(x^{1'}, \ldots, x^{m'})}\, \mathrm{d}x^{1'} \wedge \ldots \wedge \mathrm{d}x^{m'} = \int_{x'(D)} \psi_{1'\ldots m'}\, \mathrm{d}x^{1'} \wedge \ldots \wedge \mathrm{d}x^{m'} \,,$$

in which *the Jacobian, which is positive since x, x' are oriented, appears automatically* because of the anticommutativity of the $\wedge$-product.

Now, *any oriented, compact manifold with boundary, M, can be « cubulated »*, *i.e.* there exists a finite collection $\{D_1, \ldots, D_k\}$ of subsets of M such that α) $\bigcup_{j=1}^{k} D_j = M$, β) the interiors of any two sets D_j, D_l $(j \neq l)$ have empty intersection, γ) each D_j is contained in the domain of an oriented co-ordinate system x_j, and $x_j(D_j) = \tilde{D}_j$ is a closed cube in R^m. (See ref. [6], p. 103.)

That is, *M can be cut into nice pieces* of the type considered in the preceding paragraph.

We now define $\int_M \psi$ by taking a cubulation and putting

$$\int_M \psi = \sum_{j=1}^{k} \int_{D_j} \psi \,. \tag{69}$$

This integral is, in fact, independent of the cubulation. For, consider the intersection $\{D_j \cap E_l\}$ of two cubulations $\{D_1, \ldots, D_k\}$, $\{E_1, \ldots, E_s\}$, then

$$\sum_{j=1}^{k} \int_{D_j} \psi = \sum_{j=1}^{k} \sum_{l=1}^{s} \int_{D_j \wedge E_l} \psi = \sum_{l=1}^{s} \int_{E_l} \psi \,.$$

Now consider the general case of an *oriented, compact m-submanifold with boundary, M, of an n-manifold N*, and a continuous m-form field ψ on N. The imbedding of M into N is a special case of a mapping $g: M \to N$ as considered at the end of Subsect. 3˙1; hence the form ψ can be pulled back from N to M, and *we define*

$$\int_M \psi = \int_M g^* \psi . \tag{70}$$

Locally, this last step amounts to nothing else than the use of a parametrization of M to compute the integral (cf. (59)).

The important point is that $\int_M \psi$ *is an invariant determined by* M *and* ψ, although it is computed by cutting M into pieces and using co-ordinates on N and on M.

As a consequence of (70) and (66), we note that if $\psi = \mathrm{d}\varphi$ (*on* N), *then*

$$\int_M \mathrm{d}\varphi = \int_M g^*(\mathrm{d}\varphi) = \int_M \mathrm{d}(g^*\varphi) . \tag{71}$$

3˙5. – We are now in a position to state the fundamental

Stokes' theorem. Let M be an oriented, compact m-submanifold with boundary ∂M of an n-manifold N, and let φ be a continuously differentiable $(m-1)$-form field on N. Then

$$\int_M \mathrm{d}\varphi = \int_{\partial M} \varphi . \tag{72}$$

This theorem can be *reduced* immediately to the *special case where only* M *and a* $(m-1)$*-form on* M *are given*, since (72) is, in view of the definition (70) and the auxiliary formula (71), equivalent to

$$\int_M \mathrm{d}(g^*\varphi) = \int_{\partial M} g^*\varphi .$$

(We have also used the fact that the pull-back of φ to ∂M can be carried out in two steps, first pulling φ from N to M, and then from M to ∂M.)

Considering the *special case* (no N, φ given on M), assume first that M admits a global co-ordinate system x such that $x(M) = \tilde{M}$ is, say, a solid sphere in R^m. Then φ can be represented on M as

$$\varphi = L \cdot \Omega = L^1 \, \mathrm{d}x^{23\dots m} + (-1)^{m-1} L^2 \, \mathrm{d}x^{3\dots 1} + - \dots ,$$

where $\Omega = \mathrm{d}x^{1\dots m}$ is an m-form on M, and L a vector field (cf. the remark following the lemma in *a*)). Hence

$$d\varphi = L^a{}_{,a}\, dx^{1\ldots m} .$$

Formula (72) thus *reduces*, in this case, *to Gauss' theorem* applied to the vector field L and a sphere $\tilde{M}$ in R^m.

If M is not of such a simple form, one has again to cut M into simple pieces, and to prove (72) by patching the pieces together. The most elegant tools for this procedure are *decompositions of uniiy*; but we refer to the literature [16] instead of discussing these somewhat tricky, technical arguments. We also only mention that *noncompact domains* of integration can be used, provided the integrands vanish sufficiently strongly « at infinity »; and that one can replace M by a manifold with boundary and corners.

The *generalized Stokes formula* (72) contains as special cases not only *Gauss' divergence theorem*, but also the *familiar Stokes formula* ($m = 2$) and its higher-dimensional analogues.

3'6. – *Let X be the spacetime manifold.* Using an oriented co-ordinate system we define the 4-form

$$\eta = \sqrt{-g}\, dx^{1234} \qquad \left(g = \det(g_{ab}),\ \sqrt{-g} > 0\right), \tag{73}$$

η is independent of the co-ordinate system and is called the (Riemannian) *volume element.*

The components of η with respect to an oriented co-ordinate system are, according to (73) and (52), given by

$$\eta_{abcd} = \eta_{[abcd]}, \qquad \eta_{1234} = \sqrt{-g} . \tag{74}$$

If A is a vector field on X and D a 4-dimensional, oriented, compact sub-manifold with boundary of X—henceforth called a *region*—we obtain from Stokes' theorem (72)

$$\int_D d(A\cdot\eta) = \int_{\partial D} A\cdot\eta . \tag{75}$$

The integrand on the left can be rewritten as

$$d(A\cdot\eta) = \left(\sqrt{-g}\, A^a\right)_{,a} dx^{1234} = A^a{}_{;a}\,\eta ,$$

and that on the right as

$$A\cdot\eta = A^a\sigma_a , \qquad \sigma_a = \tfrac{1}{6}\,\eta_{abcd}\, dx^{bcd} . \tag{76}$$

σ_a are the components of the *vectorial hypersurface element*, which is a vector-valued 3-form.

Then (75) goes over into

(77) $$\int_D A^a{}_{;a}\eta = \int_{\partial D} A^a \sigma_a$$

the familiar *metric-dependent version of Gauss' theorem in Riemannian space.*

3·7. – Since each tangent space T_q of spacetime is itself a (flat) semi-Riemannian space, it has a volume element like (73); we write

(78) $$\pi = \sqrt{-g}\, \mathrm{d}p^{1234}\,.$$

g is to be evaluated at q, with respect to oriented co-ordinates (x^a) such that $p = p^a(\partial/\partial x^a)$. The p^a's are co-ordinates in T_q.

Physically important hypersurfaces (3-submanifolds) of T_q are the *mass shells* for masses $m \geqslant 0$. The mass shell $P_m(q)$ consists of all future-directed (4-momentum-) vectors p at q which belong to (proper) mass m, $p^2 = -m^2$. As co-ordinates on P_m we shall take p^ν, $\nu = 1, 2, 3$; p^4 is then determined by

(79) $$g_{ab}(x^c)\, p^a p^b = -m^2\,.$$

The 3-form

(80) $$\pi_m = \frac{\sqrt{-g}}{|p_4|}\, \mathrm{d}p^{123}$$

is a co-ordinate-independent *volume element on* P_m. It can be obtained from

(81) $$\pi_m = 2H(p)\,\delta(p^2 + m^2)\sqrt{-g}\, \mathrm{d}p^{1234}\,,$$

where δ is the Dirac distribution and

$$H(p) = \begin{cases} 1 \text{ if } p \text{ is future-directed,} \\ 0 \text{ otherwise}\,. \end{cases}$$

(For $m > 0$, $m\pi_m$ is the *induced Riemannian volume element of* $P_m(q)$, considered as a hypersurface in T_q.) A derivation of (80) without the use of δ will be given in Appendix II.

In orthonormal co-ordinates

(82) $$\pi_m = \frac{\mathrm{d}p^{123}}{E}\,,$$

where $E=\sqrt{m^2+\boldsymbol{p}^2}$ is the *energy*. Taking polar co-ordinates in $\boldsymbol{p}$-space, we have

$$\pi_m = \frac{\boldsymbol{p}^2}{\sqrt{m^2+\boldsymbol{p}^2}}\,\mathrm{d}|\boldsymbol{p}|\wedge(\sin\theta\,\mathrm{d}\theta\wedge\mathrm{d}\varphi)\,; \tag{83}$$

also

$$\pi_m = \sqrt{E^2-m^2}\,\mathrm{d}E\wedge(\sin\theta\,\mathrm{d}\theta\wedge\mathrm{d}\varphi)\,. \tag{84}$$

4. – Kinetic theory.

4·1. – Before studying a gas, we *consider a single charged particle with mass* m ($\geqslant 0$) *and charge* e, *which moves in a given gravitational field* g_{ab} *and electromagnetic field* F_{ab}. Its worldline $(x^a(v))$ is determined by the *Lorentz-Einstein equations of motion*

$$\frac{\mathrm{d}x^a}{\mathrm{d}v} = p^a\,, \qquad \frac{\mathrm{D}p^a}{\mathrm{d}v} = eF^a{}_b p^b\,, \tag{85}$$

if radiation reaction is neglected.

The (affine) parameter v is defined (uniquely except for its origin) by the requirement that p^a *is the* 4-*momentum*, whether $m>0$ or $m=0$. If $m>0$, $mv=s$ is proper time.

According to (85) the *instantaneous state* of a particle is given by a 4-momentum p^a at an event x^a; these data determine the motion uniquely. Therefore, we define the set

$$M = \{(x, p)\colon x\in X,\, p\in T_x,\, p^2\leqslant 0,\, p \text{ future-directed}\} \tag{86}$$

as the *one-particle phase-space for particles of arbitrary rest masses.*

M is an 8-dimensional manifold with boundary. If (x^a) are local co-ordinates in X and (p^a) the corresponding components of vectors, then (x^a, p^a) are local co-ordinates in M. The boundary ∂M consists of all states belonging to zero rest mass. (M is a part of the *tangent bundle* $T(X)$ of space-time, obtained from (86) by admitting *all* vectors p.)

The equations of motion (85) define on M a vector field

$$L = p^a\frac{\partial}{\partial x^a} + (eF^a{}_b p^b - \Gamma^a_{bc}p^b p^c)\frac{\partial}{\partial p^a}\,, \tag{87}$$

the *Liouville vector* (or *operator*). The (directed, parametrized) integral curves $(x^a(v),\, p^a(v))$ of (85) form a congruence in M, the *phase flow* generated by L.

Physically, the phase flow represents the set of all test-particle motions which are possible in the combined gravitational and electromagnetic fields occuring in L.

The rest mass m, given by

$$m^2 = -g_{ab}(x^a)p^a p^b \,, \qquad m \geqslant 0 \,, \tag{88}$$

is a scalar function on M which is constant on each phase orbit,

$$L(m) = 0 \,. \tag{89}$$

The hypersurface of M defined by $m = \text{const}$ is generated by all those phase orbits which belong to the assigned mass value; we use the symbol M_m for it. M_m is the *phase space for particles of mass* m, its dimension is 7. As co-ordinates on M_m we take (x^a, p^ν).

Because of (89) L *is tangent to* M_m. The restriction of L to M_m,

$$L_m = p^a \frac{\partial}{\partial x^a} + (eF^\nu{}_b p^b - \Gamma^\nu_{bc} p^b p^c) \frac{\partial}{\partial p^\nu} \,, \tag{90}$$

is the *Liouville operator associated with* M_m. The nonrelativistic analogue of M_m is the $(\boldsymbol{x}, t, \boldsymbol{p})$-space.

The phase spaces M, M_m *are orientable manifolds* (even if X were not orientable). For M this follows immediately from the fact that if (x^a), $(x^{a'})$ are two co-ordinate systems on X and if (x^a, p^a), $(x^{a'}, p^{a'})$ are their extensions to M, then

$$\frac{\partial(x^{a'}, p^{a'})}{\partial(x^b, p^b)} = \frac{\partial(x^{a'})}{\partial(x^b)} \frac{\partial(p^{a'})}{\partial(p^b)} = \left(\frac{\partial(x^{a'})}{\partial(x^b)}\right)^2 > 0 \,.$$

Since $M_0 = \partial M$, M_0 is also orientable (Subsect. **3**.3). For the analogous reason, M_m is orientable (for it is the boundary of the part of M given by $p^2 \leqslant -m^2$). *We orient* M *by taking* (x^a, p^a) *as oriented co-ordinate systems, and take in* M_m *the induced orientation.*

4.2. – Consider a *large number of particles* interacting through gravitational, electromagnetic and (possibly) short range (*e.g.*, nuclear) forces. The particles might be *macroscopic* (galaxies, stars) or *microscopic* (molecules, atoms, ions, nuclei, electrons, ...).

An exact relativistic description of such a system in terms of a many-particle (statistical) mechanics does not seem to be available at present; moreover, it would have to be very complicated.

However, for several important systems the following description seems to be reasonable: The particles move almost like test particles in the smoothed-out mean field generated by all the particles together, except (possibly) during very short intervals of time (collisions) during which they are thrown from one test-particle orbit into another one.

Examples for nearly *collision-free systems* are *galaxies* (as systems of stars), *the system of galaxies*, and the photon gas known as *cosmic fireball* (after decoupling from matter). The more complicated picture, where *collisions* interrupt the quasi-free motions, seems to be adequate for *gases and plasmas*, except at low temperatures and high densities. In particular, it seems to hold for most *stellar matter*.

For these reasons, we adopt the following *model of gas*: A gas is represented by a *broken complex of worldlines* [17] in space-time. The *vertices* of the complex *represent collisions*; *between collisions the worldlines satisfy the equations of motion* (85). (See Fig. 3.)

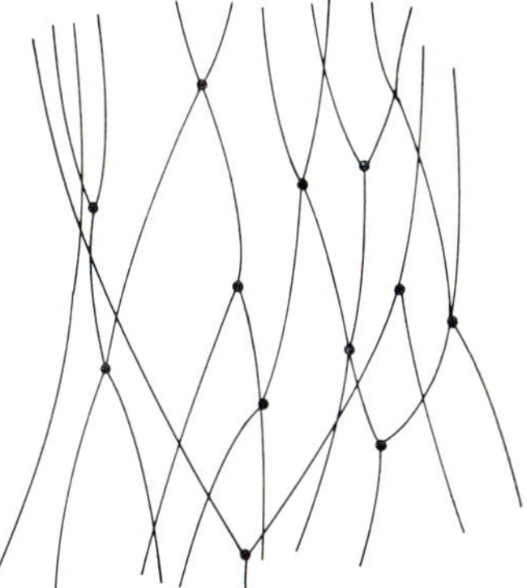

Fig. 3. – Space-time picture of a relativistic gas.

The fields g_{ab}, F_{ab} in (85) *are* given, *external fields* for a *test gas*, and for a *self-gravitating gas* they are taken to be the *average fields* produced collectively by the gas particles themselves.

The *collisions* are subject to *conservation laws*; their *space-time distribution* is governed by *statistical laws* which will be considered in Subsect. **4**·9.

The classical model of an individual gas described above is too complicated and detailed to be useful. Moreover, at least for microscopic particles, its sharply defined collision events and worldlines cannot be considered as an adequate model of reality. We therefore imagine, in one and the same space-time X and field (g_{ab}, F_{ab}), a large number of such gas histories—a *Gibbs ensemble of microstates* of a gas representing a *macrostate* of the many-particle system. *The average properties of such an ensemble are the subject of kinetic theory.* They may well provide an approximate, macroscopic description of a gas even if the particles of the latter obey quantum laws [46].

In order to translate these ideas into mathematical language, we shall make some *smoothness assumptions* about the average distributions of occupied states and the average numbers of collisions in an ensemble of gas histories. This will lead us to the (one-particle) *distribution function* f in phase space and to an expression for the average *density of collisions*. By means of f we shall define certain *flux densities* in space-time which will be used to formulate *balance equations* and *conservation laws*. We then set up the *Einstein-Maxwell field equations* for the gas and derive conservation laws from them.

Subsequently we shall study *collisions* and formulate the *Boltzmann equa-*

tion. With its help, we shall derive the *H-theorem* and the *equilibrium distributions*. Finally we shall indicate the treatment of *transport processes*, and establish the basic equations of *hydrothermodynamics*.

4.3. – Imagine a *gas* consisting of *particles of different types*. Particles of *type* j have *rest mass* m_j, *charge* e_j, and possibly further characteristics (like *baryon number* b_j, *spin* s_j, e- and *μ-lepton number* l_j^e, l_j^μ, or in case of stars *spectral types* etc.). With each particle type we associate a *phase space* $\boldsymbol{M}_j$ $(=M_{m_j})$ and a *Liouville operator* L_j. The fields g_{ab}, F_{ab} are, of course, independent of j.

A definite *microstate*, or history, of the gas is represented in $\boldsymbol{M}_j$ as a *collection of segments of phase orbits*, the states occupied by j particles between collisions. A *collision* can be symbolized by

$$(x;\, ip,\, j\bar{p} \rightarrow kp',\, lp'')\,. \tag{91}$$

This indicates that at $x \in X$ particles of types i and j with 4-momenta p, $\bar{p}$ collide to produce particles of types k, l with 4-momenta p', p''. The collision (91) gives rise, *e.g.*, in $\boldsymbol{M}_i$ to an end point (x, p) of an occupied orbit segment, an *annihilation*; and in $\boldsymbol{M}_k$ it gives rise to an initial point (x, p') of such a segment, a *creation*. (See Fig. 4.) (*)

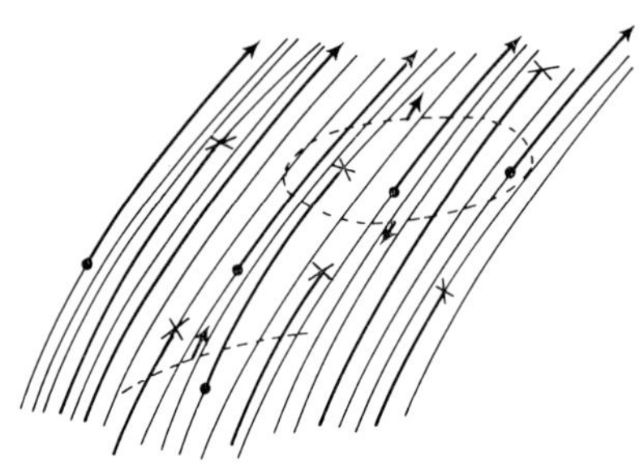

Fig. 4. – Phase-space picture of a gas. Occupied states fat; creations: •; annihilations: ×. The open dashed line represents a hypersurface intersected by 3 occupied states. The area enclosed by a dashed line represents a domain of M containing one annihilation and two creations.

We shall use the term *hypersurface of* $\boldsymbol{M}_j$ for oriented, 6-dimensional submanifolds with boundary of $\boldsymbol{M}_j$.

Let $N_j[\Sigma]$ be the *number of occupied states intersecting* the compact *hypersurface* $\Sigma \subset \boldsymbol{M}_j$. The intersection of an occupied phase orbit k with Σ is counted *positively* (*negatively*) if, at the point of intersection, the vector basis $(L, A_1, \ldots, A_6)$ has the same (opposite) orientation as the basis of an oriented co-ordinate system of $\boldsymbol{M}_j$, L being the tangent to k and $(A_1, \ldots, A_6)$ an oriented basis tangent to Σ.

The functionals $N_j\colon \Sigma \rightarrow N_j[\Sigma]$ (for all j) fully specify a *microstate* of the gas.

(*) We do not assign a phase trajectory to a particle « during » a collision, but maintain that the incoming particles are annihilated, the outgoing ones created in the collision.

If D is any region in M_j, then $N_j[\partial D]$ *is the number of collisions in* D, if creations are counted positively, and annihilations negatively. (See Fig. 4.)

For a *macrostate*, let $\bar{N}_j[\Sigma]$ be the *ensemble average of* $N_j[\Sigma]$. We need an analytical expression for $\bar{N}_j[\Sigma]$. Since $\bar{N}_j[\Sigma]$ is a *flux* through Σ of a fictitious fluid flowing in M_j with velocity L_j we expect it to be expressible as an *integral*. We thus need a volume element of hypersurfaces Σ.

A co-ordinate-independent volume element on M_j *is*

$$\Omega_j = \eta\wedge\pi_j = \frac{-g}{|p_4|}\,\mathrm{d}x^{1234}\wedge\mathrm{d}p^{123} \qquad (\pi_j = \pi_{m_j}), \tag{92}$$

the (exterior) product of the space-time volume element η (eq. (73)) and the measure π_j on the mass shell P_j (eq. (80)).

A *volume element for hypersurfaces in* M_j is obtained [18] by contracting Ω_j with the Liouville vector L_j:

$$\omega_j = L_j\cdot\Omega_j = p^a\sigma_a\wedge\pi_j + \frac{1}{2|p_4|}\,\eta_{\lambda\mu\nu4}(F^\lambda{}_b p^b - \Gamma^\lambda_{bc}p^b p^c)\,\mathrm{d}p^{\mu\nu}\wedge\eta\,. \tag{93}$$

In evaluating the contraction we have used the quantities σ_a and η_{abcd} defined in (76) and (74), respectively.

This volume element ω_j *is*, except for a numerical factor, *uniquely determined by the requirement that it assigns a nonzero volume to any hypersurface element not tangent to* L_j (see the lemma in Subsect. 3·1), a property which is obviously needed for our purposes.

The measure ω_j shares an important property with the Lebesgue measure on a Newtonian phase space: It *is invariant with respect to the phase flow*. To prove this, verify first by means of (93) that

$$\mathrm{d}\omega_j = 0\,; \tag{94}$$

this is seen most easily by using inertial co-ordinates x^a at an (arbitrary) event, so that there $\Gamma^{\cdot}_{\cdot\cdot}=0$, $p^4=(m^2+p^\nu p^\nu)^{\frac{1}{2}}$, $\partial p^4/\partial x^4=0$, and $p^4\,\mathrm{d}p^4 = p^\nu\,\mathrm{d}p^\nu$ on M_j. Next, apply Stokes' theorem to a tube T of phase orbits bounded by two cross-sections Σ, Σ' and a mantle Λ (Fig. 5):

$$0=\int_T \mathrm{d}\omega_j = \int_{\partial T}\omega_j = \int_{\Sigma'}\omega_j - \int_{\Sigma}\omega_j + \int_{\Lambda}\omega_j\,.$$

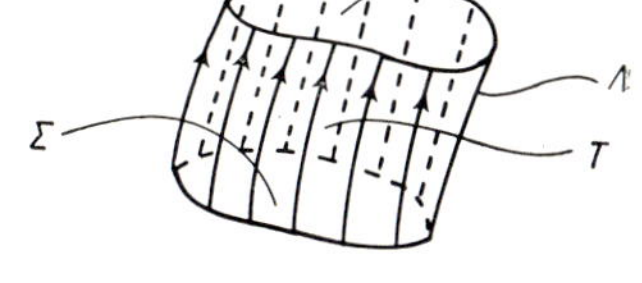

Fig. 5. – Tube T of phase orbits, bounded by cross-sections Σ, Σ' and mantle Λ.

The last term vanishes since L is tangent to Λ (see the corollary to the lemma in Subsect. 2·1).

Hence

$$\int_{\Sigma'} \omega_j = \int_{\Sigma} \omega_j ,$$

which expresses the asserted invariance.

By (94) and the previously stated property, ω_j *is uniquely characterised except for a scalar factor constant on the phase orbits*, for $0 = \mathrm{d}(f\omega_j) = \mathrm{d}f \wedge \omega_j = \mathrm{d}f \wedge (L_j \cdot \Omega_j) = L_j(f)\,\Omega_j$, by (67).

A seven-dimensional relativistic phase space M_j corresponds in the Newtonian case to the direct product of the time axis $\{t\}$ with the ordinary, six-dimensional $(p; q)$-phase space of one particle. Hence, *a spacelike hypersurface* $G \subset X$ *at each point of which a momentum region* $K_x \in P_{m_j}(x)$ *is prescribed represents a hypersurface* Σ *in* M_j *which corresponds to a part of an « instantaneous » ordinary phase space.* On such a Σ, ω_j from (93) reduces to its first part (*),

$$\omega_j = p^a \sigma_a \wedge \pi_j \quad \text{on } \Sigma . \tag{95}$$

If we choose G small and contained in the local, instantaneous rest space of an observer at x with 4-velocity u, then $\Sigma = G \times K_x$ is a part of the ordinary phase space associated with that observer, with space projection G and momentum domain $K_x \subset P_{m_j}(x)$. If K_x is also small and contains p, (95) gives

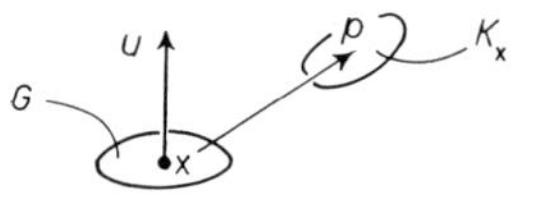

Fig. 6.

$$-\,\omega_j(\Sigma) = p^4 V(G)\pi_j(K_x) = p^4 V(\pi/p^4) ,$$

if V is the volume of G and $\pi\left(=\int_{K_x} \mathrm{d}^3 p\right)$ the « ordinary » momentum volume of K_x.

Hence, in this case

$$|\omega_j(G \times K_x)| = V\pi , \tag{96}$$

the *ordinary phase-space volume* the observer would assign to $G \times K_x$. This result shows that the normalization factor in ω_j has been chosen adequately.

After this digression on volume elements we continue to study $\bar{N}_j$. We make the following *smoothness assumptions* about a macrostate of a gas:

D_1) On any fixed hypersurface $\Sigma \subset M_j$ there exists a continuous, non-negative density function f_Σ such that, for all parts $\Sigma' \subset \Sigma$,

$$\bar{N}_j[\Sigma'] = \int_{\Sigma'} f_\Sigma \omega_j . \tag{97}$$

(*) (95) holds irrespective of whether G is spacelike, timelike or null.

D_2) For any sequence of regions in M_j which shrinks to a point, $D_l \to \xi$, there exists a positive number A such that

$$|\bar{N}_j[\partial D_l]| \leqslant A \int_{D_l} \Omega_j \,. \tag{98}$$

The first property, D_1, corresponds to the existence of a continuous one-particle distribution function in ordinary statistical mechanics. The inequality (98) asserts that the average number of collisions in D is at most of the order of the Ω-measure of D.

These assumptions imply the existence of an *invariant* (one particle) *distribution function* $f_j(x, p)$ on M_j such that

$$\bar{N}_j[\Sigma] = \int_{\Sigma} f_j \, \omega_j \qquad \text{for any } \Sigma \subset M_j \,. \tag{99}$$

According to (96) the statement (99), *i.e.* the independence of f_j of Σ, is equivalent to the assertion that *the phase-space density at* (x, p) *is the same for all observers at* x.

To prove (99), we have to show that if a point $\xi \in M_j$ is contained in two hypersurfaces Σ_1, Σ_2, then $f_{\Sigma_1}(\xi) = f_{\Sigma_2}(\xi)$. Consider the situation shown in Fig. 7. T is a tube of phase orbits, having ξ on its boundary. Σ_1', Σ_2' are the parts of Σ_1, Σ_2 cut out by T. D is the region in T between Σ_1' and Σ_2'. Since no orbits cross the mantle of T we have

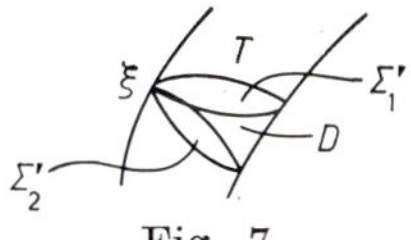

$$\bar{N}_j[\partial D] = \bar{N}_j[\Sigma_1'] - \bar{N}_j[\Sigma_2'] \,.$$

Fig. 7.

With the first assumption D_1 and the mean-value theorem for integrals this can be rewritten as

$$\bar{N}_j[\partial D] = f_{\Sigma_1}(\xi_1) \int_{\Sigma_1'} \omega_j - f_{\Sigma_2}(\xi_2) \int_{\Sigma_2'} \omega_j \,,$$

with $\xi_l \in \Sigma_l'$. But we proved above that $\int_{\Sigma_1'} \omega_j = \int_{\Sigma_2'} \omega_j$, hence

$$f_{\Sigma_1}(\xi_1) - f_{\Sigma_2}(\xi_2) = \frac{\bar{N}_j[\partial D]}{\int_D \Omega_j} \, \frac{\int_D \Omega_j}{\int_{\Sigma_1'} \omega_j} \,.$$

Now, let T shrink to the phase orbit passing through ξ. Then, according to our second assumption D_2, the first factor on the right remains bounded, whereas the second factor goes to zero since its numerator is one order smaller than its denominator. It follows, therefore, that $f_{\Sigma_1}(\xi) = f_{\Sigma_2}(\xi)$, as was claimed.

Because of physical applications, let us specialize the formula (99) to a hypersurface Σ of the type described in connection with (95). We obtain:

$$N_j = \int_G \sigma_a \left\{ \int_{K_x} f_j p^a \pi_j \right\} \tag{100}$$

is the average number of particles whose worldlines intersect the hypersurface G *and which have, at the point* x *of intersection, a 4-momentum* p *contained in the region* $K_x \subset P_j$ (*).

So far we have only postulated that the restriction of f_j to any hypersurface is continuous. Because of the interpretation of f_j as an average density, contained in (99), we now assume in addition:

D_3) $f_j(x, p)$ is a continuously differentiable function on M_j.

4·4. – The average number of collisions in D ($\subset M_j$) is, according to (99), given by

$$N_j[\partial D] = \int_{\partial D} f_j \omega_j = \int_D \mathrm{d}(f_j \omega_j)\,.$$

Because of $\mathrm{d}\omega_j = 0$ (eq. (94)) and the identity (67),

$$\mathrm{d}(f_j\omega_j) = \mathrm{d}f_j \wedge \omega_j = \mathrm{d}f_j \wedge (L_j \cdot \Omega_j) = L_j(f_j)\,\Omega_j\,,$$

whence we obtain

$$N_j[\partial D] = \int_D L_j(f_j)\,\Omega_j\,. \tag{101}$$

Hence,

$$L_j(f_j) = p^a \frac{\partial f_j}{\partial x^a} + (e_j F^\lambda{}_b p^b - \Gamma^\lambda_{bc} p^b p^c) \frac{\partial f_j}{\partial p^\lambda} = \frac{\mathrm{d}f_j}{\mathrm{d}v}\,, \tag{102}$$

where the last expression means the derivative of $f_j(x, p)$ along a phase orbit, *gives the density* (with respect to Ω_j) *of collisions in phase space.*

(*) Hence $f_j \geqslant 0$, and (96) shows that f_j equals the ordinary phase-space density for every local observer.

For applications we specialize the domain D in (101) to be

$$D := \{(x, p) : x \in G, p \in K_x \subset P_{m_j}(x)\}$$

where G is a region in space-time. Then, from (92):

$$N_j[\partial D] = \int_G \eta \left\{ \int_{K_x} L_j(f_j)\pi_j \right\} \tag{103}$$

is the average number of collisions in G in which a particle of type j is thrown into a state with 4-momentum in K_x at $x \in G$.

It follows from these considerations that the *Liouville equation* (in one-particle phase space) (*),

$$L_j(f_j) = p^a \frac{\nabla f_j}{\partial x^a} + e_j F_b^\lambda p^b \frac{\partial f_j}{\partial p^\lambda} = \frac{df_j}{dv} = 0 , \tag{104}$$

expresses the absence of collisions or, more generally, the *detailed balancing* between creations and annihilations of particles of type j with 4-momentum p at x.

4·5. – It is obvious from (100) that the vector field

$$N_j^a(x) := \int_{P_j(x)} f_j(x, p) p^a \pi_j \tag{105}$$

is the *particle 4-current density* for particles of kind j. Analogously,

$$J_j^a := e_j N_j^a , \qquad J^a := \sum_j J_j^a \tag{106}$$

are the partial and total *electric 4-current densities*. In a similar way, current densities can be associated with other quantities carried by the particles, *e.g.*, baryon number.

The physical meaning of the components of N_j^a for an observer at x is obtained by specializing (105) to an inertial frame at x, using the expression (82) for π_j, and remembering that f_j equals the ordinary phase-space density. Thus one gets that N_j^4 *is the number density n_j of j-particles in the observers's rest space, and*

$$\mathbf{N}_j = (N_j^\lambda) = n_j \langle \mathbf{v} \rangle_j , \tag{107}$$

(*) In (104), $\nabla/\partial x^a = \partial/\partial x^a - p^c \Gamma^b_{ca}(\partial/\partial p^b)$ is the covariant partial-derivative operator.

where $\boldsymbol{v}$ is the 3-velocity of a particle with respect to the observer and $\langle\,\rangle_j$ indicates averaging over j-particles near x.

One can also associate flux densities with tensorial quantities. For example,

$$T_j^{ab}(x) := \int_{P_j(x)} p^a p^b f_j \pi_j\,, \qquad T_k^{ab} := \sum_j T_j^{ab} \tag{108}$$

are the partial and total 4-momentum flux densities. They are also called *kinetic stress energy-momentum tensors*, respectively.

We assume throughout that each distribution function vanishes at infinity of the mass-shell sufficiently strongly so that the integrals (105) and (108) exist.

The physical meaning of the T_j^{ab}'s for an arbitrary local observer can be obtained as in the preceding case; one gets

$$\begin{cases} T_j^{44} = \mu = n_j\langle E\rangle_j = \textit{energy density}\,, & (p^4 = E)\,, \\ (T_j^{4\lambda}) = n_j\langle \boldsymbol{p}\rangle_j = \textit{momentum density}\,, \\ (T_j^{\lambda\mu}) = n_j\langle \boldsymbol{p}\otimes\boldsymbol{v}\rangle_j = \textit{kinetic pressure tensor}\,, \end{cases} \tag{109}$$

in accordance with (21).

We also define the *mean kinetic pressure* (with respect to the observer) by $p_j = \frac{1}{3} T^{\lambda}_{j\lambda}$, so that

$$p_j = \tfrac{1}{3} n_j \langle \boldsymbol{p}\cdot\boldsymbol{v}\rangle_j \tag{110}$$

(Bernoulli's formula), and write $\varrho_j = m_j n_j$ for the *rest mass density*. ((110) can be used, in conjunction with the virial theorem $\overline{\boldsymbol{F}\cdot\boldsymbol{x}} = -\overline{\boldsymbol{p}\cdot\boldsymbol{v}}$ of particle mechanics, to show that p_j is indeed the pressure exerted on the walls confining the gas.)

The preceding formulae, together with $E = m/\sqrt{1-v^2}$, $\boldsymbol{p} = E\boldsymbol{v}$, imply the following *inequalities between* μ, p, ϱ (we omit j, considering one component):

$$0 \leqslant 3p \leqslant \tfrac{3}{2}p + \sqrt{(\tfrac{3}{2}p)^2 + \varrho^2} \leqslant \mu \leqslant \varrho + 3p\,. \tag{111}$$

They restrict the equations of state of a gas (*).

Only the third of these inequalities, due to TAUB [20], is not obvious. To prove it, consider $(1-v^2)^{-\frac{1}{4}}$ and $(1-v^2)^{\frac{1}{4}}$ as elements of the space of square-integrable functions with respect to the measure $f\,dp^{123}$, and apply

(*) Note also: $T^a{}_a = -m^2\int_P f\pi = 3p - \mu \leqslant 0$.

Schwartz' inequality:

$$\left(\int f\,\mathrm{d}p^{123}\right)^2 \leqslant \left(\int \frac{1}{\sqrt{1-v^2}}\,f\,\mathrm{d}p^{123}\right)\left(\int \sqrt{1-v^2}\,f\,\mathrm{d}p^{123}\right).$$

On multiplication with m^2, one gets $\varrho^2 \leqslant \mu(\mu-3p)$, which is equivalent to the asserted inequality.

From (109) one infers:

$$(112)\quad \begin{cases} \text{if } p \ll \varrho, \text{ then } \mu \approx \varrho + \frac{3}{2}p \text{ (nonrelativistic « monatomic » gas)}, \\ \text{if } p \gg \varrho, \text{ then } \mu \approx 3p \text{ (ultrarelativistic gas)}, \\ \text{if } m = 0, \text{ then } \mu = 3p \text{ (photon gas, } e.g.\text{)}. \end{cases}$$

Excluding the trivial case $n = 0$ we infer from (108)

$$(113)\qquad T_{ab}v^a v^b > 0 \quad \text{for all nonspacelike } v^a\text{'s}.$$

This property implies the existence of a timelike eigenvector [47] u^a of $T^a{}_b$. Hence, *any kinetic stress energy tensor admits a unique decomposition*

$$(114)\quad \begin{cases} T^{ab} = \mu\, u^a u^b + p^{ab} \\ \text{with } u^a \text{ future-directed,} \quad u_a u^a = -1, \quad p_{ab}u^b = 0. \end{cases}$$

((113) is true provided the distribution function is, in fact, a *function*, and *not a distribution*. If one permits f to be a distribution, then there is precisely one exception to (113). It is given by $m = 0$, $f(x^a, p^a) = g(x^a, p^a)\,\delta(k_a p^a)$, where $k^a(x^b)$ is a lightlike vector field and g an (ordinary) function on M_0, which will be chosen such that

$$\pi\int_0^\infty \lambda^2 g(x^a, \lambda k^a)\,\mathrm{d}\lambda = 1$$

for all x^a. This distribution describes a *stream of « photons » without velocity dispersion*; at x^a, all particles have worldlines tangent to $k^a(x^b)$. g describes the *spectrum* of the « radiation ». One finds

$$T^{ab} = k^a k^b, \qquad N^a = \left(\pi\int_0^\infty \lambda g(x^b, \lambda k^b)\,\mathrm{d}\lambda\right) k^a.$$

Such a T^{ab} is known from electromagnetic *null fields* and the geometrical optics approximation; it does *not* permit a representation of the form (114).)

An observer who travels with the 4-velocity u^a occuring in (114) will measure a vanishing momentum density. Therefore, u^a is called the *dynamical mean 4-velocity* of the gas.

The dynamical mean 4-velocity of a gas is, in general, different from its *kinematical mean 4-velocity*, defined as the unit vector collinear with the particle 4-current (105). An observer travelling with the latter will find the mean peculiar velocity of the gas particles to vanish, according to (107). For a mixture, there are still more reasonable ways to define mean 4-velocities; *e.g.*, one can use the baryon numbers b_j as weights and put

$$B^a = \sum_j b_j N_j^a = \varrho_b u_b^a \qquad (u_b^2 = -1) \tag{115}$$

to define a mean velocity u_b^a and a mean baryon density ϱ_b.

If the velocity dispersion of the gas is negligible, we have according to (114) and (109)

$$T^{ab} = \varrho u^a u^b ; \tag{116}$$

usually such a gas is called « *dust* » in relativity theory.

((116) is obtained from (108) if one uses the singular distribution « function »

$$f(x^a, p^a) = \frac{n(x^a)}{2\pi m} \frac{\delta(m + u_a p^a)}{\left(m^2 - (u_a p^a)^2\right)^{\frac{1}{2}}},$$

where $m > 0$ and $u^a(x^b)$ is the 4-velocity field.)

A weakened form of the inequality (113) and

$$(T_{ab} - \tfrac{1}{2} g_{ab} T)\, v^a v^b \geqslant 0 \qquad \text{for } v_a v^a < 0, \tag{117}$$

which follows from (111), is important in the theory of singularities of solutions of the Einstein field equations [48].

4·6. – The current densities can be used to formulate *balance equations* and, in particular, *conservation laws*.

Thus, it follows from the meaning of the N_j^a's and Gauss' theorem (77) that *conservation of electric charge is expressed by*

$$J^a{}_{;a} = 0, \tag{118}$$

and *conservation of baryon number is similarly given by*

$$B^a{}_{;a} = (\varrho_b u_b{}^a)_{;a} = 0 , \tag{119}$$

where B^a is defined by (115).

Analogously it follows that $N_j{}^a{}_{;a}$ is the *space-time production density for j-particles, i.e.* the average number of j-particles created (by inelastic collisions) per unit time and unit volume.

To express $N_j{}^a{}_{;a}$ in terms of f_j, let D be an arbitrary region in X, and define the « cylindrical » region

$$\hat{D} = \{(x, p) : x \in D, (x, p) \in M_j\} ,$$

lying in M_j over D. Then

$$\int_{\partial\hat{D}} f_j \omega_j = \int_{\hat{D}} L_j(f_j)\,\Omega_j ,$$

as derived in Subsect. **4**·4. Since $\partial\hat{D}$ « lies over » ∂D,

$$\int_{\partial\hat{D}} f_j \omega_j = \int_{x\in\partial D} \sigma_a \left\{ \int_{P_j(x)} p^a f_j \pi_j \right\} = \int_{\partial D} \sigma_a N_j{}^a = \int_D N_j{}^a{}_{;a} \eta ,$$

where we have used (97), (105), and (77). Also, with (92),

$$\int_{\hat{D}} L_j(f_j)\,\Omega_j = \int_{x\in D} \eta \left\{ \int_{P_j(x)} L_j(f_j)\,\pi_j \right\} .$$

Since the two integrals $\int_D \ldots$ are equal *for arbitrary* D, we infer

$$N_j{}^a{}_{;a} = \int_{P_j} L_j(f_j)\,\pi_j . \tag{120}$$

This (intuitively rather obvious) result is the *balance equation for j particles.*

(120) implies

$$J^a{}_{;a} = \sum_j \int_{P_j} e_j L_j(f_j)\,\pi_j , \tag{121}$$

the *balance equation for electric charge*, and a similar equation for B^a.

In order to see how $T_j{}^{ab}{}_{;b}$ is related to f_j, we take again a region $D \subset X$, form $\hat{D}$, and apply Stokes' theorem to $v_a p^a f_j$, where $v_a(x^b)$ is a vector field on X which satisfies

$$v_{a;b} = 0 \quad \text{at } x_0 \in X. \tag{122}$$

We obtain

$$\int_{\partial\hat{D}} v_a p^a f_j \omega_j = \int_{\hat{D}} L_j(v_a p^a f_j)\Omega_j .$$

Again, both sides can be rewritten as (iterated) integrals over the arbitrary region D, whence

$$(v_a T_j{}^{ab})_{;b} = \int_{P_j} L_j(v_a p^a f_j)\pi_j .$$

Evaluating this equation at the event x_0 where (122) holds and taking into account that, at x_0,

$$L_j(v_a p^a f_j) = \frac{\mathrm{D}}{\mathrm{d}v}(v_a p^a f_j) = v_a\left(p^a L_j(f_j) + f_j \frac{\mathrm{D}p^a}{\mathrm{d}v}\right) = v_a\left(p^a L_j(f_j) + e_j F^a{}_b p^b f_j\right)$$

($\mathrm{D}/\mathrm{d}v$ = absolute derivative along the particle orbit through (x_0^a, p^a)), we obtain

$$T_j{}^{ab}{}_{;b} = F^a{}_b J_j{}^b + \int_{P_j} p^a L_j(f_j)\pi_j . \tag{123}$$

This is the 4-*momentum balance equation for type j particles.* (Balance for energy and momentum.) The two vectors of the right-hand side represent the *electromagnetic* and *collisional 4-force densities* acting on the j-th component of the gas. An example for the latter is the force exerted on an electron gas by photons due to Compton scattering.

4·7. – The preceding development was independent of field equations for g_{ab} and F_{ab}. We may apply them to a *test gas* embedded in an external Einstein-Maxwell field.

Let us now, however, require that g_{ab}, F_{ab} are the (average) fields produced by the gas itself:

$$G^{ab} + \Lambda g^{ab} = T^{ab} := T_k{}^{ab} + T_M{}^{ab} , \tag{124}$$

$$F_{[ab,c]} = 0 , \qquad F^{ab}{}_{;b} = J^a . \tag{125}$$

The source term in *Einstein's gravitational field* equation is the *total stress energy-momentum tensor*, consisting of the kinetic part defined in (108) and the Maxwellian contribution:

$$T_M^{ab} := F^{ac}F^b{}_c - \tfrac{1}{4}g^{ab}F_{cd}F^{cd}\,, \tag{126}$$

and J^a in *Maxwell's equations* is the total electric current density defined in (106).

As is well known, (124) and (125) imply the conservation laws

$$T^{ab}{}_{;b} = 0\,, \tag{127}$$

and

$$J^a{}_{;a} = 0\,, \tag{128}$$

respectively, and (127) can be rewritten (because of Poynting's identity) as

$$T_k{}^{ab}{}_{;b} = F^{ab}J_b\,. \tag{129}$$

Inserting (128) and (129) into the balance eqs. (121) and (123), respectively, we obtain

$$\sum_j \int_{P_j} e_j L_j(f_j)\pi_j = 0 \tag{130}$$

and

$$\sum_j \int_{P_j} p^a L_j(f_j)\pi_j = 0\,. \tag{131}$$

These two integral conditions which express conservation of charge and 4-momentum in collisions, represent restrictions imposed on the time evolution of the distribution functions by the Einstein-Maxwell equations. They are trivially satisfied if each f_j satisfies a Liouville equation (104).

The eqs. (124), (125), (104), coupled with the definitions (108), (105), (106), (126), *govern a collisionless charged, gravitating gas.* They define a deterministic physical system (see Subsect. **2**'12) with the basic variables g_{ab}, F_{ab}, f_j. The corresponding theory generalizes the *Vlasov-Landau approximation* of plasma physics and the usual formulation of *stellar dynamics* to a general-relativistic setting.

4'8. – Our next aim is to take over into the relativistic gas theory the covariant analogue of *Boltzmann's collision integral.* We wish to include elastic and inelastic collisions, absorption and emission processes etc., and we also want to take into account those quantum effects which are due to the indistinguishability (« statistics ») of microparticles.

For the last reason it is advisable to renorm the P_j-element π_j and the distribution function f_j [21]. Let r_j be the *spin degeneracy* of a particle of type j, *i.e.*

$$(132)\qquad r_j = \begin{cases} 2s+1 & \text{for particles with } m>0 \text{ and spin } s\,, \\ 2 & \text{for particles with } m=0 \text{ and spin } s>0\,, \\ 1 & \text{for particles with } m=0 \text{ and spin } s=0\,. \end{cases}$$

We then substitute

$$(133)\qquad \pi_j \to \frac{(2\pi)^3}{r_j}\,\pi_j\,, \qquad f_j(x,p) \to \frac{r_j}{(2\pi)^3}\,f_j(x,p)\,.$$

Since a free particle of type j with fixed 4-momentum p has r_j mutually orthogonal polarization states, and since each eigenvalue of p of a particle enclosed in a box G with 3-volume V corresponds asymptotically to a cell of size $(2\pi)^3$ in the classical (p, q) phase space, the change (133) has the effect that *the new* ω_j*-volume* (95) *of a domain*

$$D := \{(x,p) : x \in G,\ p \in K_x \subset P_j(x)\} \subset M_j$$

equals approximately the number of mutually orthogonal quantum states which « belong to » the spacelike hypersurface $G \subset X$ *with attached momentum ranges* K_x $(x \in G)$. Consequently, *the value* $f_j(x, p)$ *of the* (renormed) *distribution function equals approximately the average occupation number of simple, (quasi) p-eigenstates localized near x.* Thus,

$$(134)\qquad f_j \leqslant 1 \text{ for fermions}$$

because of Pauli's exclusion principle.

The qualification « approximately » will not (in fact, could not) be made precise here; the above statements should be understood in a semi-heuristic way, as indicating a reasonable correspondence with a (not yet existing) quantum theory in curved space-time.

The formula (100) which characterizes the distribution function, eq. (103) for the number of collisions, as well as the expressions (105), (108) for the currents remain valid if the substitution (133) is made.

4·9. – In order to obtain *time development equations for the* f_j*'s* we need an expression for $\mathfrak{C}[D; jK_1, iK_2, \ldots \to lK_3, \ldots]$, *the average number of collisions*

$(jp_1, ip_2, \ldots \to lp_3, \ldots)$ *in the (small) space-time region* D *with momenta* $p_1, p_2, \ldots, p_3, \ldots$ *in the ranges* $K_1, K_2, \ldots, K_3, \ldots$ *of the respective mass shells.*

In a rigorors many-particle theory, one would expect $\mathfrak{C}$ to be a functional of (at least) the f_j's *and* the pair distribution functions. Hence, equations for the latter would be required etc., as discussed in (nonrelativistic) statistical mechanics (BBGKY-hierarchy).

We shall follow here the simple *method of Boltzmann* in which $\mathfrak{C}$ is expressed as *a functional of the* f_j*'s alone*, the physical assumption being that in the situations to be described—gases not too far from equilibrium, and not too cold and dense—particles which are about to collide are not correlated.

In order to obtain a reasonable « Stosszahlansatz » we first consider a *special relativistic quantum gas* [22] enclosed in a cubical box G. If periodic boundary conditions are imposed on the admissable wave functions of single particles, there exists a complete orthonormal set $|p_\lambda\rangle$ of single-particle momentum eigenstates. Let a_λ be the *annihilation operator* for $|p_\lambda\rangle$, so that a^*_λ is the *creation operator*. Furthermore, let $|\{n_\lambda\}\rangle$ denote the element of the *Fock space* of the gas which is a joint eigenstate of the occupation number operators $N_\lambda := a^*_\lambda \cdot a_\lambda$ with respective eigenvalues n_λ. In the boson case, n_λ is any nonnegative integer, in the fermion case, n_λ equals 0 or 1.

These $|\{n_\lambda\}\rangle$, corresponding to all sets $\{n_\lambda\}$ with $\sum_\lambda n_\lambda < \infty$, form (with suitable phases) an *orthonormal basis of the Fock space*, and we have from the commutation rules for the a_λ's

$$(135)\qquad \begin{cases} a_\lambda|\{n_\mu\}\rangle = \sqrt{n_\lambda}\,|\{n_1, \ldots, n_\lambda - 1, \ldots\}\rangle\,, \\ a^*_\lambda|\{n_\mu\}\rangle = \sqrt{1 \pm n_\lambda}\,|\{n_1, \ldots, n_\lambda + 1, \ldots\}\rangle\,. \end{cases}$$

The dots indicate unchanged occupation numbers; the upper sign in $\pm$ refers to bosons, the lower one to fermions, here and throughout. (The signs of the square roots obey rules which we will not need.)

Let the *Hamiltonian* of the gas be of the form

$$(136)\qquad H = H_0 + H_I\,,$$

where H_0 describes the « free » part and H_I is the interaction part.

Let the gas be in the state $\psi_i = |\{n_\lambda\}\rangle$ at time $t_i = 0$ (with respect to the rest frame of the box). The probability for a collision $(p_{\varrho_1}, p_{\varrho_2}, \ldots \to p_{\sigma_1}, p_{\sigma_2}, \ldots)$ to happen during $0 \leqslant t \leqslant T$ is given by $|\langle\psi_T, \psi_f\rangle|^2$, where

$$\psi_f = |\{n_1, \ldots, n_{\varrho_1} - 1, \ldots, n_{\varrho_2} - 1, \ldots, n_{\sigma_1} + 1, \ldots, n_{\sigma_2} + 1, \ldots\}\rangle\,.$$

(We assume the ϱ's and σ's to be pairwise different.) ψ_T is the state at time T

which evolves out of ψ_i according to Schrödinger's equation. T is assumed to be large in comparison with the collision time, but small enough so that the occurence of more than one collision in $0 \leqslant t \leqslant T$ can be neglected.

Using first-order Dirac perturbation theory one obtains, as is well known:

$$|\langle \psi_T | \psi_f \rangle|^2 = T\,\delta(E_f - E_i) |\langle \psi_f | H_I | \psi_i \rangle|^2 , \tag{137}$$

where E_i, E_f are the (unperturbed) energies of ψ_i, ψ_f, respectively.

Suppose that H_I can be expanded in a series the terms of which are of the form

$$w(\mu_1, \dots, \mu_r; \lambda_1, \dots, \lambda_s)\, \overset{*}{a}_{\mu_1} \dots \overset{*}{a}_{\mu_r} a_{\lambda_1} \dots a_{\lambda_s} , \tag{138}$$

with the creation operators to the left, and with complex coefficients symmetric in the μ's as well as in the λ's. We note in passing that the self-adjointness of H_I implies

$$\overline{w}(\lambda_1, \dots, \lambda_s; \mu_1, \dots, \mu_r) = w(\mu_1, \dots, \mu_r; \lambda_1, \dots, \lambda_s) . \tag{139}$$

($\overline{w}$ = complex conjugate of w.)

If one inserts the expansion $H_I = \dots$ into (137), takes into account the definitions of ψ_i and ψ_f, and uses (135) in conjunction with the orthonormality of the vectors $|\{n_\lambda\}\rangle$, one recognizes that only the term $w(\sigma_1, \sigma_2, \dots; \varrho_1, \varrho_2, \dots) \cdot \cdot \overset{*}{a}_{\sigma_1} \overset{*}{a}_{\sigma_2} \dots a_{\varrho_1} a_{\varrho_2} \dots$ and terms with additional factors $\overset{*}{a}_\lambda a_\lambda \overset{*}{a}_\mu a_\mu \dots$ give non-vanishing contributions to the transition amplitude in (137). Neglecting the latter (higher order) terms *one obtains*

$$|\langle T_T | \psi_f \rangle|^2 = T\,\delta(E_f - E_i) |w(\sigma_1, \dots; \varrho_1, \dots)|^2 n_{\varrho_1} n_{\varrho_2} \dots (1 \pm n_{\sigma_1})(1 \pm n_{\sigma_2}) \dots \tag{140}$$

for the average number of collisions $(p_{\varrho_1}, p_{\varrho_2}, \dots \to p_{\sigma_1}, p_{\sigma_2}, \dots)$ *taking place in the box* G *during* $0 \leqslant t \leqslant T$.

If the gas contains particles of different types, the index λ must be interpreted as indicating both the particle type and a translation state; also the spin state can be included in λ.

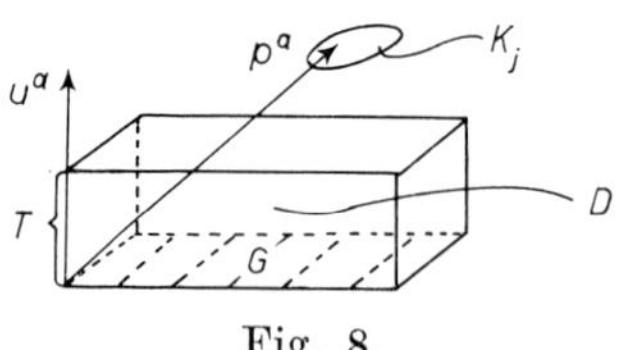

Fig. 8.

To obtain from (140) the *collision functional* $\mathfrak{C}[D; jK_1, \dots \to lK_3, \dots]$ defined at the beginning of this Subsection, we take as the space-time region D the « product » of a spacelike hypersurface G with a time interval of duration T, as indicated in Fig. 8. We assume that G and T are so small that D can be considered as flat, as far as the collisional behaviour of the gas in D is concerned, and that F_{ab} is weak enough so as not to affect the collisions. Let u^a be the unit normal of G.

Applying to the gas in D the result (140), averaging over a Gibbs ensemble of microstates $|\{n_\lambda\}\rangle$, assuming the random variables n_λ to be statistically independent, using that the number of quantum states corresponding to a small) 4-momentum range K_j surrounding p^a equals $(-u_a p^a) V \pi_j(K_j)$ (V: volume of G; $\pi_j(K_j)$: volume of K_j), and remembering that $\bar{n}_\lambda \approx f_j(x, p_\lambda)$, one obtains

$$(141) \qquad \mathfrak{C}[D; jK_1, iK_2, \ldots \to lK_3, \ldots] \approx (TV)\,\pi_j(K_1)\,\pi_i(K_2) \ldots \pi_l(K_3) \ldots$$
$$\ldots f_j(x, p_1) f_i(x, p_2) \ldots (1 \pm f_l(x, p_3)) \ldots W(jp_1, ip_2, \ldots; lp_3, \ldots)\,,$$

provided H_I is spin-independent. Some power of V, the factors $(-u_a p^a)$ etc., and $\delta(E_f - E_i)$ have been absorbed into W. This does not destroy the symmetry implied by (139); *W satisfies*

$$(142) \qquad W(jp_1, ip_2, \ldots; lp_3) = W(lp_3, \ldots; jp_1, ip_2, \ldots)$$

(see [50]).

In accordance with the derivation and with the interpretation of (141) we assume that *W is a Lorentz-invariant « function » of its momentum variables*, independent of the position and size V of the quantization volume G.

Since W vanishes whenever $E_f \neq E_i$, Lorentz-invariance requires that

$$(143) \qquad W(\ldots) = \delta(p_1 + p_2 + \ldots - p_3 - \ldots)\,R(\ldots)\,,$$

where *R is an ordinary, nonnegative, Lorentz-invariant function with the same symmetry* (142) as W.

Summarizing the result of our semi-heuristic « derivation » of (141), we state that *the space-time density of collisions* $(jK_1, iK_2, \ldots \to lK_3, \ldots)$ *at x equals*

$$(144) \qquad \int_{K_1}\pi_j \int_{K_2}\pi_i \ldots \int_{K_3}\pi_l \ldots \{f_j(x, p_1) f_i(x, p_2) \ldots (1 \pm f_l(x, p_3)) \ldots\} W(jp_1, ip_2 \ldots, lp_3, \ldots)$$

where W satisfies (142) *and* (143). It should be kept in mind that several approximations have been made in obtaining (144). (A fully satisfactory derivation of (144), with precise statements about its domain of validity, is not known to the author.)

4·10. – The function W in (144) is some measure of the *probability* with which the respective collisions occur. How is it related to the *cross-section*?

Consider collisions of the type $(ip_1, jp_2 \to kp_3, \ldots, lp_q)$ with $q-2$ outgoing particles with $p_r \subset K_r$. The *differential cross-section* $\mathrm{d}Q^u$ for collisions of the specified kind with respect to an observer having 4-velocity u *is defined by*

$$\text{(145)}\qquad dQ^u = \frac{\text{(space-time density of collisions)}}{\text{(density of target particles)(relative flux density of projectivele particles)}}$$

provided the final states are unpopulated.

For the specified observer the (spatial) density of (i, K_1)-particles is, from (100),

$$\text{(146)}\qquad n_1 = f_i(x, p_1)\, E_1 \pi_i(K_1)\,,$$

where E_1 is the energy of these particles; a similar formula holds for n_2.

The relative velocity of the incident particles, judged by the same observer, is

$$\text{(147)}\qquad |\boldsymbol{v}_2 - \boldsymbol{v}_1| = \frac{|(u\cdot p_1)p_2 - (u\cdot p_2)p_1|}{(u\cdot p_1)(u\cdot p_2)}\,,$$

as is seen by taking $u^a = \delta^a_4$.

Inserting (144), (146), and (147) into (145) one obtains

$$\text{(148)}\qquad dQ^u|(u\cdot p_1)\,p_2 - (u\cdot p_2)\,p_1| = W(ip_1, jp_2; kp_3, \dots, lp_q)\,\pi_k\wedge\dots\wedge\pi_l\,.$$

It follows from (148) that dQ^u *has the same value* dQ *for all observers whose 4-velocites are linear combinations of* p_1 *and* p_2. Since the corresponding frames of reference include the center-of-mass frame of (p_1), (p_2) and the rest frames of p_1 and p_2, *this particular value* dQ *is usually called «the» relativistic cross-section.* One finds from (148)

$$\text{(149)}\qquad [(p_1\cdot p_2)^2 - m_i^2 m_j^2]^{\frac{1}{2}}\, dQ = W(ip_1, jp_2; kp_3, \dots, lp_q)\,\pi_k\wedge\dots\wedge\pi_l\,.$$

Comparing (149) with the formula for dQ derived in (special) relativistic quantum scattering theory (*) one finds for the function R of (143):

$$\text{(150)}\qquad R(ip_1, jp_2; kp_3, \dots, lp_q) = (2\pi)^{3q-4}\left|\langle kp_3, \dots, lp_q|M|ip_1, jp_2\rangle\right|^2.$$

The function M is defined by

$$\text{(151)}\qquad \langle kp_3, \dots, lp_q|S-1|ip_1, jp_2\rangle = \delta(p_1 + p_2 - p_3 - \dots - p_q)\langle\dots|M|\dots\rangle\,,$$

where S is the *scattering operator* and initial and final configurations are ob-

(*) Ref. [23], eq. (3.18).

tained from single-particle states $|ip_r\rangle$ normalized relativistically according to

$$\langle ip_1|jp_2\rangle = E_1\,\delta_{ij}\,\delta(\boldsymbol{p}_1-\boldsymbol{p}_2)\,. \tag{152}$$

If the microscopic collision theory is invariant under the space-time reflection PT, then the symmetry (142) *of W or R*, which was obtained above only in first order of perturbation theory, *follows* rigorously *from* (150) *and* (151) (*).

Since W contains a (four-dimensional) δ-factor, dQ *is effectively a differential form of degree* $3(q-2)-4=3q-10$. Only in the case of absorption or fusion, $q=3$, dQ in (149) is of the form

$$\mathrm{d}Q = \sigma_a(-p_1\cdot p_2)\,\delta\left(\frac{m_i^2+m_j^2-m_k^2}{2}-p_1\cdot p_2\right),$$

and either dQ or the constant σ_a is called the absorption cross-section.

4·11. – In Subsect. **4·4** we deduced from the kinematics of the phase flow that $\int_{K_1} L_j(f_j)\pi_j$ is the space-time density of all collisions in which j-particles with 4-momenta in K_1 are produced. In Subsect. **4·9** we obtained the expression (144) for the contribution to this density due to collisions of a particular kind. If we know all collisions in which j-particles participate, we can combine these two results to obtain the (generalized) Boltzmann equation.

Consider the fairly general case in which *two kinds of collisions* occur between particles of several species $j=1,\ldots,N$: *Binary collisions* (elastic and/or inelastic), and *emission* and *absorption* processes (Fig. 9).

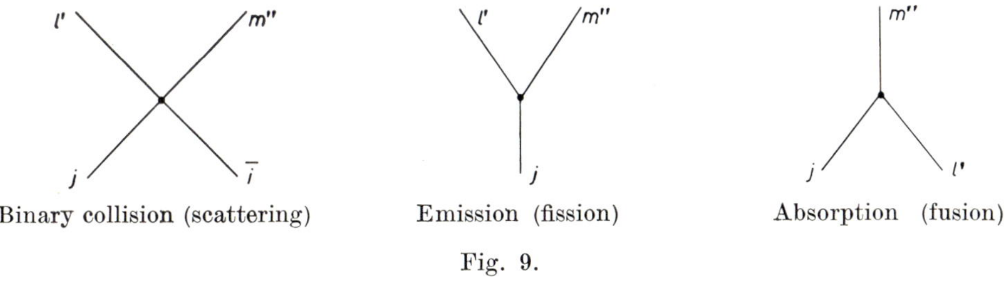

Fig. 9.

To simplify the notation we omit the common argument x and write j instead of (j, p), $\bar{i}$ instead of $(i, \bar{p})$ etc.; *i.e.* we let the index indicate the particles species *and* the four-momentum. Also, we abbreviate $\sum_{l=1}^{N}\int_{P_l}\pi'_l f_l(x, p')\ldots$ by $\int_{l'}$, and we put $\hat{f}_j = f_j^{-1}\pm 1$.

(*) See ref. [23], eq. (2.28).

Then we obtain the following *generalized Boltzmann equation* (*):

$$f_j^{-1}L_j(f_j) = \frac{1}{2}\int\limits_{\bar{i}}\int\limits_{l'}\int\limits_{m''}(\hat{f}_j\hat{f}_{\bar{i}} - \hat{f}_{l'}\hat{f}_{m''})\,W_{j\bar{i};l'm''} + \int\limits_{l'}\int\limits_{m''}(\hat{f}_j\hat{f}_{l'} - \hat{f}_{m''})V_{jl';m''} + \frac{1}{2}\int\limits_{l'}\int\limits_{m''}(\hat{f}_j - \hat{f}_{l'}\hat{f}_{m''})V_{l'm'';j}\,. \tag{153}$$

The symmetry (142) has been used. The factors $\frac{1}{2}$ are necessary since without them the respective collisions would be counted twice.)

If (153) is to be compatible with the field eqs. (124), (125), it must satisfy the two integral conditions (130) and (131), as discussed in Subsect. **4**·7. In our new notation these conditions read

$$\int\limits_j e_j f_j^{-1}L_j(f_j) = 0\,, \qquad \int\limits_j p^a f_j^{-1}L_j(f_j) = 0\,.$$

In view of the Boltzmann eq. (153) and the symmetries of the integrands, the first of these equations can be transformed into

$$\frac{1}{4}\int\limits_j\int\limits_{\bar{i}}\int\limits_{l'}\int\limits_{m''}(e_j + e_{\bar{i}} - e_l - e_m)\hat{f}_j\hat{f}_{\bar{i}}W_{j\bar{i};l'm''} + \frac{1}{2}\int\limits_j\int\limits_{l'}\int\limits_{m''}(e_j + e_l - e_m)(\hat{f}_j\hat{f}_{l'} - \hat{f}_{m''})\,V_{jl';m''}\,.$$

This condition is satisfied for arbitrary distribution functions if and only if

$$\begin{cases} W_{j\bar{i};l'm''} = 0 & \text{if } e_j + e_{\bar{i}} \neq e_l + e_m\,, \\ V_{jl';m''} = 0 & \text{if } e_l + e_l \neq e_m\,, \end{cases} \tag{154}$$

i.e. if the W or R functions or the differential cross-sections obey the charge conservation law, which we will, of course, assume.

Similarly, (131) *is implied by the Boltzmann equation and* (143).

Additional microscopic conservation laws will impose conditions like (143) and (154) on the cross-sections; there will then also be additional « macroscopic » conservation laws like $J^a{}_{;a} = 0$, $T^{ab}{}_{;b} = 0$.

The Einstein-Maxwell field equations (124), (125) *together with a set of Boltzmann equations* (153) *presumably define a deterministic physical system with the*

(*) See ref. [24].

basic variables g_{ab}, F_{ab}, $f_1, \ldots, f_N$. For, if initial data are prescribed on a spacelike hypersurface Σ: $x^4 = \text{const}$ which satisfy the *constraints* $G^4_a + \Lambda\delta^4_a = T^4_a$, $F_{[12,3]} = 0$, $F^{4a}{}_{;a} = J^4$ on Σ, then *the equations*

$$R_{\mu\nu} - \Lambda g_{\mu\nu} = T_{\mu\nu} - \tfrac{1}{2} g_{\mu\nu} T\,,$$

$$F_{[\lambda\mu,4]} = 0\,, \qquad F^{\lambda a}{}_{;a} = J^\lambda\,,$$

$$L_j(f_j) = \ldots \ (\text{see (153)})\,,$$

combined with four *co-ordinate conditions, will uniquely determine the evolution of the system*; *and* since the evolution equations imply, as we have just seen, the relations $J^a{}_{;a} = 0$, $T^{ab}{}_{;b} = 0$, *the constraints will be propagated off* Σ. (Existence and uniqueness, though plausible, have not actually been proven for the above system. See however, ref. [11] for the discussion of the Cauchy problem for relativistic magnetohydrodynamics, and ref. [25] for the relativistic Boltzmann equation in an external gravitational field.)

4·12. – Bound, isolated, macroscopic physical systems like a gas enclosed in a box or, approximately, a star, tend to relax into an equilibrium state. A formal expression of this empirical fact is the *second law of thermodynamics* which we wish to establish now as a consequence of the assumed Boltzmann equation.

Returning for a moment to the quantum model of a gas considered in Subsect. **4·9** we recall that a quantum-statistical ensemble with a *statistical operator* (density matrix) W has an *entropy* $S = -\operatorname{Tr}(W \log W)$. If the ensemble is characterized by a set $\{\overline{n}_\lambda\}$ of average occupation numbers, and if no further information is given about its state, then its statistical operator is found (*) by maximizing S under the conditions $\operatorname{Tr}(WN_\lambda) = \overline{n}_\lambda$, and its entropy is then found to be (**)

$$S[\{\overline{n}_\lambda\}] = -\sum_\lambda \{\overline{n}_\lambda \log \overline{n}_\lambda \mp (1 \pm \overline{n}_\lambda) \log (1 \pm \overline{n}_\lambda)\}\,.$$

Also, the entropy of a mixture is found to be equal to the *sum* of the entropies of its components.

According to the correspondence between the quantum and the classical gas models described in Subsect. **4·8**, we have $\overline{n}_\lambda \approx f_j(x, p_\lambda)$; and the state $|p_\lambda\rangle$ corresponds to a cell $K_\lambda \subset P_j$ whose size is given, in terms of the quantization volume G, by $\sigma_a(G) p^a_\lambda \pi_j(K_\lambda) = 1$.

(*) For a justification of this prescription see, *e.g.*, ref. [26].
(**) Compare, *e.g.*, ref. [51].

Hence, we are led to define the 4-vector field

(155) $$S^a_j(x) := -\int_{P_j(x)} \left(f_j \log f_j \mp (1 \pm f_j) \log (1 \pm f_j)\right) p^a \pi_j$$

as the *entropy flux density of the j-component of the gas,* and

(156) $$S^a := \sum S^a_j$$

as the *total entropy flux density.* The entropy flux through a hypersurface $\Sigma \subset X$ is

(157) $$S[\Sigma] = \int_\Sigma S^a \sigma_a \,.$$

If Σ is a cross-section of the world tube of a gaseous body, then $S[\Sigma]$ is the *total entropy* of that body at the «instant» Σ.

In the classical limit, $f_j \to 0$, (155) reduces to the Boltzmann expression

(158) $$S^a_j = N^a_j - \int_{P_j} p^a f_j \log f_j \pi_j \,. \qquad (f_j \ll 1)$$

We proceed to compute the entropy production density. From (155) we find, using the method used previously to compute divergences of currents (Subsect. **4˙6**):

(159) $$S^a{}_{;a} = \int_j f_j^{-1} \log \hat{f}_j L_j(f_j) \,.$$

If this result is combined with the Boltzmann equation and the symmetries of W and V are used, one obtains

$$S^a{}_{;a} = \frac{1}{8} \int_j \int_i \int_{l'} \int_{m''} \log \left(\frac{\hat{f}_j \hat{f}_i}{\hat{f}_{l'} \hat{f}_{m''}}\right) (\hat{f}_j \hat{f}_i - \hat{f}_{l'} \hat{f}_{m''}) W_{ji;l'm''} + \frac{1}{2} \int_j \int_{l'} \int_{m''} \log \left(\frac{\hat{f}_j \hat{f}_{l'}}{\hat{f}_{m''}}\right) (\hat{f}_j \hat{f}_{l'} - \hat{f}_{m''}) V_{jl';m''} \,.$$

Since $(a-b) \log (a/b) \geqslant 0$ for arbitrary positive numbers a, b, we infer

(161) $$S^a{}_{;a} \geqslant 0 \,.$$

This is the relativistic version of Boltzmann's H-theorem [24 *b*), *c*)]. It implies that the flux of S^a through any closed hypersurface is nonnegative, and for a gaseous body localized in space it implies that *the total entropy never decreases in time* (Fig. 10):

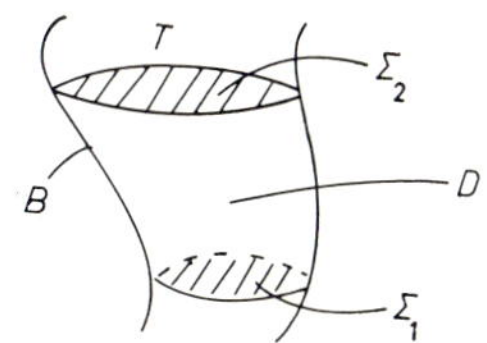

Fig. 10. – World tube T of an isolated body with boundary B and cross-sections Σ_1, Σ_2.

$$\int_{\Sigma_2} S^a\sigma_a - \int_{\Sigma_1} S^a\sigma_a = \int_D S^a{}_{;a}\eta \geqslant 0\,, \tag{162}$$

if Σ_2 is later than Σ_1.

4·13. – A gas with distribution f and fields g_{ab}, F_{ab} is in a *stationary state* in a domain D of space-time if f, g_{ab}, F_{ab} are invariant under a one-dimensional, fixed point free group of transformations of D into itself with timelike orbits.

Let a gas be in a stationary state in D. Assume that the gas occupies a world tube $T\subset D$ which is bounded by a group-invariant hypersurface B (which may be at infinity). Also, let the gas be adiabatically isolated, so that there is no entropy flux through B. Finally, let T have a spacelike cross-section Σ. Then

$$S^a{}_{;a} = 0 \tag{163}$$

in D. For, if g is a transformation of the symmetry group, the entropy flux through $g(\Sigma_1)=\Sigma_2$ equals that through Σ_1 on account of the assumed invariance. Because of the adiabatic isolation of the gas and (161,) (162) shows that the entropy production vanishes between Σ_1 and Σ_2 (see Fig. 10).

Let us now, independently of the assumption of stationarity, *investigate those distributions which have vanishing entropy production at an event* x. From (160) we infer that for those distributions the expressions

$$\log \hat{f}(j) + \log \hat{f}(\bar{i}) - \log \hat{f}(l') - \log \hat{f}(m'') \tag{164}$$

and

$$\log \hat{f}(j) + \log \hat{f}(l') - \log \hat{f}(m'') \tag{165}$$

must vanish for all admissible collisions; and this property is also sufficient for (163). That is, *the entropy production density vanishes at* x *if and only if* $\log(\hat{f}_j)$ *is an additive collision invariant.* If this result is combined with eqs. (159) and (160), a further proposition follows: *The entropy production density vanishes at* x *if and only if, at* x, $L_j(f_j)=0$ *for all* $p\in P_j(x)$. This latter condition expresses *detailed balancing*: the number of particles thrown into the state p at x by collisions equals the number of particles thrown out of that state.

The preceding considerations do not prove, but strongly indicate that an isolated gaseous body tends toward a state where (163) holds. Therefore, we define an *equilibrium distribution* (f_j) as one for which the entropy production density vanishes.

To find the equilibrium distributions we first evaluate (164), (165) at a point x of space-time, and then use $L_j(f_j)=0$ to determine the space-time dependence of the parameters occurring in the equilibrium f_j's.

4·14. – In order to find f_j from (164) and (165), we define (at x) the *collision manifold* for binary collisions to be that submanifold S of the product manifold

$$M = P_j \times P_i \times P_l \times P_m = \{m = (p_1, p_2, p_3, p_4)\}$$

on which $p_1 + p_2 = p_3 + p_4$. The points of S correspond to collisions which satisfy the law of conservation of 4-momentum.

Let us consider *elastic collisions*, so that $m_j = m_l$, $m_i = m_m$.

Any C^1-function h on $P_j \cup P_i$ gives rise to five functions on $M: h_q(m) := h(p_q)$, $q = 1, 2, 3, 4$, and $\Delta h := h_1 + h_2 - h_3 - h_4$. h is called an *additive collision invariant* if Δh vanishes on S. Trivially, the functions

$$h(p) = -\beta_a p^a - \begin{cases} \alpha_j, & \text{if } p \in P_j, \\ \alpha_i, & \text{if } p \in P_i, \end{cases} \tag{166}$$

are additive collision invariants, if β_a and α_j, α_i are six constants (*). *They are, in fact, the only additive collision invariants* [27]. To prove this, we first remark that a tangent vector T of M at a point of S is tangent to S precisely if $\mathrm{d}(\Delta p^a)\cdot T = 0$ for $a = 1, 2, 3, 4$, according to the definition of S. If h is an additive collision invariant, $\Delta h = 0$ on S, hence $\mathrm{d}(\Delta h)\cdot T = 0$ at all points of S and for all tangents T to S. Consequently, $\mathrm{d}(\Delta h) = \beta_a \mathrm{d}(\Delta p^a)$ on S with some C^1-functions β_a defined on S. The last equation implies, since the four arguments p_1, p_2, p_3, p_4 are independent on M:

$$\mathrm{d}h_r = \beta_a \mathrm{d}p^a_r, \qquad r = 1, 2, 3, 4. \tag{167}$$

We wish to show that the β_a's are constants, for then (167) yields the result (166) by integration. We therefore restrict the functions and the d-operator in (167) to S and then write those equations explicitely, using $p^4_r = \sqrt{m_r^2 + \boldsymbol{p}_r^2}$, the independence of p^ν_r on S (for each r), and $h_r(p_1, \dots, p_4) = l^r(\boldsymbol{p}_r)$, obtaining

$$l^r{}_{,\alpha}(\boldsymbol{p}_r) = \beta_\alpha(p_1, \dots, p_4) + \beta_4(p_1, \dots, p_4)\left(\frac{p_{r\alpha}}{p^4_r}\right). \tag{168}$$

(*) For a fixed pair i, j.

Subtracting two such equations (with $s \neq r$) from each other, we get

$$l^r{}_{,\alpha}(\boldsymbol{p}_r) - l^s{}_{,\alpha}(\boldsymbol{p}_s) = \left(\frac{p_{r\alpha}}{p_r^4} - \frac{p_{s\alpha}}{p_s^4}\right) \beta_4(p_1, \ldots, p_4) \, .$$

This equation shows: β_4 is constant on any subset $\{(p_1, \ldots, p_4)\}$ of S whose points have the same « projections » p_r, p_s; *e.g.* $\beta_4(p_1, p_2, p_3, p_4) = \beta_4(p_1, p_2, \overline{p}_3, \overline{p}_4)$. (This follows first for $\boldsymbol{p}_r \wedge \boldsymbol{p}_s \neq 0$, and then generally by continuity.)

Any two elastic collisions $(p_1, \ldots, p_4)$, $(\overline{p}_1, \ldots, \overline{p}_4)$ are contained in the union of four overlapping subsets of S of the type described above. This is indicated by the following sequence, in which any two consecutive members have two arguments in common:

$$(p_1, p_2, p_3, p_4), \quad (p_1, p_2, \overline{p}_1, p_2), \quad (\overline{p}_1, p_2, \overline{p}_1, p_2), \quad (\overline{p}_1, \overline{p}_2, \overline{p}_1, \overline{p}_2), \quad (\overline{p}_1, \overline{p}_2, \overline{p}_3, \overline{p}_4) .$$

Hence, $\beta_4 = \text{const}$ on S. Equation (168) then shows that the functions β_α are even constant on subsets of S whose points have *one* common projection p_r only; thus they are also constant on S.

This finishes the proof concerning additive invariants of elastic binary collisions.

4·15. – After this digression we return to the condition (164) on the distribution functions. If we assume that all elastic collisions between pairs of particles which are permitted by 4-momentum conservation occur with positive probabilities (*i.e.* that the corresponding functions R in (143) are *positive* on S), then it follows from (164) and the theorem just established that $\log \hat{f}_j = -\beta_a p^a - \alpha_j$ $(p \in P_j)$, *i.e.*

$$f_j(x, p) = (\exp[-\alpha_j(x) - \beta_a(x) p^a] \mp 1)^{-1} \, . \tag{169}$$

β_a is independent of j, as established above, whereas the α's may depend on the particle type.

In order that f_j in (169) vanishes at infinity on P_j (for large energies), it is necessary and sufficient that *β^a is a time-like, future-directed vector.* We put

$$\beta^a = \beta u^a \, , \qquad u_a u^a = -1 \, , \qquad \beta > 0 \, , \tag{170}$$

and call the timelike, future-directed unit vector u^a the *mean* 4-*velocity.*

Since $f_j \geqslant 0$ it follows that

$$\alpha_j < \beta m_j \text{ for bosons} \, , \tag{171}$$

and this inequality suffices for $f_j \geqslant 0$. (The requirement $f_j \leqslant 1$ for fermions does not restrict the values of the parameters α_j and β_a.)

If inelastic collisions, emissions and absorption occur, (164) *and* (165) *restrict the scalars* α_j *in* (169) *by the conditions*

$$\alpha_j + \alpha_i = \alpha_l + \alpha_m\,, \qquad \alpha_j + \alpha_l = \alpha_m\,, \tag{172}$$

to be satisfied for all admissible collisions.

As an *illustrative example*, consider a mixture of photons, electrons, hydrogen atoms and ions, and helium atoms and ions (a model stellar atmosphere). Different excited states of bound systems count as different species. Using l to denote the state of ionization and j to label excited states, we have the following list of species: γ, e, H_{lj} ($l=0, 1$), He_{lj} ($l = 0, 1, 2$). Taking into account all charge-conserving binary collisions (ionizations, *e.g.*), and emission and absorption of γ's, one obtains from (172): $\alpha_\gamma = 0$, $\alpha_{\mathrm{H}_{lj}} = \alpha_\mathrm{H} - l\alpha_\mathrm{e}$, $\alpha_{\mathrm{He}_{lj}} = \alpha_\mathrm{He} - l\alpha_\mathrm{e}$. α_e, α_H, α_He are three disposable parameters. Consequently *$\log \hat{f}$, considered at a fixed event x as a function of 4-momentum and species, is a linear combination of the conserved quantities*

4-momentum, charge, « hydrogen number », « helium number »,

with coefficients $-\beta_a$, α_e/e, $-\alpha_\mathrm{H}$, $-\alpha_\mathrm{He}$ ($e =$ charge of proton) which might depend on x. This result contains the law for the distributions of atoms and ions over excited states, that for the relative abundances of atoms and ions (Saha equation), the dependence of the latter on the electron density, and Planck's law for the thermal distribution of photons.

Similar results hold for other mixtures in equilibrium, *e.g.* the (approximate) equilibrium presumed to have existed in the early phases of a big-bang universe. In that case, the species are the « elementary » particles and nuclei formed out of them, and the conserved quantities are 4-momentum, charge, baryon number, e-lepton number, μ-lepton number.

4·16. – The result

$$f_j(x, p) = \left(\exp\left[\beta(x)E - \alpha_j(x)\right] \mp 1\right)^{-1}, \tag{173}$$

of the last Subsection, in which we have put $E = -u_a(x)p^a$, shows that the equilibrium distributions are isotropic in the momentum space of an observer travelling with the 4-velocity u^a. This justifies the name « mean velocity » for u^a, and leads us to call E the « thermal energy » of a particle.

The isotropy of f_j implies that the currents introduced in eqs. (105) and

(108) have the forms

$$N^a_j = n_j u^a \,, \tag{174}$$

$$T^{ab}_j = (\mu_j + p_j)\, u^a u^b + p_j g^{ab} \,. \tag{175}$$

Hence, u^a *is identical with the kinematical as well as the dynamical mean velocity of each component* (Subsect. **4·5**).

The scalars n_j, μ_j, and p_j are the *number density*, *energy density*, and *pressure*, respectively, of the j-component of the gas with respect to a local frame with time axis u^a.

Also, eq. (155) reduces for an equilibrium distribution to

$$S^a_j = s_j u^a \,, \tag{176}$$

where s_j is the *entropy density* of component j.

Evaluation of the quantities n_j, μ_j, p_j, and s_j (in the mean rest frame where $u^a = \delta^a_4$) by means of eqs. (84), (105), (108), (155), (169), (174), (175), and (176) gives, if we consider one component and temporarily omit the index j:

$$n = \frac{r}{2\pi^2}\int_m^\infty \frac{\sqrt{E^2 - m^2}\, E\,\mathrm{d}E}{\exp[-\alpha + \beta E] \mp 1}\,, \tag{177}$$

$$\mu = \frac{r}{2\pi^2}\int_m^\infty \frac{\sqrt{E^2 - m^2}\, E^2\,\mathrm{d}E}{\exp[-\alpha + \beta E] \mp 1}\,, \tag{178}$$

$$p = \frac{r}{6\pi^2}\int_m^\infty \frac{(E^2 - m^2)^{\frac{3}{2}}\,\mathrm{d}E}{\exp[-\alpha + \beta E] \mp 1}\,, \tag{179}$$

$$s = \frac{r}{2\pi^2}\int_m^\infty \left\{\frac{-\alpha + \beta E}{\exp[-\alpha + \beta E] \mp 1} \mp \log\left(1 \mp \exp[\alpha - \beta E]\right)\right\} \cdot \sqrt{E^2 - m^2}\, E\,\mathrm{d}E\,. \tag{180}$$

(As to (180), write $-f\log f \pm (1 \pm f)\log(1 \pm f) =$

$$= f\left(\log(1 \pm f) - \log f\right) \pm \log(1 \pm f) = f\log\left(\frac{1}{f} \pm 1\right) \mp \log\frac{1}{1 \pm f}\Big)\,.$$

These formulae show that *the two parameters* α *and* β *fully determine the thermodynamic state of a simple gas* which is specified by the mass and spin of its particles; the spin determines the sign $\pm$ as well as the degeneracy index r (*).

The *thermodynamical meaning of α and β* can be found as follows: We first observe that

$$s = -\alpha n + \beta\mu \mp \frac{r}{2\pi^2}\int_m^\infty \log\left(1 \mp \exp[\alpha - \beta E]\right)\sqrt{E^2 - m^2}\, E\,\mathrm{d}E\,.$$

We transform the last term by partial integration and recognize, with eq. (179), that

$$s = -\alpha n + \beta\mu + \beta p\,. \tag{181}$$

Next, we use eq. (179) to compute, using again integration by parts,

$$\mathrm{d}p = \frac{n}{\beta}\,\mathrm{d}\alpha - \frac{\mu + p}{\beta}\,\mathrm{d}\beta\,. \tag{182}$$

Combining the last two equations we obtain

$$\mathrm{d}\mu = \beta^{-1}\,\mathrm{d}s + \alpha\beta^{-1}\,\mathrm{d}n\,. \tag{183}$$

According to thermostatics, *$\mu(s, n)$ is a thermodynamic potential*, and the relation $\mathrm{d}\mu = T\,\mathrm{d}s + \tilde{\mu}\,\mathrm{d}n$ is one way of characterizing the temperature T and the chemical potential (per particle), $\tilde{\mu}$. We therefore conclude that

$$\beta = \frac{1}{T}\,, \qquad \alpha = \beta\tilde{\mu} = \frac{\tilde{\mu}}{T}\,; \tag{184}$$

i.e. β is the inverse temperature, and *α is the ratio* (*chemical potential/temperature*). (Equation (171) implies: $\tilde{\mu} \leqslant m$ for bosons.)

Equation (119), when rewritten as

$$\mu + p = Ts + \tilde{\mu} n\,, \tag{185}$$

is a well-known thermostatic identity.

The preceding relations hold for each component of a mixture. As an example for the analogue of the Gibbs relation (183) for a heterogeneous system, consider again the mixture of Subsect. **4**·15. Summing the relations (183)

(*) For details concerning the evaluation of these formulae see ref. [52].

over all components and using the expressions for the α's one gets

$$\beta d\mu - ds = -\frac{\alpha_e}{e} d\varrho + \alpha_H dn_H + \alpha_{He} dn_{He}, \tag{186}$$

where ϱ is the total charge density and n_H, n_{He} are the number densities of H and He particles, respectively, including all ionization and excitation states. $-\alpha_e/e_\beta$ is the electrochemical potential (per unit charge), and α_H/β, α_{He}/β are the chemical potentials (per particle) for hydrogen and helium, respectively. The analogue of (185) is

$$\mu + p = T\left(s - \frac{\alpha_e}{e}\varrho + \alpha_H n_H + \alpha_{He} n_{He}\right). \tag{187}$$

(For a charged system, the total energy density is, of course, the sum of the particle contribution considered in (186), (187) and the field contribution $T_{Mab} u^a u^b$.)

Similar formulae govern the thermostatics of other mixtures; instead of ϱ, n_H, n_{He} the densities of a set of independent, conserved quantities will appear.

4˙17. – Having studied some consequences of the form of the equilibrium distributions at an event, we now consider the *space-time dependence of the parameters* α *and* β_a in (169). According to the last remark of Subsect. **4˙13** this dependence is determined by the requirement $L(f) = 0$ for each component. This equation demands that the function $\alpha + \beta_a p^a$ be constant on any particle trajectory. That is, we must have $(\alpha_{,a} + e\beta_b F^b{}_a)p^a + \beta_{a;b} p^a p^b = 0$ for all 4-momenta satisfying $g_{ab} p^a p^b = -m^2$. We split p^a according to

$$p^a = Eu^a + \pi e^a, \qquad u_a e^a = 0, \qquad e_a e^a = 1, \qquad E^2 - \pi^2 = m^2, \tag{188}$$

and obtain, with $A_a := \alpha_{,a} + e\beta_b F^b{}_a$:

$$A_a(Eu^a + \pi e^a) + \beta_{(a;b)}(Eu^a + \pi e^a)(Eu^b + \pi e^b) = 0.$$

For fixed x, E, and π, the left-hand side is a function on the unit sphere $\{e^a\}$ which may be decomposed into spherical harmonics of degrees 0, 1, and 2. Since spherical harmonics of different degrees are linearly independent, it follows that

$$\left\{\begin{aligned} &E^2\beta_{a;b} u^a u^b + A_b u^b E + \tfrac{1}{3}(E^2 - m^2) h^{ab}\beta_{a;b} = 0, \\ &(A_a + 2Eu^b\beta_{(b;a)}) h^{ac} = 0, \\ &h^{ac}h^{bd}\beta_{c;d} = \tfrac{1}{3} h^{ab}h^{cd}\beta_{c;d}. \end{aligned}\right. \tag{189}$$

Here

$$h^a_b = \delta^a_b + u^a u_b \tag{190}$$

is the tensor which projects 4-vectors into the 3(vector) space orthogonal to u^a.

Since these equations hold, at a given event, for all energies, we get

$$\left\{\begin{array}{lll} m h^{ab}\beta_{a;b} = 0\,, & A_a u^a = 0\,, & \beta_{a;b} u^a u^b = -\tfrac{1}{3} h^{ab}\beta_{a;b}\,, \\ A_a h^{ab} = 0\,, & u^a \beta_{(a;b)} h^{bc} = 0\,. & \end{array}\right. \tag{191}$$

Consequently,

$$A_a = \alpha_{,a} + e\beta_b F^b{}_a = 0\,, \tag{192}$$

and with (189) and the identity

$$\beta_{(a;b)} = \beta_{(c;d)}\,\delta^c_a\,\delta^d_b = \beta_{(c;d)}(h^c_a - u^c u_a)(h^d_b - u^d u_b):$$

$$\beta_{(a;b)} = \lambda g_{ab}\,;\ \lambda = 0 \text{ if } m > 0\,. \tag{193}$$

We summarize the results in the following *theorem* [24 *b*)]. A gas is in equilibrium in a space-time domain D if and only if:

a) its distribution function is of the form (169) over D,

b) $\beta^a\left(= \dfrac{u^a}{T}\right)$ is a $\left\{\begin{array}{l}\text{Killing} \\ \text{conformal Killing}\end{array}\right\}$ vector in D, if $\left\{\begin{array}{l} m > 0 \\ m \geqslant 0 \end{array}\right\}$,

c) α is, in D, related to the electric field intensity

$$E_a := F_{ab} u^b \tag{194}$$

and the temperature T by

$$T\,\mathrm{d}\alpha = eE\,. \tag{195}$$

We consider several *consequences* of this theorem.

If $m > 0$ (or if this holds for one component of a mixture), space-time must be *stationary* in D in order that an equilibrium can exist there.

In a stationary space-time the quantity U defined by

$$U := \tfrac{1}{2}\log(-\xi\cdot\xi)\,, \tag{196}$$

where ξ is a dimensionless timelike Killing vector, has several properties in common with the sum of the Newtonian gravitational potential and the centrifugal potential, if the frame of reference defined by the ξ-orbits is considered as corresponding to a Newtonian rotating frame. If, in an equilibrium state, there is only one timelike Killing vector, except for a factor of proportionality (and this will be the generic case), then we must have $T_0\beta^a = \xi^a$ with a constant T_0, and we obtain *Tolman's relation*

$$e^U = \frac{T_0}{T} \tag{197}$$

between potential and temperature.

If there is no (macroscopic) electromagnetic field or *if* at least $E = 0$—as in an infinitely conducting plasma—, then α *is a constant* in an equilibrium domain. Hence, from (184), *each chemical potential* $\tilde{\mu}$ *depends on the gravitational potential like the temperature.*

In the general case with $m > 0$, the stationarity of space-time, Einstein's field equation and (192) imply that the quantity α, the temperature T, the mean 4-velocity u^a, and the kinetic and field stress-energy tensors are invariant under the group which expresses the stationarity. If the additional assumption is made that F_{ab} itself is invariant (*), one can also find an invariant 4-potential A_a, and then one deduces from (192) that $\alpha - e\beta^a A_a = \gamma = \text{const}$, *i.e.* one has

$$f(x, p) = (\exp[-\gamma - \beta_a(p^a + eA^a)] \mp 1)^{-1} \tag{198}$$

with a constant γ.

It is instructive to derive the *nonrelativistic limit* of eq. (198). For this purpose we re-introduce temporarily the velocity of light c, maintaining our convention $k = 1$. Specializing to a static, asymptotically flat space-time we recall that the metric can be written such that

$$g_{\lambda\mu} = \delta_{\lambda\mu} + \mathcal{O}(c^{-2}), \qquad g_{\lambda 4} = 0, \qquad g_{44} = -1 - \frac{2U}{c^2} + \mathcal{O}(c^{-4}), \tag{199}$$

where $U(x^\lambda)$ is the scalar potential defined in (196), since the time-translating Killing vector ξ is chosen as $\partial/\partial x^4$. Then

$$\beta^a = \frac{1}{T_0}\,\delta^a_4\,. \tag{200}$$

By definition

$$p^a = mc\,\frac{dt}{ds}\,\frac{dx^a}{dt}\,, \tag{201}$$

(*) This is almost, but not quite implied by the invariance of T^{ab}_M.

where t is defined by $ct = x^4$, and s is proper time. If we put $\bar{\gamma} = \gamma - mc^2/T_0$ we find from the preceding three equations that

$$\lim_{c\to\infty} (\gamma + \beta_a(p^a + eA^a)) = \bar{\gamma} - \frac{1}{T_0}\left(\frac{mv^2}{2} + mU + e\phi\right), \tag{202}$$

where $\phi = A^4$ is the electrostatic potential and v the Newtonian speed of the particle. (In the limit $c \to \infty$, (x^λ, t) are inertial co-ordinates.) Since Tolman's law gives

$$\lim_{c\to\infty} T = T_0 , \tag{203}$$

we obtain the familiar formulae for equilibrium in a static gravitational and electric field.

In a similar way, one can pass to the nonrelativistic limit in all formulae we have deiived.

4·18. – A strict equilibrium solution is rarely a good model of a real situation. Much more frequently one wishes to describe *quasi-equilibria*, *i.e.* states which deviate slightly from equilibrium. Then the formulae of Subsect. **4·15** and **4·17** have to be amended by « transport terms ».

A delicate point is the *definition of a mean 4-velocity u^a in a nonequilibrium state.* We have seen in Subsect. **4·5** that any distribution function defines a kinematic and a dynamic mean velocity and, particularly for a mixture, there are even more reasonable possibilities. Whereas in equilibrium these mean velocities coincide (Subsect. **4·16**), there is no reason that they should do so off equilibrium. So, a choice has to be made.

Once a mean velocity u^a has been chosen, one can decompose $N_j{}^a$, S^a, T^{ab} (and higher order moments of f_j) uniquely:

$$\left\{\begin{aligned} &N_j{}^a = n_j u^a + i_j{}^a , \qquad S^a = su^a + s^a , \\ &T^{ab} = \mu u^a u^b + 2u^{(a}q^{b)} + ph^{ab} + \pi^{ab} , \\ &\text{where } u_a q^a = u_a \pi^{ab} = \pi^a{}_a = 0 . \end{aligned}\right. \tag{204}$$

(h_{ab} is defined in eq. (190).) In accordance with eq. (21), these quantities are called:

n_j: mean particle density (of component j),

i_j^a: diffusion current (of component j),

s: mean entropy density,

s^a: entropy diffusion current,

μ: mean energy density,

q^a: heat flow,

p: mean kinetic pressure,

π^{ab}: shear stress.

(All these quantities depend on u^a.)

One would like to choose u^a such that the thermodynamic scalars in (204) obey, at least approximately, *equations of state*, and such that the other quantities obey simple *transport equations*. In simple cases these equations of irreversible thermohydrodynamics should define, together with conservation laws like (119) and the Einstein field equation, a deterministic physical system.

Within relativistic kinetic theory of gases, approximation methods for quasi-equilibria seem to have been devised so far only for a simple Boltzmann gas with elastic binary collisions. It is assumed that there exists a *local equilibrium distribution*

$$g = \exp\left[\alpha(x) + \beta_a(x)\, p^a\right] \tag{205}$$

such that $f/g - 1$ is small in the sense that physically important moments like N^a, T^{ab} are nearly the same for g as for the actual distribution f. One can partly or fully determine g in terms of f by requiring certain moments to be exactly the same for g and f; such conditions will be called *matching conditions*.

The choice of matching conditions is a matter of convenience, dictated by the same considerations as the choice of u^a. In order to avoid confusion, one should express physically important quantities in principle in terms of f, and not in terms of the auxiliary function g.

Israel (*) has adapted the *Chapman-Enskog method* to the relativistic case. For the sake of simplicity, I shall specialize his matching conditions to the (« Eckart »)-form

$$N^a = N^a_0\,, \qquad (T^{ab} - T_0^{\,ab})\, u_a u_b = 0\,, \tag{206}$$

where u^a is the kinematic mean velocity; a subscript 0 indicates a quantity computed with g. The Boltzmann equation

$$L_m(f_1) = \tfrac{1}{2}\int (f_3 f_4 - f_1 f_2)\, W_{12;34}\, \pi_2 \wedge \pi_3 \wedge \pi_4 \tag{207}$$

is *linearized* by inserting $f = g(1 + h)$ and neglecting nonlinear terms in h, and omitting on the left-hand side even h itself and its derivatives. The mo-

(*) See [24*d*)] and [53].

mentum-dependence of the solutions $h(x, p)$ of the resulting linear inhomogeneous integral equation

$$L_m(g_1) = \tfrac{1}{2} g_1 \int g_2(h_3 + h_4 - h_1 - h_2)\, W_{12;34}\, \pi_2 \wedge \pi_3 \wedge \pi_4 \tag{208}$$

is then determined; ISRAEL finds that h can be expressed in terms of the parameter functions α, β, u_a of g, the vector $h^a{}_b p^b$, and three functions of the energy $u_a p^a$ and the temperature β^{-1} which themselves have to satisfy integral equations similar to (208). From h the quantities defined in (204) can be computed. The result is

$$\left\{\begin{aligned} & p = p_{\text{th}} + \bar{p}\,, && \bar{p} = -\xi\theta\,, \\ & q^a = \lambda h^{ab}\left(\frac{\alpha}{m}\right)_{,b}\,, && \pi^{ab} = -2\eta\sigma^{ab}\,, \\ & p_{\text{th}} = p_{\text{th}}(\mu, \varrho)\,, && N^a = n u^a\,, \qquad \varrho = mn\,. \end{aligned}\right. \tag{209}$$

θ and σ^{ab} are the *dilation rate* and the *rate of shear* associated with u_a,

$$\theta = u^a{}_{;a}\,, \qquad \sigma_{ab} = h^c_a h^d_b u_{(c;d)} - \tfrac{1}{3}\theta h_{ab}\,. \tag{210}$$

p_{th}, the *thermostatic pressure*, is computed from μ and ϱ as in equilibrium. It differs from the *kinetic pressure* p by a viscous pressure $\bar{p}$. The coefficients ζ, λ, η are *positive*. Their computation is tedious; it has been performed (for ζ and λ) by ISRAEL for a special « Maxwellian » cross-section.

The result $\zeta > 0$ means that *a relativistic gas has a bulk viscosity*; this may be taken to « explain » the result of Subsect. **4**˙17 that a relativistic equilibrium distribution is incompatible with nonrigid mean motion.

The *heat conduction law* can be transformed (in sufficient approximation) into the more familiar form

$$q^a = -\bar{\lambda} h^{ab}(T_{,b} + T\dot{u}_b) \tag{211}$$

($\dot{u}_a = u_{a;b} u^b =$ mean acceleration) by using (182) and the relativistic Euler equation $h^a_b T^{bc}{}_{;c} = 0$. *These results establish the relativistic theory of a linearly viscous fluid* (gas) *and contain the irreversible thermodynamics of a simple gas.*

ANDERSON and STEWART have developed a *relativistic version of Grad's method of moments* (*). This method is described separately by STEWART in this volume, so I only mention, for the sake of comparison, that these authors

(*) See [54] and MARLE [53].

also adopt the first of the two matching conditions (206) and use the kinematic mean velocity, too, but take as their last matching condition $T^a{}_a = T_0{}^a{}_a$. They also obtain the results (209), (211). One advantage of their method is that *the transport coefficients are obtained directly as* certain *integrals*; no integral equations have to be solved.

No treatment of multicomponent gases seems to exist until now, and no inelastic processes have been included.

If one accepts eqs. (204) and an equation of state

$$\mu = F(s, n_j) \tag{212}$$

for a relativistic fluid independently of kinetic theory, one can use $T^{ab}{}_{;b} = 0$, $B^a{}_{;a} = 0$ (eq. (119)) and possibly further conservation laws to compute the entropy production density $S^a{}_{;a}$. The resulting expression suggests, in the manner of irreversible phenomenological thermodynamics, transport equations like (209), with additional diffusion laws, reaction laws, and couplings between the various processes. *One thus obtains*, in fact, *the scheme of relativistic thermo-hydrodynamics due to* STUECKELBERG *and* WANDERS [28]. This scheme is more general than kinetic theory, but also less definite, since one can say almost nothing about the transport coefficients. One can, however, define equilibrium solutions by $S^a{}_{;a} = 0$, and thus obtain some of the results of Subsect. **4'17**, *e.g.* (197) and the statements about the mean motion, independently of kinetic theory.

Sometimes it is useful to *combine hydrodynamics and kinetic theory*. Thus one may describe the stellar plasma as a fluid and the radiation traversing it as a photon gas. We shall not describe this method here, however.

4'19. – This Subsection is devoted to a few simple *applications of kinetic theory to photons*, and serves to hint at some more complicated applications the details of which can be found in the cited papers.

From the defining equation (100) for the distribution function, the formula (84) for the measure π_0 and Planck's formula $2\pi\nu = -u_a p^a$ one obtains the *relation*

$$I_\nu = 4\pi\nu^3 f \tag{213}$$

between the distribution function f and the specific intensity I_ν measured by an observer with 4-velocity u^a who « sees » frequency ν. Since f is observer-independent, so is I_ν/ν^3; this result contains several kinematic effects important in cosmology.

If the photons do not interact with matter between source S and observer O,

Liouville's equationss (104) and (213) give the relation

$$I_{\nu_0} = \frac{I_{\nu_s}}{(1+z)^3} \tag{214}$$

between I_ν near the source and at the observer, $z = (\lambda_0 - \lambda_s)/\lambda_s$ being the red-shift. This equation is basic for the derivation of the *m, z relation* and the theory of *angular sizes* (half power diameters) of galaxies not only in Robertson-Walker universes, but in general models. Using the Boltzmann rather than the Liouville equation, or an approximation to it (*equation of radiative transport*, see [29]), one can derive *scattering and absorption corrections* to eq. (214).

If the radiation is *thermal*, with a temperature T and a mean velocity u^a, we infer from (169) (with $\alpha = 0$) and (213) that *an* (arbitrary) *observer will in each direction measure a Planckian energy distribution* with an effective temperature T_e depending on the velocity v which the observer has relative to the radiation and the angle ϑ between the direction of observation and the direction with which he moves through the radiation field:

$$T_e = T\,\frac{\sqrt{1-v^2}}{1-v\cos\theta}\,. \tag{215}$$

This effect is now being used to find the velocity of the Earth relative to the 3 °K fire-ball radiation (*).

If one assumes that the 3 °K radiation was emitted thermally from the « recombination hypersurface » $T \approx 3000$ °K, one obtains from (214) the observed intensity distribution in each direction in an arbitrary model universe, provided one can compute z from the null geodesics. This idea was used by SACHS and WOLFE [30] to estimate the *influence of material lumps* (super-clusters) *on the radiation*. The same method yields the *optical appearance of a collapsing star* for a distant observer [31].

The power of Liouville's theorem is further illustrated by the following theorem (proven in [32]): *A collisionless photon gas with a distribution function which is isotropic with respect to a geodesic mean 4-velocity field* u^a, *i.e.*, $f(x^a, p^a) = g(x^a, u_b(x)\,p^b)$, $\dot u_a = 0$, *can exist only in a Robertson-Walker space-time.* Since the microwave background radiation is observed to be highly isotropic, *this theorem gives a much better motivation for using such models than arguments based on* still rather poor *galaxy counts* [33].

4·20. – Finally, I should like to make a few remarks about *solutions of the coupled Einstein-Liouville equations* (124), (104) (with $F_{ab} = 0$).

Very few exact solutions are known so far.

(*) See the lectures of SCIAMA in this volume.

There is, firstly, an old *model of Einstein*'s in which particles move in concentric circular orbits under the influence of their joint mean field [34].

Much more general *static, spherically symmetrical models of* such *star clusters* have been constructed by ZEL'DOWICH and PODURETS [35], and by FACKERELL [36].

For the *construction of solutions of the Liouville equation* it is useful to remember that, in the case of neutral particles, *f solves this equation if and only if it is constant on all geodesics.* Now, if ξ^a is a Killing vector, then $\xi_a(x)\,p^a$ is constant along geodesics, since

$$\frac{\mathrm{D}}{\mathrm{d}v}(\xi_a p^a) = \xi_{a;b}p^b p^a = \xi_{(a;b)}p^a p^b = 0\,.$$

Hence, *if a space-time has symmetries,* generated by Killing vectors ξ^a_A $(A = 1, \ldots, r)$, *then any positive function g of r variables defines a solution* $f(x, p) = g(\xi_A(x)\,p)$ *of the Liouville equation.*

Suppose a symmetry mapping is given by

$$\bar{x}^a = i^a(x^b)\,, \qquad \bar{p}^a = \frac{\partial i^a}{\partial x^b}\,p^b\,, \qquad \begin{pmatrix} \bar{x} = i(x)\,, \\ \bar{p} = \mathrm{d}i(p)\,. \end{pmatrix}.$$

A distribution function f is *invariant under* i if $f(i(x), \mathrm{d}i(p)) = f(x, p)$. If this is to hold for all symmetry mappings of the group generated by the ξ_A's, consideration of the *infinitesimal transformations* yields

$$\frac{\partial f}{\partial x^a}\,\xi^a + \frac{\partial f}{\partial p^\lambda}\,\xi^\lambda{}_{,a}p^a = 0\,. \tag{216}$$

For solutions of the special kind $f = g(\ldots)$ considered above the last equation is equivalent to

$$\frac{\partial g}{\partial y_A}\,C^D_{AB}\,y_D = 0\,, \tag{217}$$

as is seen by inserting $f = g(\ldots)$ into (216) and using the Lie-commutation relations

$$[\xi_A, \xi_B] = C^D_{AB}\,\xi_D \tag{218}$$

for the group generators. (C^D_{AB} are the structure constants of the group.) Equations (217), being linear, can be solved for given C's, and thus one obtains solutions of the Liouville equation invariant under groups of space-time isometries. The distribution functions in the solutions referred to above are of this type. This method corresponds to the construction of solutions of Liouville's equation in ordinary Newtonian mechanics by means of constants of the motion associated with symmetries.

Investigations on the *stability* of such relativistic star clusters, performed by IPSER and THORNE, are dealt with in the lecture by IPSER.

A rederivation and new characterization of Robertson-Walker models has been given in ref. [32]. In fact, these models have been shown to be *the only solutions of the Einstein-Liouville equations which have a locally isotropic distribution function, are nonstationary, and are populated by particles with* $m > 0$. In the case of particles with vanishing mass such a solution also exists (Tolman universe), but it is still not known whether there exist, in addition to these, $m = 0$–solutions with isotropic distributions and *with rotation*, although RIENSTRA [37] has shown that this is not the case under various special kinematical conditions.

The *kinetic Robertson-Walker solutions* are given by the usual metric,

$$\mathrm{d}s^2 = R^2(t)\,\mathrm{d}\sigma^2 - \mathrm{d}t^2\,, \tag{219}$$

and a distribution function

$$f(x, p) = g\big(R^2(t)\,h_{ab}\,p^a p^b\big) \tag{220}$$

with h_{ab} as in (190) and $u = \partial/\partial t$. (This f is not of the form discussed above.) The evolution of the model is given by

$$t = \frac{\sqrt{3}}{2}\int_0^{R^2}\left(-3ky + \int_0^{\infty} x^2 g(x^2)\sqrt{m^2 + y^2}\,\mathrm{d}x\right)^{-\frac{1}{2}}\mathrm{d}y\,. \tag{221}$$

$m = 0$ gives the (Tolman) *radiation universe*, and

$$g(x^2) = \frac{4M}{m}\,\frac{\delta(x^2)}{x} \qquad (m > 0) \tag{222}$$

gives the (Friedmann) *dust universe* with mass constant M. $k = \pm 1$, 0 is the *curvature index* of $\mathrm{d}\sigma^2$. In the general case $m > 0$, those models which expand indefinitely start out at $t = 0$ like a radiation universe and go over asymptotically into a dust universe for $t \to \infty$, with a smooth transition in between.

APPENDIX I

Example of a 4-manifold which is not time-orientable.

Let X be Minkowski space. The transformations

$$\bar{x}^\lambda = x^\lambda + n^\lambda\,, \qquad \bar{x}^4 = (-1)^{n^1} x^4\,, \qquad \lambda = 1, 2, 3,\ n^\lambda \text{ integers},$$

leave $\mathrm{d}x^\lambda \mathrm{d}x^\lambda - (\mathrm{d}x^4)^2$ invariant, hence are isometries. As the n^λ vary, we obtain a group G of isometries. G is free of fixed points and is properly discrete; *i.e.* if we apply to a point p all the mappings of G except the identity, then the images of p all lie outside of some neighbourhood of p. Identify each point with all of its images under G, obtaining the quotient manifold $\hat{X} = X/G$ of X modulo G. Continuous vector fields on $\hat{X}$ correspond to continuous vector fields on X which are invariant with respect to G. Time-orientability is equivalent to the existence of a continuous timelike vector field. We now show that $\hat{X}$ is not time-orientable. For if it were, let $\hat{u}$ be a timelike continuous vector field of $\hat{X}$ and let u be the corresponding continuous vector field on X. Suppose $u^4(0, 0, 0, 0) > 0$. Then

$$u^4(1, 0, 0, 0) = -u^4(0, 0, 0, 0) < 0\,.$$

Since u is continuous, u^4 must vanish somewhere. Thus u is not timelike everywhere.

This example originated in a discussion between TRAUTMAN and SCHÜCKING. For quotient spaces see F. KLEIN, *Lectures on Non-Euclidean Geometry*, Berlin, 1926.

APPENDIX II

Geometrical construction of the mass-shell volume element π_m.

Let v_1, v_2, v_3 be linearly independent tangent vectors to $P_m(q)$ at p. Then the map $t \to \pi(t, v_1, v_2, v_3)$ is a linear form in t (t an arbitrary vector of T_q) which vanishes if and only if t is tangent to $P_m(q)$ at p. Hence this form must be proportional to the form $t \to p \cdot t$, with a proportionality factor depending on v_1, v_2, v_3 in the manner of a 3-form. Thus, the equation

$$t \cdot \pi|_{P_m(q)} = (t \cdot p)\pi_m$$

(the bar | indicates restriction to $P_m(q)$) defines a 3-form π_m on $P_m(q)$. Explicitely, the last equation reads

$$\sqrt{-g}(t^1\,\mathrm{d}p^{234}-t^2\,\mathrm{d}p^{341}+t^3\,\mathrm{d}p^{412}-t^4\,\mathrm{d}p^{123})|_{P_m(q)}=(t^a p_a)\pi_m\,.$$

Since t is arbitrary, we can take $t^a=\delta^a_4$ to obtain (82) of Subsect. **3**.7.

REFERENCES

[1] P. G. ROLL, R. KROTKOV and R. H. DICKE: *Ann. of Phys.*, **26**, 442 (1964). See also the review by B. BERTOTTI, D. BRILL and R. KROTKOV, chap. 1 in ref. [10], and ref. [38].

[2] WEYL has particularly stressed that only the totality of physical and space-time geometrical laws can be empirically verified. If one wants to construct Physics *ab* initio, the first laws of Physics and of spacetime geometry, and their motivations, cannot be separated. This speaks, in principle, against the flat-space approaches where Riemannian geometry is introduced, so to speak, as an afterthought. See particularly [8], Ch. I, Sect. **9**. Presently, the flat-space approaches are nevertheless important, particularly as a means to clarify the relation between Einstein's general theory and quantum physics. See [9] and the references cited there.

[3] Orientability of X is often not assumed. We do assume it in order to have available the co-ordinate–independent forms, or volume-elements, η and σ defined in Subsect. **3**.6, which are frequently used in kinetic theory (Sect. **3**). Orientability of spacetime follows from the observed violation of C-invariance, if CPT-invariance is assumed.

[4] We do not specify the differentiability class here. The reader may think of a C^∞-manifold always, or count the number of differentiations needed in a particular context.

[5] Time-orientation will be needed later in kinetic theory since « in » and « out » states enter into the Boltzmann collision integral (144) in a noninterchangeable way. Time-orientability of spacetime can be inferred from the experimentally established fact that some weak interactions violate CP-symmetry, but obey CPT.

[6] A very useful reference for the differential-geometrical concepts needed in general relativity is N. J. HICKS: *Notes on Differential Geometry* (Princeton, N. J., 1965).

[7] The idea of this « derivation » of Einstein's equation is due to F. A. E. PIRANI: *Phys. Rev.*, **105**, 1089 (1957).

[8] H. WEYL: *Space, Time, Matter* (London, 1922).

[9] W. E. THIRRING: *Ann. of Phys.*, **16**, 96 (1961).

[10] L. WITTEN (Ed.): *Gravitation: An Introduction to Current Research* (New York, 1962).

[11] A. LICHNEROWICZ: *Relativistic Hydrodynamics and Magnetohydrodynamics* (New York, 1967).

[12] *Recent Developments in General Relativity* (New York, 1962).

[13] W. G. DIXON: *Nuovo Cimento*, **34**, 317 (1964). See also a forthcoming paper by the same author in *Proc. Roy. Soc.*

[14] For a proof see, *e.g.*, J. L. SYNGE: *Relativity: The General Theory* (Amsterdam, 1960).
[15] R. H. DICKE and H. M. GOLDENBERG: *Phys. Rev. Lett.*, **18**, 313 (1967).
[16] M. SPIVAK: *Calculus on Manifolds* (New York, 1965).
[17] See J. L. SYNGE: *The Relativistic Gas* (Amsterdam, 1957).
[18] This elegant way of introducing the measures Ω_j, ω_j and the characterization of ω_j which follows, have been given by K. BICHTELER in his unpublished Ph. D.-thesis, Hamburg University (1965).
[19] Equation (104) for $F_{ab}=0$ has first been established by A. G. WALKER: *Proc. Edinb. Math. Soc.*, **4**, 238 (1936).
[20] A. H. TAUB: *Phys. Rev.*, **74**, 328 (1948).
[21] Henceforth the formulae are written for particles obeying quantum statistics. The corresponding formulae for classical particles can formally be obtained by taking $f \ll 1$, as in the transition (155)$\to$(158).
[22] For the following compare, *e.g.*, S. S. SCHWEBER: *An Introduction to Relativistic Quantum Field Theory* (New York, 1962).
[23] W. BRENIG and R. HAAG: *General Quantum Theory of Collision Processes, Fortschr. d. Phys.*, **7**, 183 (1959); reprinted in M. ROSS (Ed.): *Quantum Scattering Theory* (Bloomington, 1963).
[24] *a*) A. LICHNEROWICZ and R. MARROT: *Compt. Rend.*, **210**, 759 (1940); *b*) G. E. TAUBER and J. W. WEINBERG: *Phys. Rev.*, **122**, 1342 (1961); *c*) J. EHLERS: *Akad. Wiss. Mainz, Abh. math.-nat. Kl.*, Nr. 11 (1961); *d*) W. ISRAEL: *Journ. Math. Phys.*, **4**, 1163 (1963); *e*) N. A. CHERNIKOV: *Acta Phys. Polon.*, **23**, 629 (1963); *Dokl. Akad. Nauk SSSR*, **144**, 89, 314, 544 (1962) (reprinted in *Sov. Phys. Doklady*, **7**, 397, 414, 428 (1962)).
[25] K. BICHTELER: *Commun. Math. Phys.*, **4**, 352 (1967).
[26] E. T. JAYNES: *Phys. Rev.*, **106**, 620 (1957).
[27] *a*) N. A. CHERNIKOV: *Acta Phys. Polon.*, **26**, 1069 (1964); *b*) K. BICHTELER: *Zeits. Phys.*, **182**, 521 (1965); *c*) R. H. BOYER: *Amer. Journ. Phys.*, **33**, 910 (1965); *d*) C. MARLE: *Ann. Inst. Henri Poincaré*, A **10**, 67 (1969).
[28] E. C. G. STUECKELBERG and G. WANDERS: *Helv. Phys. Acta*, **26**, 307 (1956).
[29] R. W. LINDQUIST: *Ann. of Phys.*, **37**, 487 (1966).
[30] R. K. SACHS and A. M. WOLFE: *Astrophys. Journ.*, **147**, 73 (1967).
[31] W. L. AMES and K. S. THORNE: *Astrophys. Journ.*, **151**, 659 (1968).
[32] J. EHLERS, P. GEREN and R. K. SACHS: *Journ. Math. Phys.*, **9**, 1344 (1968).
[33] W. KUNDT: in G. HÖHLER (Ed.), *Springer Tracts in Modern Physics*, **47**, 111 (1968).
[34] A. EINSTEIN: *Ann. of Math.*, **40**, 922 (1939).
[35] YA. B. ZEL'DOVICH and M. A. PODURETS: *Astr. Žurn.*, **42**, 963 (1965) (Engl. transl. in *Sov. Astronomy-AJ*, **9**, 742 (1966)).
[36] E. D. FACKERELL: unpublished Ph. D. thesis, University of Sydney (1966).
[37] W. RIENSTRA: unpublished Ph. D. thesis at the University of Texas at Austin (1969).
[38] R. H. DICKE: *The Theoretical Significance of Experimental Relativity* (New York, 1964).
[39] For critical remarks concerning this assumption, see R. PENROSE: *An Analysis of the Structure of Space-Time*, Adams Prize essay, distributed by Princeton University (1966).
[40] See R. P. GEROCH: *Singularities in the space-time of general relativity: their definition, existence and local characterisation*, Ph. D. Thesis (Princeton, 1967).
[41] H. WEYL: ref. [8]. The proof is simple: Knowing the timelike geodesics through p

means to know the null cone, hence, the metric g except for a conformal factor. It is easy to check that geodesics are preserved under $g_{ab} \to f\, g_{ab}$ if and only if $f = \text{const}$. Thought experiments designed to measure g_{ab} without standard clocks or rulers have been devised by B. HOFFMANN and W. KUNDT, see ref. [12], and by R. F. MARZKE, see ref. [42].

[42] H. CHIU and W. HOFFMANN (Ed.): *Gravitation and Relativity* (New York, 1964).

[43] See the lectures of G. ELLIS in this volume.

[44] « *Reduction Theorem* », see *e.g.*, J. SCHOUTEN: *Ricci-Calculus* (Berlin, 1954).

[45] D. LOVELOCK: to appear in *Aequationes Mathematicae* (Waterloo).

[46] (Nonrelativistic) classical and quantum-statistical mechanics often give nearly the same results. For a derivation of the Boltzmann equation from quantum-statistical mechanics, see L. P. KADANOFF and G. BAYM: *Quantum-Statistical Mechanics* (New York, 1962).

[47] See J. L. SYNGE: *Relativity: The Special Theory* (Amsterdam, 1956), p. 292.

[48] See ref. [39, 40]; and S. W. HAWKING: *Proc. Roy. Soc.*, A **294**, 511 (1966); A **295**, 490 (1966); A **300**, 187 (1967).

[49] D. W. SCIAMA in [12], p. 415; F. HEHL and E. KRÖNER: *Zeits. Phys.*, **187**, 418 (1965).

[50] According to the derivation given here, the symmetry (142) follows from the selfadjointness (139) of H_I alone, *provided* the interaction is so weak that a first-order perturbation treatment is sufficiently accurate. If the formula (144) is (tentatively) accepted even beyond this weakness range, then there is a priori no reason for requiring (142). From (149), (150), it then follows, however, that (142) *is* true if the microscopic collision process (S-operator) is PT-invariant, as pointed out below eq. (152).

[51] H. KOPPE: *Werner Heisenberg und die Physik unserer Zeit*, edited by F. BOPP (Braunschweig, 1969), p. 182.

[52] *a*) S. CHANDRASEKHAR: *An Introduction to the Study of Stellar Structure*, esp. Chap. X (Chicago, 1939); *b*) P. T. LANDSBERG and J. DUNNING-DAVIES: *Statistical Mechanics of Equilibrium and Nonequilibrium*, edited by J. MEIXNER (Amsterdam, 1965), p. 36.

[53] C. MARLE: *Ann. Inst. Henri Poincaré*, A **10**, 127 (1969).

[54] J. L. ANDERSON: *Relativity*, edited by M. CARMELI, S. I. FICKLER and L. WITTEN (New York, 1970), p. 109.

Space-Time Structure from a Global Viewpoint (*).

R. GEROCH (**)

Department of Mathematics, Birkbeck College - London

Introduction.

Global methods have, in the past ten years or so, begun to make a significant impact on thinking in general relativity. These methods consist of a collection of more or less unrelated tools which can provide greater insight into a given (and often local) problem and which occasionally suggest new and interesting problems of their own. Global techniques are *not* 1) a highly mathematical alternative to local differential geometry, 2) a fixed body of knowledge, developed to carry out some specific program in general relativity, or 3) difficult.

The effective application of global methods often depends to a large extent on intuition. It is extremely difficult to understand global arguments (much less to discover and prove new theorems) unless one already has a good qualitative idea of what is to be done. Unfortunately, it is hard to develop such an intuition, for most papers either ignore global techniques completely or else assume that the requisite background material has already been covered. In this paper, we shall present some of these intuitive concepts on which so much of the global work is based.

What are global techniques good for? Of course, the principal application is to proving theorems about space-times. To this end, local and global methods are often used together. What one is able to do—even with local tensor calculus—is often determined to some extent by the global properties of the particular space-time. Furthermore, global methods permit one to isolate and

---

(*) Based in part on a series of lectures given by the author at Birkbeck College in January, 1968. I wish to thank M. WALKER for making available his notes on those lectures.

(**) National Science Foundation Postdoctoral Fellow. Present address: Department of Physics, University of Texas, Austin, Tex.

study in detail certain aspects of a space-time, for example, its causal structure and topology. Since such properties are, perhaps, more «fundamental» than the differentiable structure on which tensor analysis is based, it is important to know which results about a space-time involve only its causal structure, only its topology, etc. Finally, it is often necessary, in both local and global arguments, to set bounds on how pathological are the space-times being considered. Practically all pathologies are of a global nature. Global techniques, therefore, are used to classify pathological situations with a view toward deciding which situations are to be ruled out by physical considerations. This role cannot be dismissed as mere hair-splitting. How, after all, is one to identify the space-times which are too pathological to be of physical interest unless he has at least examined the possibilities which can arise?

Each global technique consists of a collection of constructions, theorems, and counterexamples. Fortunately, given a general idea of what constructions are available and how they are used, it is fairly easy to reconstruct the proofs of most of the theorems. Thus, statements of theorems are not a very important part of the subject. It is far more useful to have a good qualitative idea of what is available than a long list of results. The counterexamples, on the other hand, are extremely important. They provide not only a testing ground for new ideas, but also models to serve as a guide in checking and discovering theorems.

The purpose of this paper is to indicate in general terms what global techniques are available and what each one is good for. While it has been necessary to omit not only a number of minor theorems, but also many technical details in the proofs, we have tried to include a reasonably complete survey of the constructions and counterexamples used in global work. Furthermore, in order to give some idea of how the techniques are used, we have given some of the more important applications in each case. With a few exceptions, all definitions and theorems are stated precisely. A number of exercises are scattered throughout the paper.

1. – Topology and manifolds.

Very little topology has so far found application in general relativity. Essentially all that is needed is the notion of a topological space, some definitions (continuous, compact, closed), and a few elementary theorems. Unfortunately, most text books in topology contain a great deal more material than is normally required for applications in relativity. A good introductory discussion of general topology is contained in the first three chapters of Wallace's book [1].

In this Section we shall discuss briefly what topology is and where it is useful in relativity. We then introduce certain global properties—mostly of a negative character—about manifolds.

Let X denote the Euclidean plane. Given two points, (x, y) and (x', y'), in the plane, we define their separation,

$$d = [(x - x')^2 + (y - y')^2]^{\frac{1}{2}} . \tag{1.1}$$

The distance function (1.1) leads to the concept of an open set. Intuitively speaking, a set O in the plane is open if O contains all points « sufficiently close » to each point of O. More precisely, O is open if, for every point $p \in O$, there is a disk centered at p which lies entirely in O (the disk representing points sufficiently close to p). These open sets satisfy the following three conditions:

T1) The union of an arbitrary collection of open sets is open.

T2) The intersection of two open sets is open (*).

T3) The entire set X and the empty set are both open.

We may interpret these open sets as defining the notion « sufficiently close ». That is, intuitive statements of the form « All points sufficiently close to p have the property ... » are to be replaced by precise statements of the form « There is an open set containing p all of whose points have the property ... ». On the other hand, the original distance function (1.1) defines a stronger notion: « how close ».

The idea of topology is to retain the concept « sufficiently close », but in such a form that we do not have the additional information of « how close ». More precisely, a *topological space* [1] is a set X along with a collection of subsets of X, called *open sets*, subject to the conditions T1)-T3). A set C is then *closed* if its complement, $X - C$, is open, *i.e.* if all points sufficiently close to any point outside of C are outside of C. Note that a set need not necessarily be either open or closed. Two or three other definitions concerning topological spaces will be introduced later when they are needed.

Exercise: Show that the usual topology on Minkowski space is Lorentz-invariant (*i.e.* that Lorentz transformations take open sets to open sets). Show that this topology cannot arise from any Lorentz-invariant distance function.

(*) Therefore, the intersection of any finite collection of open sets is open. Note, however, that the intersection of an infinite collection of open sets need not be open. For example, the intersection of all disks centered at the origin of X consists of the origin alone, and this single point is certainly not an open set.

There are two situations in which topology makes an appearance in general relativity. The first involves the topology of the underlying manifold of a space-time (*). Of course, a space-time has a great deal more structure than merely this notion of « sufficiently close », but it turns out to be more appropriate to express certain arguments purely in terms of the topology. The reason for introducing topology is primarily one of convenience: an argument, stated topologically, often becomes more transparent. The second place in which topologies are used is in the discussion of certain « large » sets which arise in relativity, notably the set of all Lorentz metrics on a manifold [2, 3] and the set of all causal curves joining two points [4, 5]. Again, it would certainly be possible to avoid all terminology associated with topology, but it is more conveninent not to impose such arbitrary a rule.

Every space-time is based on a manifold, and every manifold has a topology [6]. Thus, certain very coarse features of the space-time are determined by its topology. We are already familiar with two examples of topological spaces. Euclidean n-space R^n and the n-dimensional sphere S^n (**). We can obtain other manifolds from these by taking *products* of manifolds [6]. For example, the product of R^1 (the line) and R^1 is R^2 (the plane), the product of R^1 and S^1 is the cylinder ($R^1 \times S^1$), and the product of S^1 and S^1 in the torus, $S^1 \times S^1$ (*not* S^2). Many of the manifolds encountered in relativity can be constructed from R^n and S^n by taking products.

Consider, for example, the (extended) Reissner-Nordström solution [7]. The 2-spheres $r = \text{const}$ are surfaces of spherical symmetry. It is convenient, therefore, to draw a picture of this space-time by assigning a single point to represent each 2-sphere. The result is shown in Fig. 1 *a*). In order to make the diagram easier to draw, we have brought the asymptotic regions in from infinity to a finite distance [8]. Such a transformation is permissible because it does not affect the situation from the point of view of the underlying manifold. Therefore, the manifold of the Reissner-Nordström solution is the product of the 2-sphere S^2 and the plane Figure in Fig. 1 *a*). But this Figure is topologically the same as the plane, R^2. To prove this fact, we have redrawn the plane Figure in Fig. 1 *b*), but now with a set of co-ordinate lines. The correspondence between the Figure and the plane, obtained by mapping the

---

(*) By a *space-time* we shall understand a 4-dimensional manifold with a C^∞ metric g_{ab} of Lorentz signature, $(+, -, -, -)$. (We occasionally use the word space-time to refer to lower-dimensional examples.) All manifolds will be assumed to be Hausdorff and paracompact (see Appendix).

(**) The topology on R^n is given by the distance function analogous to (1.1). The topology on S^n may be obtained by embedding the sphere in the usual way in Euclidean $(n+1)$-space, and taking for the distance function the Euclidean distance between points of the sphere. We shall use the symbols R^n and S^n to mean either the topological spaces or the differentiable manifolds.

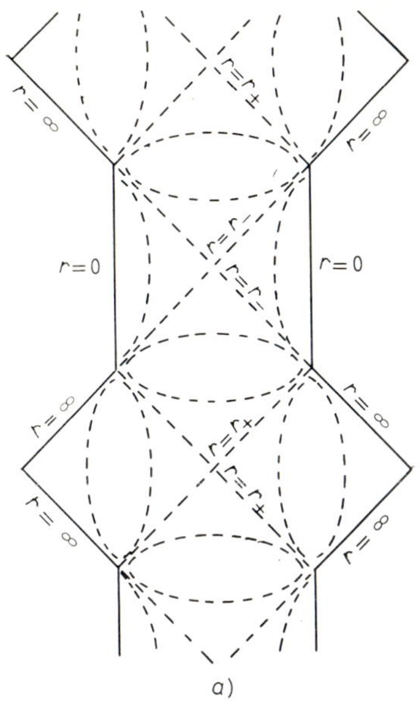

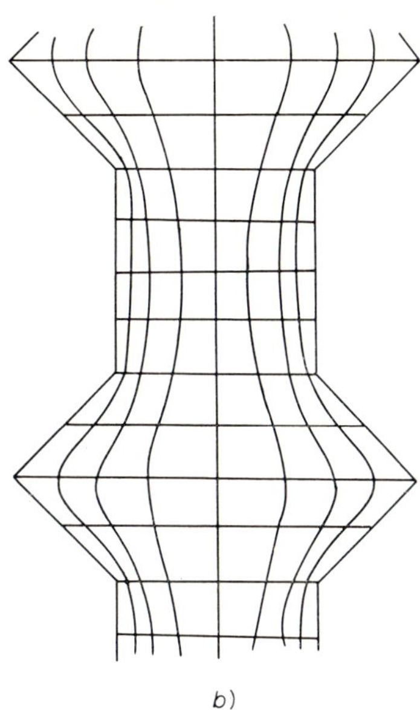

Fig. 1. – The Reissner-Nordström solution. Each point of the figure represents a 2-sphere of spherical symmetry. *a*) The curves of constant r are shown. $r=0$ represents the « singularity », and $r=\infty$ « asymptotic infinity ». Of course, neither of these regions are represented as points of the manifold itself. *b*) A co-ordinate grid which, when mapped to a rectangular grid in the plane, shows that the Reissner-Nordström manifold is $S^2\times R^2$.

co-ordinate lines into a rectangular grid on the plane, is continuous (*), both from the figure to the plane and from the plane to the figure. (Such a mapping is called a *homeomorphism*, or topological equivalence.) We conclude that the Reissner-Nordström manifold is $S^2\times R^2$.

Exercise: Verify that the Schwarzschild manifold is also $S^2\times R^2$.

What information does this conclusion reveal about the Reissner-Nordström solution? The answer is, very little indeed! It does not tell us about the location or even the existence of singularities. It does not tell us that the Reissner-Nordström solution has no closed timelike curves. It does not tell us whether or not the Reissner-Nordström solution has a Cauchy surface. It does not tell us that the Reissner-Nordström solution has asymptotic regions, or a « wormhole » representing a charge. What the statement « The Reissner-Nord-

(*) A mapping from one topological space to another is said to be *continuous* if the inverse image of each open set in the second space is an open set in the first space. *Exercise*: Verify that, for functions of a real variable (maps from R^1 to R^1), this definition reduces to the usual $\varepsilon-\delta$ definition of continuity.

ström manifold is $S^2 \times R^2$. » does say is precisely what the underlying manifold—the arena for the geometry—is.

This point—that the underlying manifold gives practically no information about the spacetime—is an important one. We illustrate it with a few examples.

Consider the open disk D_1 in the plane (Fig. 2 a)). (Topologically, the disk is equivalent to the entire plane.) One can see at a glance from Fig. 2 a) how to put a boundary on our disk. One might imagine, therefore, that if this disk

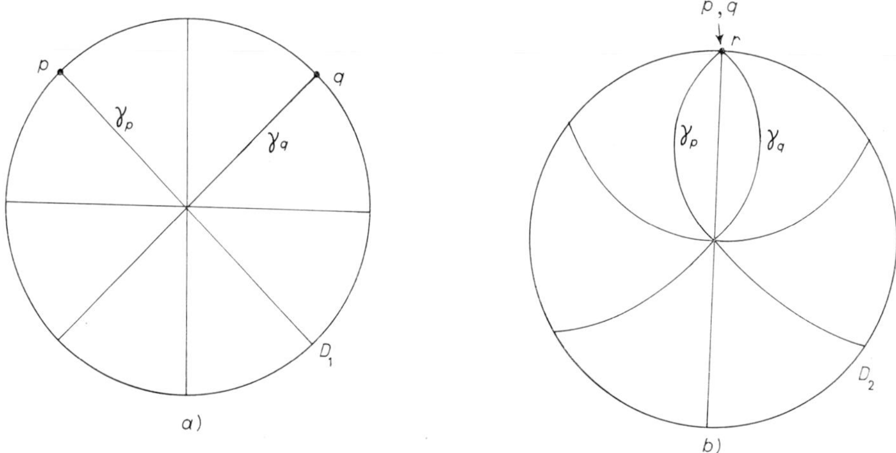

Fig. 2. – Two disks in the plane. The mapping from D_1 onto D_2, defined by mapping the spokes as shown (and mapping concentric circles about the center into themselves), takes the « boundary points » p and q of D_1 into a single « boundary point » r of D_2. A disk has no natural boundary.

were to represent a space-time, then the « singular points » (if any) would be represented by certain points on the boundary. Other points on the boundary might be interpreted as « points at infinity » as in Fig. 1 a)). In fact, in placing this particular boundary on the disk we have used more than just the manifold structure of the disk. Consider the second disk D_2 in Fig. 2 b). We map the disk D_1 onto D_2 by mapping the radial spokes as shown (a homeomorphism). Although D_1 and D_2 represent exactly the same manifold, the « natural » boundary for D_2 is quite different from that for D_1. For example, the boundary points p and q of D_1 are mapped into the single boundary point r of D_2. We see that the notion of a « natural boundary » is simply not defined for a manifold: some additional structure, such as a Lorentz metric or a causal structure, is needed. Although one might, intuitively, feel that certain statements concerning « singular surfaces » etc. for space-time should be meaningful, the precise formulation of such statements may turn out to be a difficult problem.

Consider the two curves γ_p and γ_q in D_1, consisting of spokes radiating from the center to the boundary points p and q, respectively. In D_1, these curves do not appear to « meet at the boundary », yet the appearance is just the reverse in D_2. That is to say, the notion of two curves « meeting at infinity » is also not defined by the manifold itself.

Finally, let ξ^a and η^a be two vector fields on D_1. What do the following mean: « ξ^a approaches η^a at the boundary » and « ξ^a approaches zero at the boundary »? Using mappings similar to that above, it can easily be shown that neither of these statements can be given any reasonable meaning. Two vector fields whose components, in one co-ordinate system, approach each other as we approach the boundary will not in general have this property in another co-ordinate system.

Exercise. Verify that the following statements have no obvious meaning.

1) The Friedmann models have a point singularity; the Schwarzschild solution has a line singularity.

2) A spacelike surface is asymptotically null.

3) A timelike curve is asymptotically null.

4) The metric approaches the Minkowski metric as $r \to \infty$.

5) Two geodesics approach the singularity from the same direction.

We next attempt to construct a space-time whose underlying manifold is the same as that of the Reissner-Nordström solution—$S^2 \times R^2$—but which has a very different geometrical structure. There is a general and extremely useful method for constructing examples to illustrate global properties: « cutting holes and patching together ».

The method of cutting holes is based on the following fact: if M is a space-time and C is any closed subset of M, then M, with the region C removed, is also a space-time. Removing selected regions C from a known space-time (usually Minkowski space) provides a powerful technique to find space-times having certain properties. Consider Minkowski space with the t-axis (in the usual co-ordinates) removed. This space-time is topologically $S^2 \times R^2$ (*Exercise*: Prove it.). Although this space-time has the same topology as the Reissner-Nordström solution, it is very different in most other ways. It has only a single « asymptotic infinity » rather than an infinite number, its « singularity » has an entirely different structure from that of the Reissner-Nordström solution, etc.

The second technique is that of « patching together ». Suppose for example that we wanted to decide whether or not there is a space-time M which is without singularity, satisfies Einstein's equations (source-free), and has a compact (*)

(*) We may define a manifold S as *compact* if every sequence of points of S has a point of accumulation. Intuitively speaking, compactness means that the manifold has no « open edges » or « edges which go off to infinity ». For example, a sphere, a torus, and a closed disk are compact while an open disk, a plane, and a cylinder are noncompact.

spacelike 3-surface (*i.e.* M is to represent a nonsingular closed universe). It is very easy to construct such a model. Begin with a timelike tube a Minkowski space whose cross-section is a cube (Fig. 3). Now identify opposite faces of each cross-section to form compact cross-sections. (Of course, one can carry out such identifications only when the metric can be joined smoothly.) The resulting space-time M satisfies all our conditions and, in fact, one more: it is flat. (Topologically, M is $S^1 \times S^1 \times \times S^1 \times R^1$.) It is an open question whether or not there are nonflat models which satisfy our three conditions [2].

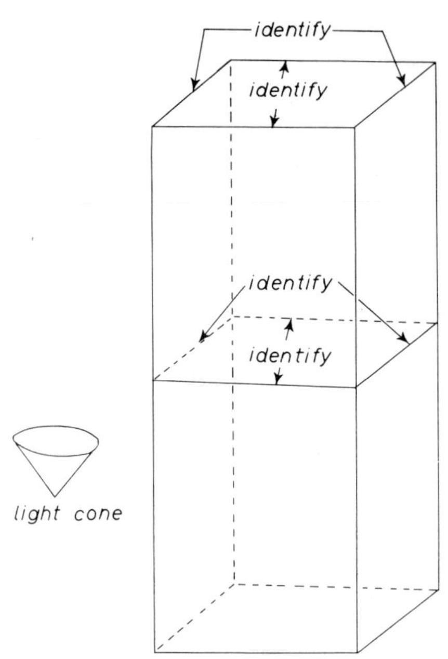

Fig. 3. – A closed, flat universe, obtained from Minkowski space by cutting and patching (one spatial dimension suppressed). Opposite vertical planes are identified. The spacelike sections in the resulting space-time are toruses, $S^1 \times S^1$. ($S^1 \times S^1 \times S^1$ in the 4-dimensional case).

One occasionally hears the view that space-times constructed by cutting and patching are « artificial » and should be ruled out. The problem here is that it is not enough to say merely that certain space-times leave one uncomfortable unless he can pin down precisely which space-times are to be accepted and which will not. No suitable criterion has yet been found. (It would not do to say « space-times constructed by cutting and patching are unacceptable », for any space-time can be represented in this way.) It is felt that such space-times, while they do appear rather artificial, must be accepted, at least until a suitable criterion for their rejection is discovered. Much work on global properties consists of trying to formulate just such intuitive ideas.

The space-times obtained by cutting and patching are not normally considered as serious models for our universe. However, the mere existence of a space-time having certain global features suggests that there are many models —some perhaps quite reasonable physically—with similar properties. For example, Minkowski space with the line removed is very similar in its global properties to certain Weyl solutions. The idea of the method is not so much to construct realistic models of our universe as to show what is possible within the context of space-times. Its power stems from the fact that one is able to mold the global properties of a space-time without having to worry about details of the metric.

Exercise. Find a flat space-time any two of whose points can be joined by a timelike curve. Find a flat spacetime which is topologically $S^1 \times R^3$.

2. – Direction fields, orientation.

A number of questions in relativity are dealt with most easily by introducing a timelike direction field. Such fields carry certain of the information contained in the light-cone structure, but in a more convenient form. In this Section we shall see how direction fields arise in relativity and what they can be used for. In particular, direction fields lead to the notions of space and time orientation.

Let M, g_{ab} be a space-time. Choose an abitrary positive-definite metric h_{ab} on M (*). This h_{ab} defines a *direction field* (*i.e.* a vector field which is nonzero at each point, and which is defined at each point only up to an arbitrary nonzero factor) on M as follows [4]. Let p be a point of M. At the point p, choose, from among all vectors whose norm with respect to h_{ab} is unity, that vector whose norm with respect to g_{ab} is maximum. Since the unit vectors with respect to h_{ab} define a sphere while the vectors having a given norm with respect to g_{ab} define a hyperboloid (Fig. 4), it is clear that there will always exist such a maximum, unique up to sign. The timelike direction field so defined depends, of course, on the choice of h_{ab}. However, certain important properties of the field turn out to be independent of this choice.

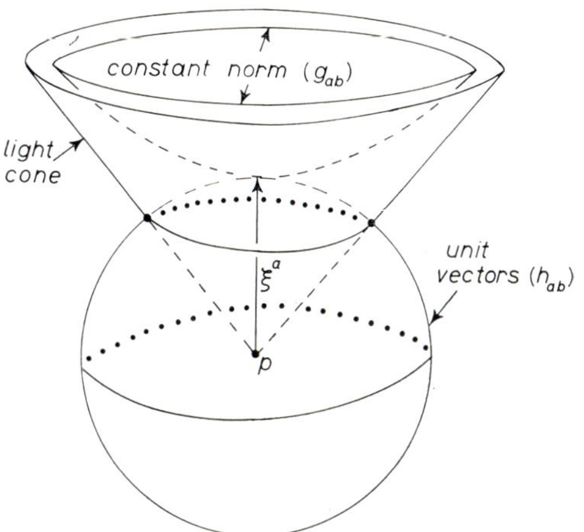

Fig. 4. – Construction of a timelike direction field. The unit vectors (with respect to a positive-definite h_{ab}) define a sphere, while the set of vectors with a given norm μ (with respect to g_{ab}) define a hyperboloid. For an appropriate μ, the sphere and hyperboloid intersect in a point, thus defining the direction ξ^a.

Conversely, let M be a 4-dimensional manifold, and let $\alpha\xi^a$ be a direction field on M (**). We shall construct a Lorentz metric g_{ab} on M for which $\alpha\xi^a$

(*) It is well known [33] that such a metric exists on every (Hausdorff, paracompact) manifold.

(**) We shall use this notation « $\alpha\xi^a$ » for a direction field. However, it cannot be assumed, as we shall see later, that the direction field consists of all nonzero scalar multiples of some fixed vector field.

is timelike. Choose an arbitrary positive-definite metric h_{ab} on M and set [4]

$$g_{ab} = -h_{ab} + 2h_{ap}\xi^p h_{bq}\xi^q(h_{cd}\xi^c\xi^d)^{-1}. \tag{2.1}$$

(Note that (2.1) is independent of the scaling of ξ^a.) This g_{ab} has Lorentz signature, and $\alpha\xi^a$ is timelike with respect to g_{ab}. Of course, the direction field does not define a Lorentz metric uniquely.

As an example of the use of direction fields, let us find the condition for a given manifold M (*) to admit a Lorentz metric. As we have seen, the existence of a Lorentz metric is equivalent to the existence of a direction field. It turns out that there are two separate cases, depending on whether or not M is compact.

Suppose first that M is not compact. Choose on M an arbitrary direction field which is regular everywhere except, possibly, at certain isolated singular points (Fig. 5 *a*)). (Such a field always exists [9].) The idea is to « push the

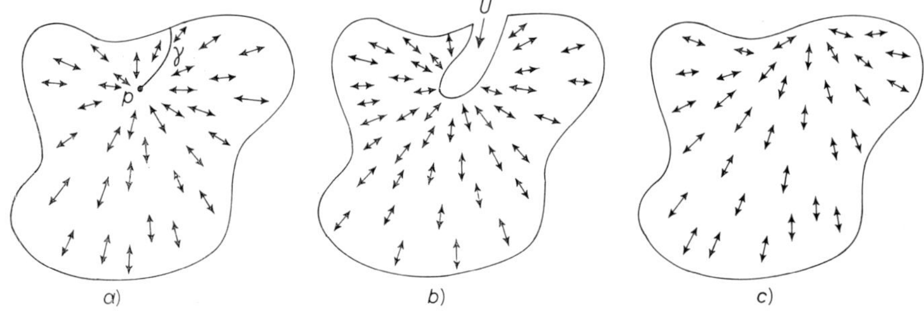

Fig. 5. – How to push a point off a noncompact manifold. Step *a*): Draw a curve γ from the point which goes off to infinity. Step *b*): Remove a neighborhood U of the curve. Step *c*): Restore the manifold to its original shape. (The double arrows represent a direction field.)

singular points off to infinity » and thus obtain a nonsingular direction field. Let p be a singular point. Choose a curve γ from p which « goes off to infinity » (**), and let U be a small tubular neighborhood of γ (***) (Fig. 5 *a*)). The manifold consisting of M with the tube U removed is equivalent (diffeomorphic) to M (Fig. 5 *b*) and 5 *c*)). We thus obtain a direction field on M for which p is no longer a singular point (Fig. 5 *c*)). Continuing in this way, we obtain a nonsingular direction field, and thus a Lorentz metric, on M.

(*) We assume in this discussion that M is connected.

(**) More precisely, γ must eventually leave and remain outside of every compact set.

(***) The *interior* of a set A is the « largest » open set contained in A or, more precisely, the union of all open sets contained in A (see axiom $T1$). A *neighborhood* of a set B is any set whose interior contains B.

Exercise. Show that the above argument applies even in the case in which there are an infinite number of isolated singular points.

We have shown that every noncompact manifold carries at least one (and therefore an infinite number) of Lorentz metrics. That is to say, every non-noncompact manifold is a *possible* arena for space-time physics. PENROSE [10] has obtained a stronger version of this result: the Lorentz metric can always be so chosen that there are no closed timelike curves.

The situation is quite different in the compact case. It is known [11] that a necessary and sufficient condition for a compact manifold M to carry a Lorentz metric is the vanishing of a certain integer, called the Euler characteristic, associated with M. However, there is a more interesting and much simpler result along these lines: every compact space-time has closed timelike curves [12, 13]. Let M be a compact manifold with a Lorentz metric, and let $\alpha\xi^a$ be a timelike direction field on M. Choose any point $p \in M$. Draw the timelike curve γ which begins at p and is everywhere tangent to $\alpha\xi^a$. Choose a sequence of points p_i on γ such that every point of γ lies beyond some p_i. Since M is compact, the p_i have an accumulation point (*), say at q (Fig. 6). Choose a cylindrical neighborhood N of q such that every trajectory of $\alpha\xi^a$ which passes through N can be distorted, in N, so that it remains timelike and passes through the point q. Since $\alpha\xi^a$ is timelike, every trajectory of $\alpha\xi^a$ which enters N must reemerge from N. Since the p_i accumulate at q, our trajectory γ must continually re-enter N. By distorting γ on two of its successive passages through

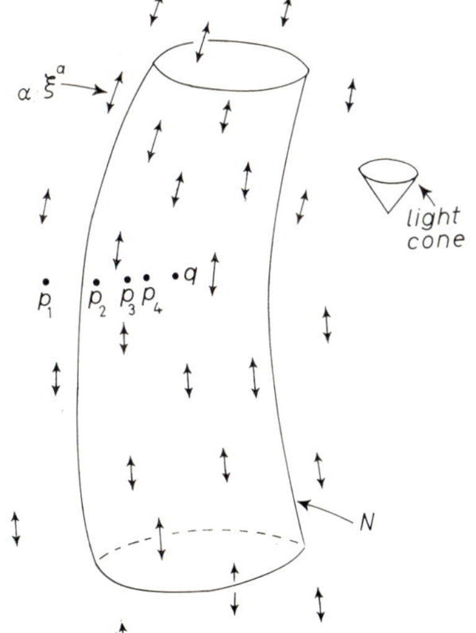

Fig. 6. – Constructing a closed timelike curve in a compact space-time. The sequence of points p_i on a trajectory of $\alpha\xi^a$ accumulate at q. Every trajectory which enters N can be distorted in N so that, while remaining timelike, it intersects q. By distorting our given trajectory on two successive passages through N, we obtain a closed timelike curve.

(*) An *accumulation point* of a sequence is a point every neighborhood of which contains an infinite number of points of the sequence. A *limit point* of a sequence is a point every neighborhood of which contains all points of the sequence beyond a certain member.

Exercise: Show that a sequence of points on a manifold can have many accumulation points, but no more than one limit point.

N, we obtain a closed timelike curve, beginning and ending at q. That is to say, if one wishes to avoid closed timelike curves, then *no* compact manifold is a suitable arena for space-time physics.

Exercise. Find an example of a compact space-time containing two points which cannot be joined by a timelike curve.

Using similar arguments, one can prove the following result: Let S and S' be two compact 3-manifolds. Then there exists a space-time M having S and S' as spacelike sections, and such that the region between S and S' is compact. If, however, S and S' have different topologies, then M will always have closed timelike curves [13]. That is, the topology of spacelike sections in a space-time can change in an arbitrary way from one epoch to another, but such changes are accompanied by closed timelike curves.

Exercise. Show that, given two 3-manifolds S and S', there exists a space-time M, without closed timelike curves, having S and S' as spacelike sections.

Another property of space-times which can be described in terms of direction fields is time and space orientation. Let $\alpha\xi^a$ be a timelike direction field in the space-time M. With any closed curve γ (not necessarily timelike) beginning and ending at a point p, we associate three integers as follows.

1) Choose one of the two possibilities for the sign of $\alpha\xi^a$ at the point p. Carry this choice continuously around γ. On returning to p, we will obtain either the same or the opposite sign as that with which we began. That is, γ may be either time-preserving $(+1)$ or time-reversing (-1).

2) Choose three spacelike vectors at p which are linearly independent, and such that the 3-plane they span does not contain $\alpha\xi^a$. Carry these three vectors around γ in an arbitrary continuous way, always keeping the vectors spacelike and independent of each other and of $\alpha\xi^a$. On returning to p, our vectors will have either the same or opposite orientation as that with which they began. That is, γ may be either space-preserving $(+1)$ or space-reversing (-1).

3) Choose a sign of charge which is to be called «positive» at p, and carry this choice continuously around γ. On returning to p, our choice of «positive» charge may either agree or disagree with that originally made at p. That is, γ may be either charge-preserving $(+1)$ or charge-reversing (-1).

Thus, with every closed curve we associate three integers, each either $+1$ or -1. The first two integers depend only on the geometry of the space-time (and, in particular, are independent of the direction field and the mode of transport), while the third involves also the physical properties of the space-time. The integers are clearly invariant under continuous deformations of the curve γ (*).

A space-time M is said to be *time* (respectively *space, charge*) *orientable* if every closed curve in M is time (respectively space, charge) preserving [11]. For example, that a space-time be time-orientable means that we may speak of the « future » and « past » light cones at each point. These three notions are completely independent, *i.e.* a space-time can be orientable in any combination of the three senses. An example of a nontime-orientable space-time is given in Fig. 7. Finally, a space-time is said to be *orientable* if every curve is both space and time preserving or neither.

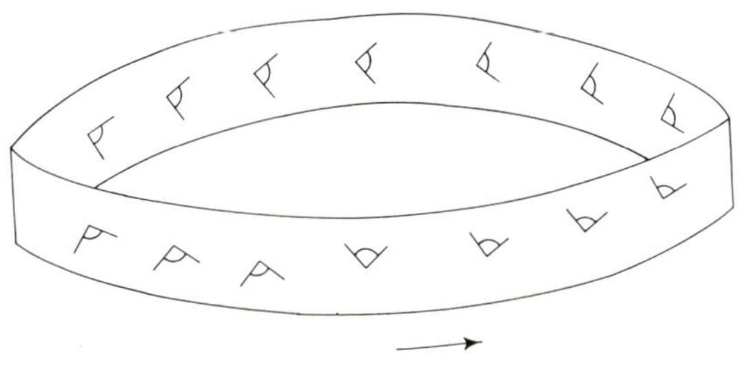

Fig. 7. – A 2-dimensional space-time (topologically $S^1 \times R^1$) which is not time-orientable. This example is not space-orientable either.

Exercise: Using a Möbius strip, construct an example which is space-orientable but not time-orientable.

Exercise. Prove that whether or not a space-time is orientable depends only on the underlying manifold, not on the metric.

Exercise. Prove that a space-time is time-orientable if and only if it carries a timelike vector field.

Exercise. Do electrodynamics in a space-time which is not charge-orientable. (The electromagnetic field will be double-valued).

Exercise. Show that every space-time which is topologically R^4 is time, space, and charge orientable.

(*) That is, the integers depend only on the homotopy class of the curve (in fact, only on its homology class) [1]. If we combine curves (in the sense of homotopy [1]), then the corresponding integers for the two curves are multiplied. Thus, we obtain a homomorphism from the group $\pi_1(M)$ into the group of all triples of integers, each either $+1$ or -1.

How can one decide what type of orientability our own universe satisfies? It turns out that this can be accomplished with local experiments, provided we assume that the results hold « universally », *i.e.*, provided we assume the strong principle of equivalence [14]. Suppose for example that, in every local region, one can, by means of statistical mechanics, determine a « future » time direction. Then our universe must certainly be time-orientable. By a similar argument, one can show [15] that if, in each local neighborhood, one has the following results: 1) an experiment demonstrating noninvariance under separate charge and parity reversal [16, 17], 2) an experiment demonstrating noninvariance under simultaneous charge-parity reversal [18], and (3) the *CPT* theorem [19], then every closed curve must reverse either all or none of time, space, and charge. (In particular, our universe must be orientable.) If, in addition, time-orientability is assumed, then our universe must be orientable in all three senses.

3. – Casual structure I. The domain of influence.

We have seen in Sect. **1** that there is associated a topology with the underlying manifold of any space-time. We have also seen that this topology gives very little information about the physics of the space-time. In particular, it does not involve the metric in any way. Is there any structure on our space-time which refers to the metric, but which does not involve all the complications inherent in the metric and differentiable structure? One of the most important of these is the causal structure. (Can event p influence event q by means of a signal?) It turns out [20] that a surprising number of global arguments concerning space-times can be formulated in terms of this particular —and very simple—aspect of the metric structure. The advantage of using causal structure is that arguments are thereby often simpler and more transparent. Furthermore, causal structure appears, physically, to be more « fundamental » than, for example, differentiable structrure, and so it is useful to know which results depend only on that structure.

Let M, g_{ab} be a space-time. We assume throughout this and the following Section that M is time-oriented, *i.e.* that there is given a continuous choice of a future light-cone at each point. We write $p \ll q$ to mean that there is a smooth, future-directed timelike curve (*) which begins at the point p and ends at the point q [20].

This is the fundamental relation of causal space theory. We now derive two of its properties.

(*) Since « timelike curve » means that the tangent vector to the curve must have positive norm, the « zero curve » (which remains always at p) is not counted as timelike.

CS1) If $p \ll q$ and $q \ll r$, then $p \ll r$. See Fig. 8.

The condition that M have no closed timelike curves may be expressed in the following form:

CS2) For no point p is $p \ll p$. The proof is obvious (*).

We shall not automatically assume the absence of closed timelike curves in what follows.

Exercise. Prove that if a space-time has one closed timelike curve, then it has an infinite number.

Let $p \in M$. We denote by $I^-(p)$ (respectively, $I^+(p)$) the set of all points q such that $q \ll p$ (respectively, $q \gg p$). This set is called the *past* (respectively, *future*) of p. For example, the past of a point p in Minkowski space is the interior of the past light cone of p.

Exercise. Show that, if $p \ll q$, then $I^-(q) \supset I^-(p)$. Show that the converse is false (*e.g.* in Minkowski space).

There is a relation between the causal structure and the topology of M. For any point p, $I^-(p)$ is open. To prove this result, we must show that arbitrary sufficiently small variations of any point $q \in I^-(p)$ result in points again in $I^-(p)$. Let γ be a timelike curve from p to q (Fig. 9), and let r be a point on this curve which is near to q. Since r is close to q and timelike related to q, $I^-(r)$ certainly includes a neighborhood of q. But $r \ll p$, and so $I^-(r) \subset I^-(p)$. Since $I^-(p)$ contains a neighborhood of each of its points, $I^-(p)$ is open. It follows from axiom T1 for a topological space that, for any set A, $I^-[A]$ (defined as the union of the pasts of the points of A) is open.

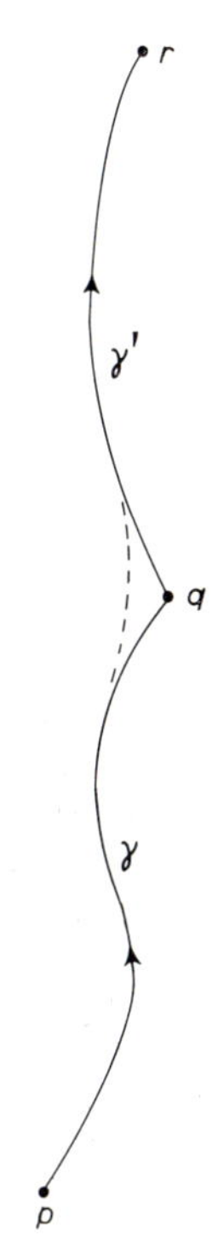

Fig. 8. – If $p \ll q \ll r$, then $p \ll r$. Let γ and γ' be timelike curves from p to q and q to r, respectively. Take the combined curve and « round off its corner » at q (the dashed line). The result is a smooth timelike curve from p to r.

To illustrate this interplay between topology and causal structure, consider the example of Fig. 10. The sequence p_i approaches p as a limit. The point q is such that $q \ll p_i$ for each i. However, q cannot be joined to p by a timelike (or even null) curve. Is the converse true? That is, is it true that, if the p_i approaches p and $q \ll p$, then $q \ll p_i$ for sufficiently large i? The answer is yes.

(*) See footnote on preceding page.

Proof. $I^+(q)$ is open and contains p. Therefore, $I^+(q)$ contains the p_i for sufficiently large i. That is, $q \ll p_i$ for sufficiently large i.

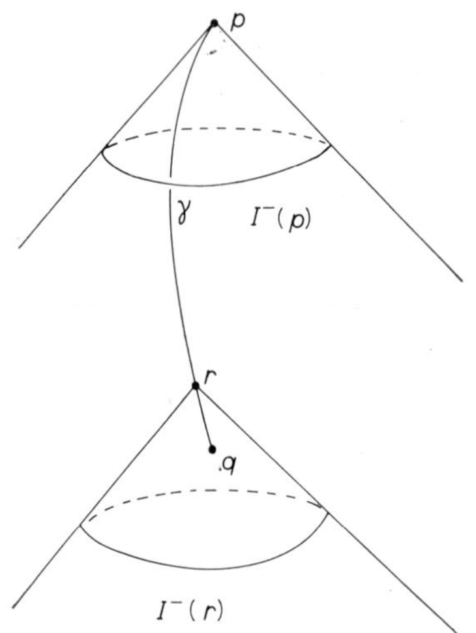

Fig. 9. – $I^-(p)$ is open. Let γ be a timelike curve from p to q. Then $I^-(r)$, where r is on γ and near q, contains a neighborhood of q. Hence, $I^-(p)(\supset I^-(r))$ contains a neighborhood of q.

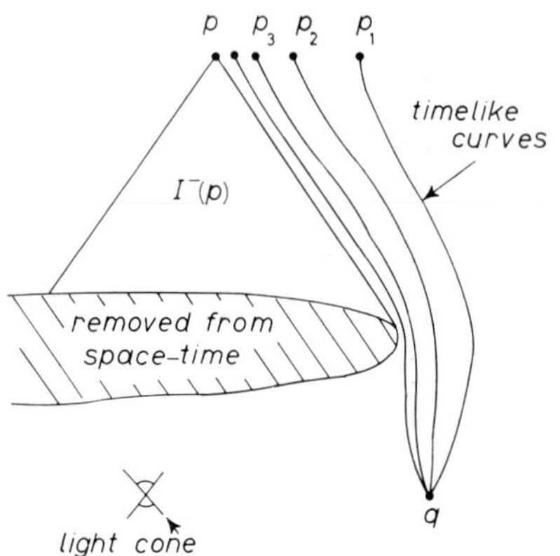

Fig. 10. – Two-dimensional Minkowski space with a region removed. Although $q \ll p_i$ for each i, and $p_i \to p$, q and p cannot be joined by a timelike or null curve.

This relationship between the pasts of the points of a sequence and the past of their limit is not completely obvious. Yet it is easy to rederive the result when needed using the constructions and examples of causal space theory.

Since sets of the form $I^-(p)$ and $I^+(q)$ are open, it follows from axiom T2 that sets of the form $I^-(p) \cap I^+(q)$ are open. Therefore, by axiom T1, arbitrary unions of sets of this form are open. It is an interesting question whether the converse is also true. That is, if U is a given open set in M, is it true that U can be written as a union of sets of the form $I^-(p) \cap I^+(q)$? An affirmative answer would imply that the topological structure of M follows already from its causal structure.

In general, the answer to our question is « no ». Let us construct a counterexample. Consider 2-dimensional Minkowski space (*). Take a horizontal strip of Minkowski space lying between the lines $t = 0$ and $t = 1$ (Fig. 11). The idea is to violate causality so badly that sets of the form $I^-(p) \cap I^+(q)$ will

(*) Two-dimensional examples are frequently used because they are easier to draw and to visualize. Why is this not an unphysical procedure, since our own universe is 4-dimensional? It is always necessary to check with such examples that the same features can appear in a 4-dimensional space-time.

be « too large » for their unions to comprise all open sets. To achieve this end, identify the edge $t = 0$ with the edge $t = 1$. The result is a cylindrical space-time in which « time » goes around the cylinder (Fig. 11). There are closed timelike curves. Furthermore, *any* two points may be joined by a timelike curve (*e.g.*, by a spiral on the cylinder whose pitch is less than one). Thus, for any point p, $I^-(p)$ is the entire manifold M, and so there is only one set of the form $I^-(p) \cap I^+(q)$, namely, M. Clearly, no open set, except M itself, can be written as a union of sets of the form $I^-(p) \cap I^+(q)$. The point here is that, in a space-time as severely acausal as that of Fig. 11, the causal structure gives practically no information: $p \ll q$ for any two points. It could hardly be expected that such a uninteresting causal structure would be capable of defining the topology of M.

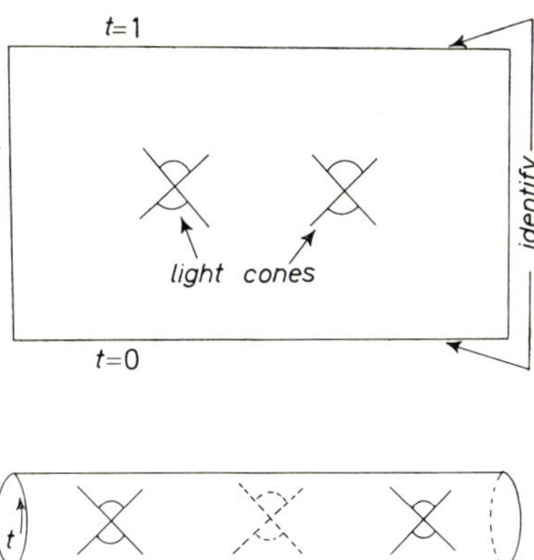

Fig. 11. – A 2-dimensional space-time with closed timelike curves. Identify the top and bottom edges of a horizontal strip from 2-dimensional Minkowski space. In the resulting cylindrical space-time, any two points may be joined by a timelike curve. Hence, for each p, $I^-(p)$ is the entire manifold.

Exercise: Construct an example with closed timelike curves, but for which certain pairs of points cannot be joined by a timelike curve.

Exercise. Show that any open set in Minkowski space can be written as a union of sets of the form $I^-(p) \cap I^+(q)$.

We shall see later that it is possible to infer the topology of M from its causal structure provided we assume that the latter is « reasonable » in a certain sense.

By far the most important element of the causal structure is the relation « $\ll$ ». However, there is a second causal relation which can be defined on a space-time. A smooth curve γ will be called a *causal curve* if its tangent vector is either timelike or null at each point. (The tangent vector may be timelike at some points and null at others (*)). We write $p \prec q$ (*p causally precedes q*) if there is a future-directed causal curve from p to q. Finally, $J^-(p)$ is the set of all points which causally precede p. For example, if p is a point in Minkowski space, then $J^-(p)$ is the closed past light-cone at p (*i.e.* including $I^-(p)$, the null cone, and the point p itself). Clearly, we have

CS3) If $p \ll q$, then $p < q$.

(*) We consider the zero vector as null and both future and past-directed. Thus, the « zero curve », which remains entirely at a point p, counts as a causal curve.

Exercise. Show that, if $p \prec q$, then p and q may be joined by a future-directed null curve.

The fundamental relation determining when points may be joined by a timelike curve is the following: Let γ_1 and γ_2 be future-directed causal curves joining p to q and q to r, respectively. Suppose $p \not\ll r$. Then γ_1 and γ_2 are null geodesics, and their tangent vectors coincide at q. The proof is essentially a generalization of the argument of Fig. 8. If γ_2 were not a null geodesic, then γ_2 would enter the past of r. Hence, q, the entire curve γ_1, and the point p would all be in $I^-(r)$. Similarly for γ_1. If γ_1 and γ_2 had different tangents at q, then we could round off the corner of the combined curve at q, and thus obtain a causal curve from p to r which is timelike near q. This curve would therefore enter $I^-(r)$ in the «rounding-off region», and thereafter remain in $I^-(r)$.

Exercise. Using the above result, establish the following properties of «$\ll$» and «$\prec$»:

CS4) If $p \prec q \prec r$, then $p \prec r$.

CS5) If $p \prec q \ll r$, then $p \ll r$.

CS6) If $p \ll q \prec r$, then $p \ll r$.

The condition that there be no closed causal curves cannot be written in a form analagous to CS2. In fact, it follows from the definition that (*)

CS7) $p \prec p$ for every point p.

However, this requirement can be expressed in the following way:

CS8) If $p \prec q \prec p$, then $p = q$.

That is, there are defined on every space-time two causal relations subject to CS1), CS3), CS4), CS5), CS6) and CS7). If, in addition, there are no closed causal curves, then these relations also satisfy CS2) and CS8. A general *causal space* [20] is a set X with two relations, subject to CS1)-CS8), but not assumed to come from any manifold or metric structure on X. A number of results normally associated with relativity theory can, in fact, be proven for general causal spaces.

Having discussed some properties that «$\ll$» and «$\prec$» *do* satisfy, let us next examine some properties that are not satisfied.

1) Sets of the form $J^-(p)$ are closed in Minkowski space, but this is not true in general (Fig. 12).

2) If the past of p contains the past of q in Minkowski space, then $q \prec p$. This is not true in general (Fig. 12).

(*) See footnote on preceding page.

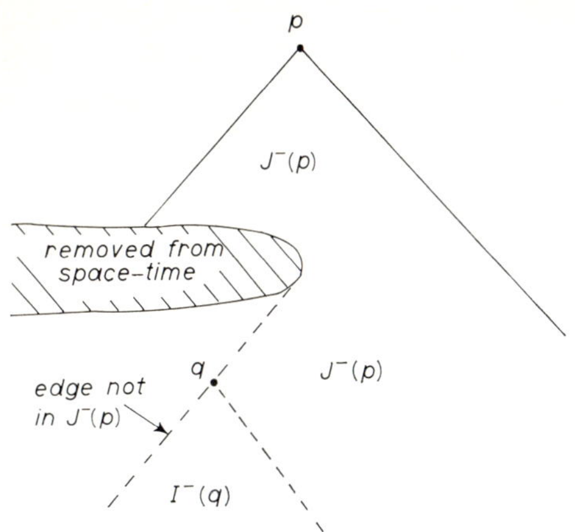

Fig. 12. – The same 2-dimensional example as Fig. 10. $J^-(p)$ includes the two solid lines from p, but not the dashed line through q. $I^-(p)$ is (as always) the interior of $J^-(p)$. Thus, $I^-(p) \supset I^-(q)$ and, furthermore, q is in the closure of $J^-(p)$, although q and p cannot be joined by a causal curve.

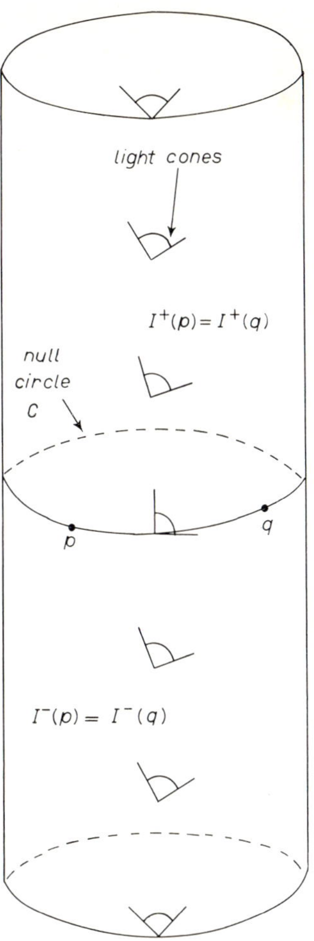

Fig. 13. – A 2-dimensional cylindrical space-time. The metric is invariant under rotations of the cylinder about its axis. The light cones « tip over » near the center of the cylinder to create a closed null circle C. There are no closed timelike curves, $I^-(p) = I^-(q)$ (= lower half of cylinder) and $I^+(p) = I^+(q)$ (= upper half of cylinder), although $p \neq q$.

3) In Minkowski space, two points with the same past and the same future coincide. This, too, is not true in general (Fig. 13). The above examples are pathological, but that is precisely the point! Should one of these properties be required in the course of a proof, then some further condition must be imposed on the space-time.

Many of the above examples involve closed timelike curves. What is wrong with such space-times [21, 22]? One would like to argue that they are artificial because our own universe does not have closed timelike curves, or, at least, none which pass through the earth. Such a strong assertion might appear to be difficult to verify, for if there is a closed timelike curve which passes through the earth, it would probably wander a great distance from the earth before coming back to itself. One could, nonetheless, explore such possibilities. If we program a rocket ship to traverse a suspected curve, we should not have to wait millions of years for its return. The ship would return the instant we lit the fuse. The experiment, then, consists of programming the ship to follow such a curve, lighting the fuse, and looking over our shoulder for its return. If the ship simply flies off, then our suspected closed timelike curve is not such an object.

As nobody has tried such an experiment, why do we suspect no closed timelike curves? There is a much easier experiment. We ask the following question: how large is the class of solutions of, say, Maxwell's equations in a space-time with closed timelike curves? We would expect, in general, to find very few. Give initial data for Maxwell's equations on a spacelike surface S. If there are closed timelike curves through S, we expect the radiation to be reflected off the « bumps » in the curvature and eventually to be re-registered on S. Thus, our initial data might, in general, be incompatible with the space-time geometry. Since the only restriction on electromagnetic fields that we observe in our local neighborhood is Maxwell's equations, there is at least a reason to expect that causality violation is not too severe in our neighborhood.

Let us try to formulate, mathematically, this additional causality condition on models of our universe. The most obvious choice is to require that there be no closed timelike curves. But is this the « right » condition: could a space-time have no closed timelike curves but nonetheless display acausal behaviour? The answer is « yes »: the space-time of Fig. 13 has no closed timelike curves, but a single closed null curve C.

Even to rule out closed causal curves is not the strongest possible condition. Remove from Fig. 13 the single point p, thus breaking the only closed causal curve in the space-time. This space still appears causally anomalous, for there are « almost closed » timelike curves through q. More precisely, given any sufficiently small neighborhod U of q, there is a timelike curve which begins in U leaves U, and then re-enters U. Such a space-time is said to violate *strong causality* [23]. This is precisely the type of causality violation one normally wants to rule out in proving theorems on the existence of singularities [24, 23].

Exercise. Prove that in a strongly causal space-time the topology follows from the causal structure [20].

Exercise. Prove that in a strongly causal space-time distinct points have distinct pasts [20].

There are even higher levels of causality violation than this strong causality (Fig. 14). What, then, is the ultimate causality condition? A space-time is said to be stably causal [3] if, when the metric is varied by an arbitrary but sufficiently small amount, there results a space-time with no closed timelike curves. The words « sufficiently small » suggest that the precise formulation of stable causality will require the notion of a topology on the set of all Lorentz metrics on a manifold. The appropriate topology is called F^0 [2]: a neighborhood of a metric g_{ab} consists of all metrics whose value at each point is within a certain range of g_{ab}, where this « range » varies continuously but otherwise arbitrarily over M. Most common space-times—Minkowski space, the Schwarzschild, Reissner-Nordström, and Friedmann solutions—are stably causal.

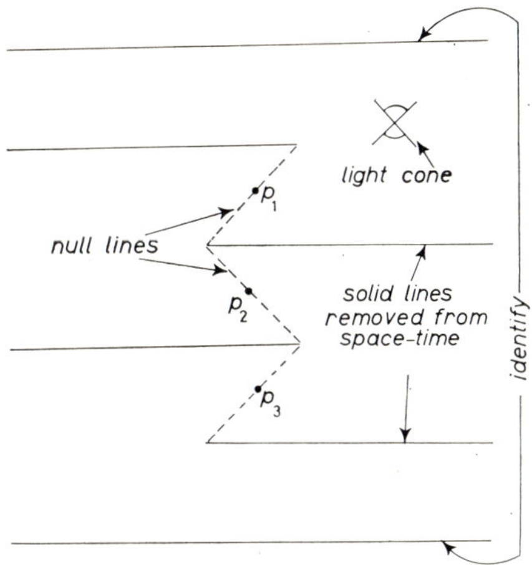

Fig. 14. – A space-time which violates a « higher-order » causality condition. Take a horizontal strip of 2-dimensional Minkowski space, identify the top and bottom edges, and remove from the manifold four (carefully chosen) horizontal half-lines (solid in the Figure). This space is strongly causal. However, given any neighbourhoods U_1, U_2 and U_3 of p_1, p_2 and p_3, respectively, there are past-directed timelike curves as follows: γ, from U_1 to U_2, γ', from U_2 to U_3, and γ'' (which crosses the identified edges) from U_3 to U_1. This situation never occurs in a stably causal space-time.

In fact, it is possible to give a general criterion for when a space-time is stably causal. Suppose that a space-time M has a time function, *i.e.* a scalar field t whose gradient is everywhere timelike. Then M is certainly stably causal. (Vary the metric at each point in such a way that $\nabla_a t$ remains timelike. The « varied » space-times also admit t as a time function, and so, since t increases along timelike curves, have no closed timelike curves.) It turns out that the converse is also true: every stably causal space-time has a time function [23]. Thus stable causality—the ultimate causality condition—is completely equivalent to the existence of a cosmic time (*). (The are, of course, many time functions, so there is no notion of simultaneity in a stably causal space-time.)

4. – Causal structure II. The domain of dependence.

In the previous Section we studied one aspect of causal structure: can event p influence event q by means of a signal? We shall now be concerned with a second aspect of causal structure: is the situation at an event p completely determined by that on a set S? Thus, whereas the domain of influence is defined by a rela-

(*) The equal-time surfaces need not, however, have all the properties usually associated with spacelike surfaces. In particular, they will, in general, be disconnected.

tion between points and points, the domain of dependence is defined by a relation between points and sets. The assumption of Sect. **3**—that all space-times are time-oreinted—remains in force in the present Section.

Let $\boldsymbol{M}$, g_{ab} be a space-time, and let S be a subset of $\boldsymbol{M}$. We assume that S is *achronal*, *i.e.* that no two points of S may be joined by a timelike curve (*). It is convenient to think of S as a spacelike surface on which to give initial data for, for example, Maxwell's equations. Since signals in relativity travel along causal curves, we expect the situation at a point p to be completely determined by that on S provided every such curve from p intersects S. More precisely, the future *domain of dependence* [5, 10, 23, 25] of S, $D^+(S)$, is the collection of all points p such that every past-directed timelike curve (without past end-point) from p intersects S. (It is convenient to include S in its own domain of dependence.) The past domain of dependence, $D^-(S)$, is defined in a similar way by interchanging the roles of past and future.

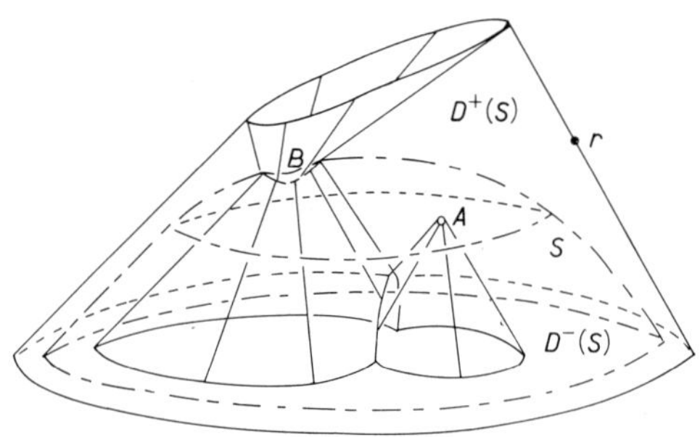

Fig. 15. – The domains of dependence of a surface S. The space-time is 3-dimensional Minkowski space with a small hole A removed. The surface S is a hyperboloid of revolution with a hole B removed. Where it not for the hole B in S, $D^+(S)$ would be the region between S and the null cone containing r. The effect of the hole is to delete from this set all points to the future of B. Were it not for the holes A and B, $D^-(S)$ would be just the past of S. The effect of these holes is to delete from this set the pasts of A and B. Many of the important features of the domain of dependence can be seen in this Figure.

The typical behaviour of the domain of dependence is illustrated in Fig. 15. The space-time is 3-dimensional Minkowski space with a hole, marked « A », removed. The achronal set S is a hyperboloid of revolution, also with a hole

(*) This assumption has two purpose. Firstly, it ensures that S represents the type of region on which we might give initial data, *i.e.* a surface which is locally space-like or null. Secondly, it automatically rules out causality violations in which data given on S may later be re-registered on S. For example, the space-time of Fig. 11 has no achronal sets at all.

removed. The sets $D^{\pm}(S)$ are shown in the Figure. We see from Fig. 15 that holes in either the space-time itself or in the surface S create « shadows » in the domain of dependence of S. Intuitively, information can enter or leave the space-time through these « holes » without its being registered on S.

We see from Fig, 15 that a past-direct *causal* curve from a point (*e.g.* r) of $D^+(S)$ need not necessarily intersect S. However, if p lies in the interior of $D^+(S)$, then every past-directed causal curve from p does intersect S [5]. To show this, suppose that p is in the interior of $D^+(S)$, and let Γ be a past-directed causal curve from p which fails to intersect S. Choose a point $q \in D^+(S)$ which lies to the future of p. It is fairly clear intuitively that one can always find a past-directed timelike curve γ from q which remains always to the future of Γ. Therefore, γ does not intersect S either, which contradicts our assumption that $q \in D^+(S)$.

Another property of the domain of dependence is that none of its points can lie on a closed timelike curve. The proof is easy. Suppose $p \in D^+(S)$, and let p lie on a closed timelike curve γ. If γ does not intersect S, then the curve obtained by continually describing γ has no past endpoint and does not intersect S, thus contradicting our assumption that $p \in D^+(S)$. On the other hand, if the closed timelike curve γ intersects S, then S cannot be achronal.

Exercise. Show that S is closed if and only if $D^+(S)$ is closed. Find an example for which $\overline{D^+(S)} \neq D^+(\bar{S})$.

Exercise. Let $p \in D^+(S)$. Show that $I^-(p) \cap I^+(S) \subset D^+(S)$.

The first important concept derived from the domain of dependence is that of a Cauchy surface [5, 10, 24]. An achronal set S is said to be a *Cauchy surface* for the space-time M if every point of M lies in either $D^+(S)$ or $D^-(S)$. (*Exercise.* Prove that a point lies in both $D^+(S)$ and $D^-(S)$ if and only if it lies in S.) That S be a Cauchy surface for M means, intuitively, that the entire evolution of M, past and future, is completely determined by S.

Does our own universe have a Cauchy surface? It is very difficult to gather evidence on this question. Consider, for example, the Reissner-Nordström solution (Fig. 16). In the immediate neighborhood of the surface S, it looks as though S might be a Cauchy surface. (The geometry in a neighborhood differs by a small amount from that in a neighborhood of a similar surface in the (extended) Schwarzschild solution, and in that case S is a Cauchy surface.) However, as the geometry evolves in the Reissner-Nordström solution, the « throat » of the wormhole first contracts and then re-expands, allowing points (p in the Figure) which are not in $D^+(S)$. Thus, S turns out not to be a Cauchy surface. However, it would have been difficult to predict this result in advance —*i.e.* by examining only a neighborhood of S. (*Exercise.* Show that the Friedmann models have a Cauchy surface.) It is a very strong condition on a space-time that it admit a Cauchy surface. We shall now derive some implications of this condition.

We have seen above that no point in the domain of dependence can lie on a closed timelike curve. Therefore, a space-time with a Cauchy surface cannot have any closed timelike curves.

We can also obtain information about the topology of the space-time. Choose an arbitrary timelike direction field $\alpha\xi^a$ on M. Since S is a Cauchy surface, every trajectory of $\alpha\xi^a$ intersects S, and, since S is achronal, no trajectory intersects S more than once. That is to say, we may describe each point p of M by giving i) the point at which the trajectory through p intersects S, and ii) the distance from p to S along that trajectory. Consequently, the space-time M, is topologically, $S\times R$ [5, 10].

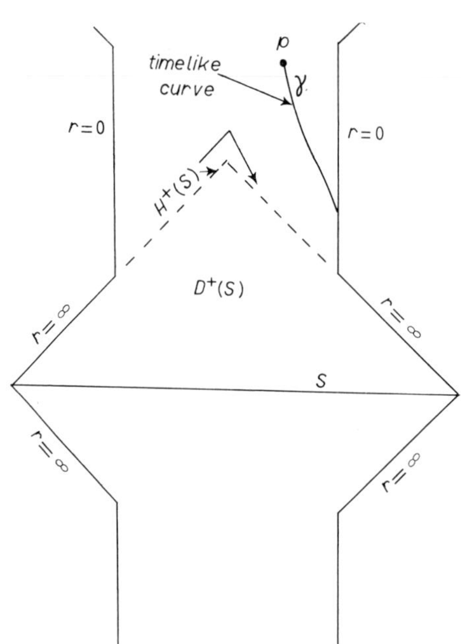

Fig. 16. – The domain of dependence of a surface S in the Reissner-Nordström solution. S is, topologically, $S^2\times R^1$. The point p is not in $D^+(S)$, for there is a past-directed timelike curve γ from p which does not meet S. $D^+(S)$ includes its future boundary, marked « $H^+(S)$ ».

In fact, the above is true in an even stronger form [5]: not only can M be written in the form of a product, $S\times R$, of a 3-manifold and the real line, but this product can be so chosen that, for each real number in R, the corresponding surface S is a spacelike Cauchy surface. (*E.g.*, a spacelike hyperplane is a Cauchy surface for Minkowski space, and Minkowski space can be filled with a one-parameter family of such hyperplanes.) It now follows that every space-time with a Cauchy surface is stably causal. (Choose the permitted variations in the metric at each point to be those under which the S-surfaces remain spacelike. Then each varied metric is also free of closed timelike curves.)

Exercise. Show that any two Cauchy surfaces in a space-time have the same topology.

Exercise. Find an example of a space-time which can be written as a product $S\times R$, with each embedded S spacelike, but which has no Cauchy surface.

One of the most useful tools for studying the domain of dependence is the Cauchy horizon [5, 10, 23]. The future *Cauchy horizon*, $H^+(S)$, of an achronal set S is defined as the «future boundary» of the domain of dependence of S, or, more precisely, as the set of all points of $D^+(S)$ no point to whose future is in $D^+(S)$. The future Cauchy horizon of a surface in the Reissner-Nordström solution is indicated in Fig. 16. (*Exercise.* Find the Cauchy horizons in Fig. 15.) The Cauchy horizon in these examples is a null-surface. This is a general and very important property of the Cauchy horizon. More precisely, we have the following result [5, 23]: let p be a point of $H^+(S)$, such that $p \notin S$. Then there is a past-directed null geodesic segment from p which lies in $H^+(S)$. This geodesic either remains entirely in $H^+(S)$, or else emerges from $H^+(S)$ at a point of the closure of S.

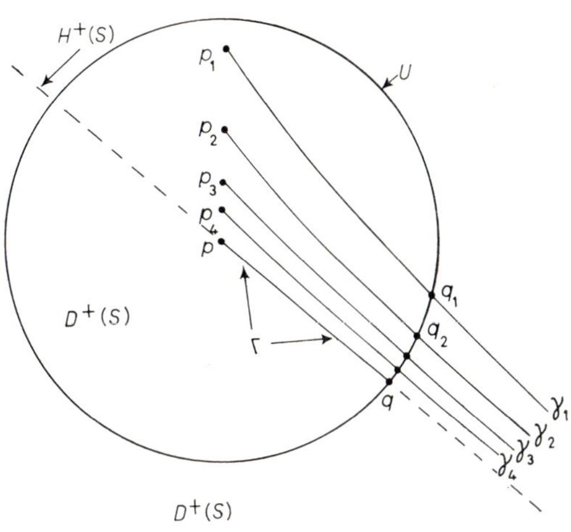

Fig. 17. – Constructing a null geodesic on the Cauchy horizon. The sequence p_i approaches $p \in H^+(S)$ from above. The timelike curves γ_i cannot enter $D^+(S)$. The points q_i, at which the γ_i exit from the neighborhood U, converge to a point $q \in H^+(S)$. Then p and q are joined by a null geodesic segment $\Gamma \subset H^+(S)$.

The proof is illustrated in Fig. 17. Let p be a point in $H^+(S)$. Draw a small neighborhood U of p which does not intersect S, and let p_i $(i = 1, 2, ...)$ be a sequence of points which lie in $I^+(p)$ and in U, and which approach p in the limit. By definition of $H^+(S)$, each of the p_i lies on a timelike curve γ_i which does not intersect S. The points q_i at which the γ_i leave the neighborhood U cannot lie to the past of p (for points in both U and $I^-(p)$ are in $D^+(S)$, while the γ_i, since they do not intersect S, cannot enter $D^+(S)$.) Therefore, the q_i have an accumulation point q which is in $H^+(S)$, and which is on the null cone of p. Draw the segment Γ of a null geodesic from p to q. We now repeat the same construction beginning from q. The resulting null-geodesic segment Γ' from q must have the same tangent as Γ at q, for otherwise Γ' would enter $I^-(p)$. Continuing in this way, we construct a past-directed null geodesic from p which lies in $H^+(S)$. This construction can only come to an end if, at some stage, we reach a point every neighborhood of which intersects S. That is, the null geodesic can emerge from $H^+(S)$ only at a point of the closure of S.

Exercise Let $p \in D^+(S)$. Show that any future-directed timelike curve from p which emerges from $D^+(S)$ does so at a point of $H^+(S)$.

Exercise. Find an example in which $H^+(S)$ has a closed null curve. Show that $H^+(S)$ is always achronal.

The above property of null geodesics is often the most convenient way to establish the existence of a Cauchy surface. One uses some condition (such as an energy condition) to show that null geodesics of the type required above cannot exist. It follows that S cannot have a Cauchy horizon and must, therefore, be a Cauchy surface.

Exercise. Show that if an achronal set S has no Cauchy horizons, then S is a Cauchy surface.

We conclude this Section with one further property of Cauchy surfaces. If S is a Cauchy surface, and p a point of M, then there is a timelike geodesic from p to S whose length is greater than (or equal to) that of every other causal curve from p to S. The proof [5, 24] is rather technical, and so we can only give it in outline. We have seen that every causal curve from p intersects S. This set of causal curves (*) is first given a topology. It is then shown that 1) this set of curves is compact, and 2) the length of a curve is a continuous function (**) on this set. It then follows that there is a curve of maximal length. It is clear that the maximal curve must be a geodesic (for otherwise we could lengthen it by straightening it out), and that it must intersect S orthogonally (for otherwise we could lengthen it by slightly moving its point of intersection with S). The existence of a maximal geodesic is frequently used in proving theorems on the existence of singularities [10, 24].

Problem. The domain of dependence is normally derived from the domain of influence. (This program has been carried out for certain causal spaces [26].) Do the reverse. That is, begin with the domain of dependence as the basic structure, subject to certain axioms, and derive from it the domain of influence.

5. – Asymptotic properties.

The purposes of this Section are to briefly describe Penrose's definition [8, 27] of asymptotic simplicity and to see what are the global implications of this definition. The idea of asymptotic simplicity is to avoid entirely the use of limits of co-ordinate-dependent quantities « as $r \to \infty$ ». Instead, one introduces a conformal transformation to bring « infinity » in to a finite boundary surface on the manifold. Asymptotic properties are then to be described as local geometrical properties of this surface $\mathscr{I}$.

Let M, g_{ab} be a space-time. Our first assumption is that there is given a

(*) It is necessary for this argument to include all continuous causal curves, whether or not they are differentiable.

(**) In fact, the length is only upper semi-continuous, but this suffices for the result.

boundary surface $\mathscr{J}$ for M (*). This manifold, consisting of M with the boundary $\mathscr{J}$ attached, will be denoted by $\tilde{M}$. It is the conformal geometry we wish to carry over from M to its boundary. We assume, therefore, that there is a metric $\tilde{g}_{ab}$ on $\tilde{M}$ such that $\tilde{g}_{ab} = \Omega^2 g_{ab}$ on M, where Ω is a nonegative scalar field defined on $\tilde{M}$.

We must impose some further conditions on this system to ensure that the definition will reflect our intuitive idea of asymptotic flatness. In the first place, we want $\mathscr{J}$ to represent « infinity », and not some finite hypersurface. We assume, therefore, that the conformal factor Ω approaches zero on $\mathscr{J}$, *i.e.* that there is an infinite « stretching » in passing from $\tilde{g}_{ab}$ to g_{ab}. Secondly, we want this stretching factor to behave « like r^{-1} » as we approach $\mathscr{J}$. This assumption requires that $\nabla_a \Omega$ be nonzero on $\mathscr{J}$ (**). Finally, we need some assumption to ensure that $\mathscr{J}$ represents the « entire surface at infinity », and not merely some portion of it. We therefore assume that every null geodesic in M intersect $\mathscr{J}$ in two points, one to the future and one to the past.

To summarize, a space-time M is *asymptotically simple* [27] if M can be given a boundary $\mathscr{J}$ and $\tilde{M}$, defined as M with $\mathscr{J}$ attached, can be given a non-singular metric $\tilde{g}_{ab}$, such that

1) $\tilde{g}_{ab} = \Omega^2 g_{ab}$ on M, where $\Omega > 0$ on M.
2) $\Omega = 0$ and $\nabla_a \Omega \neq 0$ on $\mathscr{J}$.
3) Every null geodesic in M intersects $\mathscr{J}$ in just two points.

It follows from these conditions, and the further assumption that Einstein's source-free equations (with respect to g_{ab}) hold in a neighborhood of $\mathscr{J}$, that $\mathscr{J}$ must be null [27]. We shall assume hereafter that $\mathscr{J}$ is null.

Minkowski space, for example, is asymptotically simple. The conformal factor can be chosen to be [8]

$$\Omega = (1 + x^2 + y^2 + z^2)^{-\frac{1}{2}}$$

in the usual co-ordinates. The resulting manifold $\tilde{M}$ is shown in Fig. 18. We see that $\mathscr{J}$ is divided into two parts, $\mathscr{J}^+$ and $\mathscr{J}^-$, consisting of points reached by future-directed and past-directed null geodesics, respectively.

Asymptotic simplicity, it turns out, implies practically every criterion of

(*) The resulting manifold with boundary will *not* be compact. That is, the « points » $I^0, I^{\pm}$ (in Penrose's notation) will not be considered here.

(**) Note that there is no conflict here with our earlier assertion that the limit of a vector field « at infinity » is not defined. $\nabla_a \Omega$ is a well-defined vector field on the manifold $\tilde{M}$, and we simply ask that it be nonzero at certain points of $\tilde{M}$, namely, at points of $\mathscr{J}$.

« niceness » we have discussed so far. As a review of Sect. **1-4**, we shall now derive some of these results.

The first and most important consequence of asymptotic simplicity is the existence of a Cauchy surface. To prove this result, first choose an area element $\mathrm{d}A$ on $\mathscr{I}$ such that the total area of $\mathscr{I}^+$ and the total area of $\mathscr{I}^-$ are both one. For each point p of M we define

$$A^{\pm}(p) = \int\limits_{I^{\pm}(p)\cap\mathscr{I}} \mathrm{d}A$$

(Fig. 18). Along any future-directed causal curve, A^+ is strictly decreasing and A^- is strictly increasing. Let S denote the surface given by $A^+ = A^-$. Since A^+ is decreasing and A^- increasing along every future-directed timelike curve, S is certainly achronal. Furthermore, A^+ (respectively, A^-) approaches zero along every future-directed (respectively, past-directed) timelike curve without endpoint (*). Therefore, every such curve intersects S. That is to say, S is an achronal set intersected by every timelike curve, and hence S is a Cauchy surface. The location of the Cauchy surface is shown in Fig. 18.

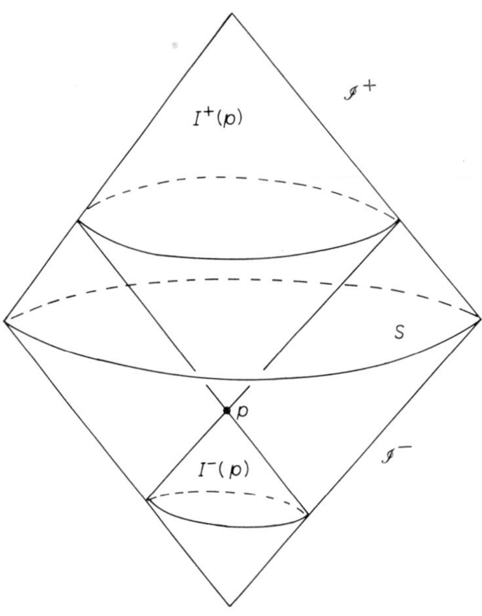

Fig. 18. – The asymptotic simplicity of Minkowski space (one spatial dimension suppressed). Every null geodesic intersects $\mathscr{I}^+$ in the future and $\mathscr{I}^-$ in the past. Each of $\mathscr{I}^\pm$ is a null surface, topologically $S^2\times R^1$. The light cones are roughly vertical, but become distorted near the boundary (« spacelike infinity ») between $\mathscr{I}^+$ and $\mathscr{I}^-$. (This « boundary » is not included in $\tilde{M}$.) The past and future of a point p are shown.

Every space-time with a Cauchy surface is stably causal. Therefore, every asymptotically simple space-time is stably causal and, in particular, has no closed timelike curves.

Since M has a Cauchy surface, M is topologically the product of that surface and the real line. To obtain further information about the topology of M, we use a method of counting null geodesics (**). Let N denote the collection of all null geodesics in M. This set is a manifold. The idea is to describe N

(*) If the curve γ reaches $\mathscr{I}^+$, then the limit of $I^+(p)$ $(p\in\gamma)$ is a single null generator of $\mathscr{I}^+$. If the curve fails to reach $\mathscr{I}^+$ (*i.e.* goes to « I^+ »), then the limit is empty.

(**) The argument outlined below arose in the course of a conversation with R. Penrose.

in terms of S and again in terms of $\mathscr{I}^+$ (since all null geodesics intersect both S and $\mathscr{I}^+$).

Let p be a point of S. The set of null geodesics emanating from p is just the set of null directions at p, and hence is, topologically, a 2-sphere S^2. Therefore, the topology of N is $S \times S^2$ (*). Let q be a point of $\mathscr{I}^+$. The set of null geodesics emanating from q is a 2-sphere. However, one of these null geodesics lies in $\mathscr{I}^+$, and so does not enter M. Hence, the null geodesics from q which do enter M are, topologically, a 2-sphere minus a single point, *i.e.* a plane R^2. Hence, N is topologically $\mathscr{I}^+ \times R^2$. But $\mathscr{I}^+$ is the union of its null generators, and so we have $\mathscr{I}^+ = R \times K$, where K is some 2-manifold. Therefore, $N = K \times R^3$.

We have shown that $S \times S^2 = K \times R^3$. This equation can hold only if K is topologically a 2-sphere and S is topologically R^3 (**). Since K is an S^2, $\mathscr{I}^+$ (and similarly $\mathscr{I}^-$) is topologically $S^2 \times R$. This is a well-known result of PENROSE [27].

Since S is topologically R^3, M itself is topologically R^4. That is, every asymptotically simple spacetime is topologically the same as Minkowski space. In particular, M is time, space, and charge orientable.

APPENDIX

Hausdorff, connected and paracompact manifolds.

In Sect. 3 we discussed some of the pathological situations which can occur in the causal structure of a space-time. It is natural to ask whether any similar anomalies can arise in the underlying manifold itself. There are, roughly speaking, three classes of pathological manifolds. A non-Hausdorff manifold contains points which cannot be properly separated by open sets. A disconnected manifold can be divided into two or more disjoint manifolds. A nonparacompact manifold is « too large ». One normally assumes that the manifolds dealt with in relativity are Hausdorff, connected, and paracompact. Furthermore, it has not so far been useful to relax any of these assumptions. Since these three assumptions often appear in the statements of theorems, however, it is of interest to know just what situations are thereby being ruled out.

A manifold (or, more generally, a topological space) is said to be *Hausdorff* if any two distinct points are contained in disjoint open sets. Although

(*) Strictly speaking, we have only shown that N is the direction bundle of S. However, the direction bundle of every orientable 3-manifold is trivial [33].

(**) To prove this statement, take the second homotopy group [1] of each side. It follows thet $\pi_2(K)$ must be nonzero. Since K is a 2-manifold, it must be topologically S^2. It then follows that S is topologically R^3.

the property of being Hausdorff does not follow from the usual definition of a manifold (*), almost all manifolds encountered in relativity are Hausdorff.

An example of a non-Hausdorff manifold is given in Fig. 19. The points p and p' have the property that every open set containing p intersects every open set containing p'. What is pathological about Fig. 19? Consider the vector field indicated in the Figure. We ask for the trajectory of this field which passes through the point r. This trajectory is unique until we reach the edge of the half-plane A. However, the curve can then continue, as a trajectory of

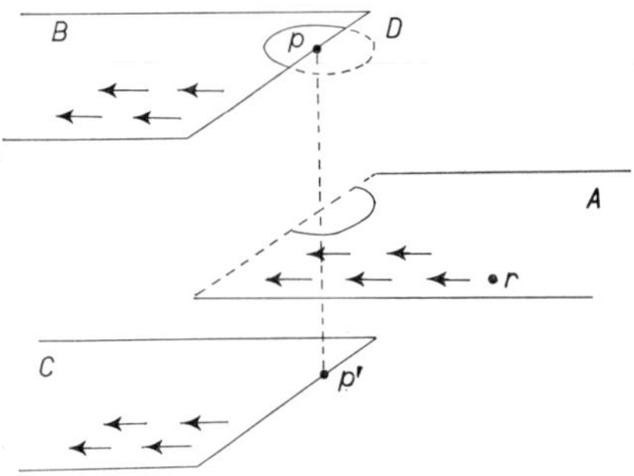

Fig. 19. – A non-Hausdorff 2-manifold. The topology in the interiors of the half-planes A, B and C is the usual plane topology. To construct a neighborhood of a point p on the edge of B, draw a horizontal disk D centered on p. The intersection of D with B plus the projection of the overhang of D onto A, defines a neighborhood of p. Similarly for each point p' on the edge of C. (A itself has no edge.) Every neighborhood of p intersects every neighborhood of p'.

our vector field, into either B or C. It is normally in situations in which one wishes to have uniqueness of such trajectories (or, similarly, of geodesics) that one must impose the condition of Hausdorffness. Some further examples of non-Hausdorff manifolds are given in [28]. In fact, the most natural extension of Taub space [29], in which both NUT extension [30, 31] are carried out simultaneously, results in a non-Hausdorff manifold [32].

A manifold is said to be *connected* if any two points on the manifold can be joined by a continuous curve. For example, the manifold consisting of two disjoint planes is not connected. Physically, a disconnected space-time represents two or more separate universes: no matter what the metric is, observers on different components of the space can never communicate with each other. It is more convenient, therefore, to study each connected component separately, that is, to impose the condition that models of our universe are connected.

The final condition to be placed on manifolds is that they be not « too large ». This condition can be stated in a number of equivalent ways.

Theorem. Let M be a connected Hausdorff manifold. The following conditions on M are equivalent (**).

1) A countable number of co-ordinate patches suffice to cover M.

(*) The situation with regard to the other separation axioms is as follows. Every manifold is T_0 and T_1. Every Hausdorff (T_2) manifold is T_3 (regular). To ensure that a Hausdorff manifold be T_4 (normal), it is necessary to impose some further condition, for example, the equivalent conditions of the theorem below.

(**) A space is said to be *separable* if it possesses a countable dense subset. For manifolds, the equivalent conditions of the theorem imply separability, but the converse is not true.

2) M may be given a nonsingular positive-definite metric h_{ab} [33].

3) M may be given a nonsingular connection [34].

4) If U_α is any collection of open sets whose union is M, then there is a countable number of U_α whose union is also M.

5) M way be written as the union of a countable collection of compact sets.

6) M is triangulable [35].

7) M has a countable basis for its topology [36].

8) M is paracompact [36].

It is a good exercise in topology to prove that 1), 4), 5), 7), and 8) are equivalent. It follows in particular from condition 3) that every space-time is based on a paracompact manifold, for the metric defines a nonsingular connection.

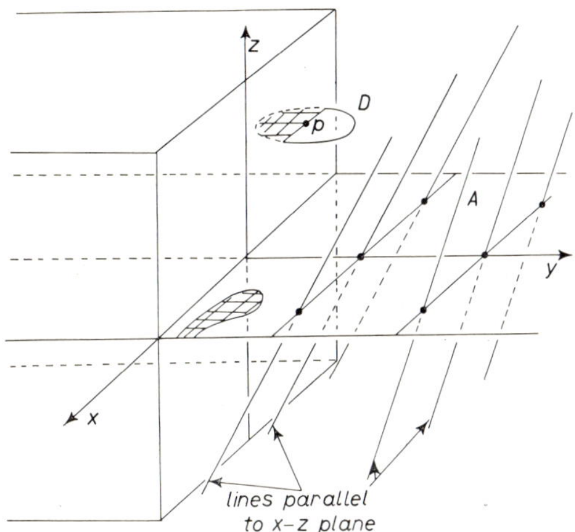

Fig. 20. – A nonparacompact 2-manifold [37]. The manifold consists of the (edgeless) half-plane A and the one-parameter stack of (edged) half-planes on the left. The straight lines are all parallel to the x-z plane and have a slope depending only on y. This slope approaches 1 for $y \to \infty$ and 0 for $y \to 0$. The topology is the usual plane topology in the interiors of the half-planes. To construct a neighborhood of a point p on an edge, draw a horizontal disk D through p. The intersection of D with its half-plane, plus the points of A intersected by the straight lines through D, defines a neighborhood of p (the hatched region). Although connected and Hausdorff, this 2-manifold is « too large » to be paracompact.

We conclude this Appendix with an example [37] of a manifold which is not paracompact. Let A denote the half-plane $z = 0$, $y > 0$ in 3-dimensional space. The manifold M consists of the half-plane A along with the family of half-planes $z =$ const, $y \leqslant 0$ (Fig. 20). In the interiors of these half-planes, the topology is the usual topology of the plane. It is only necessary to state what are neighborhoods of points on the edges of the half-planes, *i.e.* at $y = 0$.

For this purpose draw, for $y > 0$, a collection of straight lines which are parallel to the x-z plane, and whose slope depends only on y. The slope is to approach 1 as $y \to \infty$, and 0 as $y \to 0$. A neighborhood of a point p is constructed as follows. Draw a horizontal open disk D through p. An open neighborhood of p consists of all points in the left half-plane intersected by D and all points of A which are intersected by lines which pass through D. (*Exercise*: Verify that Fig. 20 is a connected Hausdorff manifold.) This manifold M is not paracompact, for there exists an uncountable collection of points of M (namely, the points $x = y = -1$) without any accumulation point, which violates, for example, condition 7) above.

REFERENCES

[1] A. H. WALLACE: *Algebraic Topology* (New York, 1963).
[2] R. GEROCH: *Singularities*, article in: *Relativity*, edited by M. CARMELI, S. FICKLER and L. WITTEN (New York, 1970).
[3] S. W. HAWKING: *Proc. Roy. Soc.*, **308** A, 433 (1969).
[4] A. AVEZ: *Ann. Inst. Fourier*, **13**, 105 (1963).
[5] R. GEROCH: *Journ. Math. Phys.*, **11**, 437 (1970).
[6] J. R. MUNKRES: *Elementary Differentiable Topology*, in *Ann. Math. Studies*, No. 54 (Princeton, 1954).
[7] J. C. GRAVES and D. R. BRILL: *Phys. Rev.*, **120**, 1507 (1960).
[8] R. PENROSE: *Conformal treatment of infinity*, article in *Relativity, Groups and Topology*, edited by C. DEWITT and B. DEWITT (New York, 1964).
[9] R. ABRAHAM and J. ROBBIN: *Transversal Mappings and Flows* (New York, 1967).
[10] R. PENROSE: *Structure of space-time*, article in *Battelle Rencontres*, edited by C. DEWITT and J. A. WHEELER (New York, 1968).
[11] L. MARKUS: *Ann. Math.*, **62**, 411 (1955).
[12] R. W. BASS and L. WITTEN: *Rev. Mod. Phys.*, **29**, 452 (1957).
[13] R. GEROCH: *Journ. Math. Phys.*, **8**, 782 (1967).
[14] R. H. DICKE: *Science*, **129**, 621 (1959).
[15] R. GEROCH: *Singularities in the space-time of general relativity*, Ph. D. Thesis, (department of Physics, Princeton University, (June 1967).
[16] J. H. CHRISTENSON, J. W. CRONIN, V. L. FITCH and R. TURLEY: *Phys. Rev. Lett.*, **13**, 138 (1964).
[17] C. S. WU: *Phys. Rev.*, **105**, 1413 (1957).
[18] R. L. GARWIN, L. M. LEDERMAN and M. WEINRICH: *Phys. Rev.*, **105**, 1415 (1957).
[19] R. F. STREATER and A. S. WIGHTMAN: *PCT, Spin and Statistics and All That* (New York, 1964).
[20] E. H. KRONHEIMER and R. PENROSE: *Proc. Camb. Phil. Soc.*, **63**, 481 (1967).
[21] P. HAVAS: *Synthèse*, **18**, 75 (1968).
[22] H. REICHENBACH: *The Philosophy of Space and Time* (New York, 1958).
[23] S. W. HAWKING: *Proc. Roy. Soc.*, **300** A, 187 (1967).
[24] S. W. HAWKING: *Proc. Roy. Soc.*, **249** A, 511 (1966).
[25] R. COURANT and D. HILBERT: *Methods of Mathematical Physics*, Vol. **2** (New York, 1965).

[26] R. GEROCH, E. H. KRONHEIMER and R. PENROSE: *Ideal points in space-time* in *Proc. Roy. Soc.* (submitted for publication).
[27] R. PENROSE: *Proc. Roy. Soc.*, **284** A, 159 (1965).
[28] R. GEROCH: *Journ. Math. Phys.*, **9**, 450 (1968).
[29] A. H. TAUB: *Ann. of Math.*, **53**, 472 (1951).
[30] C. W. MISNER: *Journ. Math. Phys.*, **4**, 924 (1963).
[31] E. T. NEWMAN, L. TAMBURINO and T. UNTI: *Journ. Math. Phys.*, **4**, 915 (1963).
[32] R. GEROCH: *The structure of singularities*, article in *Battelle Recontres*, edited by C. DEWITT and J. A. WHEELER (*New York*, 1968).
[33] N. STEENORD: *The Topology of Fibre Bundles* (Princeton, 1951).
[34] R. GEROCH: *Journ. Math. Phys.*, **9**, 1739 (1968).
[35] J. H. C. WHITEHEAD: *Ann. Math.*, **41**, 809 (1940).
[36] J. L. KELLY: *General Topology* (New York, 1955).
[37] G. SPRINGER: *Riemann Surfaces* (Reading, Mass., 1957), p. 56.

Relativistic Cosmology.

G. F. R. ELLIS

Department of Applied Mathematics and Theoretical Physics
University of Cambridge - Cambridge

1. – Introduction.

The aim of cosmology is to determine the large-scale structure of the physical universe.

One rather vague assumption is at the basis of virtually all present-day work on the subject. We shall call this (following BONDI [1]) the *Copernican Principle.* It is the assumption:

« We do not occupy a privileged position in space-time ». We may apply this to two different kinds of experiment. First, considering laboratory experiments, we may interpret the principle as implying that local physical laws are the same everywhere in the universe. Secondly, considering astronomical observations, we may assume that our view of the universe is not a preferred picture. (We can, if we wish, proceed without this second assumption; the result is to obtain, beside the picture of the universe discussed in Sect. **7**, more general situations which can be regarded as rather unlikely models of the universe.)

The over-all picture we observe is well known (see BONDI [1], MCVITTIE [2]): on a moderately large scale ($\geqslant 3\cdot 10^8$ light years) the distribution of clusters of galaxies is approximately isotropic about us. The general motion is an over-all expansion, random velocities relative to this general motion being rather small. Thus we are able to determine an average velocity vector which represents to a good approximation the over-all motion of matter in our « local » vicinity; the Copernican principle then suggests that we can assume the existence of such a vector at each point of space-time. The principle will also be taken to imply that corresponding to the very nearly isotropic distribution of matter and background radiation about us, any observer moving with the average velocity vector will see a very nearly isotropic distribution of matter and radiation on his celestial sphere.

Our aim will be to determine a cosmological model which will correctly predict the results of astronomical observations, and whose behaviour is determined by those physical laws which describe the behaviour of matter on scales up to those of clusters of galaxies. To find such laws, we remember that within the solar system the dominant long-range force is gravity. (We shall assume over-all electrical neutrality of heavenly bodies.) Accepting (with some reservations until the consequences of Dicke's solar oblateness measurements are completely understood) that general relativity is the best theory of gravitation available for phenomena up to the scale of the solar system, we shall extrapolate by assuming it is also, to a good approximation, the law valid for gravity acting on scales about that of clusters of galaxies. If this eventually proves to lead to predictions inconsistent with observations, we may have to modify or abandon general relativity; however we would like to avoid *ad hoc* modifications introduced solely for cosmological convenience! In these lectures we shall therefore assume that general relativity is valid. (The alternative theories so far proposed may be studied by much the same methods as will be used here to study the conventional theory.)

On this view, the geometry of space-time will be determined by its energy content. We may describe the matter and radiation content of the universe in two convenient ways: by using a particle distribution function (as discussed in Ehlers' lectures) or by using a fluid approximation. It is the latter course (first introduced by EINSTEIN in 1917 [3], and followed in nearly all the classical cosmological discussions) we follow in these lectures: a continuum approximation is used, and the average velocity vector may be thought of as representing the velocity of fluid « particles », which are in this case clusters of galaxies.

In Sect. **2-5**, we shall obtain dynamical relations valid in any cosmological model, and illustrate these relations by applying them to some particular cosmological models. There is a very close parallel between fluid dynamics in Newtonian theory and in general relativity, and we shall emphasize this by giving corresponding relations in parallel. The basis of this close correspondence is that the fluid equations we shall use are based on *relative* motion. Corresponding to the applications we have in mind, the Newtonian fluid will always be a self-gravitating fluid.

In Sect. **6**, we shall derive observational relations valid for any cosmological model and give as an illustration of the methods used, their application to the standard isotropic cosmological models. (The optical relations valid in special relativity are a special case of these observational relations.)

In Sect. **7**, we shall present a broad over-all picture of the observable universe, on the basis of recent observations. This over-all picture serves as a background for more detailed observational discussions.

Conventions used are as follows:

in *general relativity*,

arbitrary coordinates x^a (Latin letters run from 1 to 4) are used in the 4-dimensional space-time with metric tensor g_{ab} (signature $(-+++)$). A semi-colon (;) denotes the covariant derivative with respect to the metric g_{ab}; so $g_{ab;c}=0$.

in *Newtonian theory*,

unless otherwise stated, fixed curvilinear co-ordinates x^ν (Greek letters run from 1 to 3) and a natural time co-ordinate t is used in the (3+1)-dimensional space-time. A comma (,) denotes the covariant derivative with respect to the flat 3-space metric $h_{\mu\nu}$ (signature $(+++)$); so $h_{\mu\nu,\sigma}=0$.

Square brackets denote skew-symmetrization and round brackets denote symmetrization (SCHOUTEN [4]), so $T_{(ab)}=\frac{1}{2}(T_{ab}+T_{ba})$,

$$S_{[abc]}=\tfrac{1}{6}(S_{abc}+S_{cab}+S_{bca}-S_{acb}-S_{bac}-S_{cba})\,.$$

The general-relativity notation and units we use are the same as those used by EHLERS in this volume; in particular, note that we use units in which the speed of light is unity. As a notational convenience, we shall often use the general-relativity name of a quantity to refer to its Newtonian analogue as well.

This article is a revised and corrected version of lectures given at Varenna. It presupposes a knowledge of the basic ideas and observations of cosmology (see for example BONDI [1], RINDLER [5]). It should be supplemented by review articles written from other view-points, such as those by HECKMANN and SCHÜCKING [6, 7], ZEL'DOVICH [8], DAVIDSON and NARLIKAR [9], RINDLER [10], SCHÜCKING [11] and SANDAGE [12]; it is intended to be complementary to the articles by EHLERS and SCIAMA in this volume. I should like to thank S. W. HAWKING, J. EHLERS and M. A. H. MACCALLUM for many useful discussions, M. A. H. MACCALLUM for assistance in preparing these notes, and M. STRANGLEMAN for patiently and carefully typing them.

2. – Kinematical quantities.

2·1. *The velocity vector field.* – We assume the existence of a unique vector field representing the average velocity of matter (or of some particular component of matter) at each point of space-time. Thus there is defined,

in *general relativity*:

the normalized 4-velocity field u^a; so

(2.1) $$u_a u^a=-1\,.$$

in *Newtonian theory*:

the 3-velocity

$$v^\nu\,.$$

To express these vectors in simple form we use

normalized comoving co-ordinates (y^ν, s),

Lagrangian co-ordinates (y^ν, t),

defined locally as follows. In some arbitrarily chosen space-section (in Newtonian theory, one of the natural space sections) of space-time, label the fluid particles by co-ordinates y^ν. At all later times, label the same particles by the same co-ordinate values, so that the fluid flow lines in space-time are the curves $\{y^\nu = \text{const}\}$. Determine the time co-ordinate by measuring proper time, from the initial space section, along the flow lines. In terms of these co-ordinates, the velocity vector takes the form

$$u^a \overset{*}{=} \delta^a{}_0 .$$

$$v^\nu \overset{*}{=} 0 .$$

In terms of

general co-ordinates x^a,

co-ordinates (x^ν, t), where x^ν are fixed curvilinear co-ordinates,

the velocity vectors are given by

$$u^a = \left.\frac{\mathrm{d}x^a}{\mathrm{d}s}\right|_{y^\nu=\text{const}}$$

$$v^\nu = \left.\frac{\mathrm{d}x^\nu}{\mathrm{d}t}\right|_{y^\sigma=\text{const}} .$$

2·1.1. The projection tensor.

A $\{3+1\}$ splitting of space-time into $\{\text{space} + \text{time}\}$ is

determined at each point by u^a. The tensor

$$h_{ab} := g_{ab} + u_a u_b \tag{2.2}$$

is, at each point, a projection tensor into the rest space of an observer moving with 4-velocity u^a.

given and absolute. The Euclidean metric tensor

$$h_{\mu\nu}$$

is given in local orthonormal axes by

$$\delta_{\mu\nu} .$$

These tensors are 3-dimensional projection tensors:

$$h_{ab}u^b = 0, \quad h_a{}^c h_c{}^b = h_a{}^b, \quad h_a{}^a = 3 .$$

$$h_\mu{}^\nu h_\nu{}^\sigma = h_\mu{}^\sigma, \quad h_\mu{}^\mu = 3 .$$

By (2.2), the expression for ds^2 in general relativity is:

$$ds^2 := g_{ab}\,dx^a\,dx^b = h_{ab}\,dx^a\,dx^b - (u_a\,dx^a)^2;$$

an observer at the space-time point x^a moving with 4-velocity u^a assigns to the event $x^a + dx^a$ a spatial separation $(h_{ab}\,dx^a\,dx^b)^{\frac{1}{2}}$, and a time separation $|u_a\,dx^a|$ from himself. Note that the 3-planes defined in each tangent space by h_{ab} do *not* in general mesh together to form 3-surfaces in space-time; the condition that they do so is given in Subsect. **4**.4.1.

2.1.2. Volume elements. Effective volume elements in the instantaneous rest spaces of the co-moving observers are given by

$\eta_{abcd}\,u^d$	$\eta_{\mu\nu\sigma}$

where the totally-skew pseudotensors η are defined by

$\eta^{abcd} = \eta^{[abcd]}, \quad \eta^{1234} = (-g)^{-\frac{1}{2}},$ $g := \det(g_{ab})\,.$	$\eta^{\mu\nu\sigma} = \eta^{[\mu\nu\sigma]}, \quad \eta^{123} = (h)^{-\frac{1}{2}},$ $h := \det(h_{\mu\nu})\,.$

(In terms of an orthonormal frame, these are the alternating quantities ε^{abcd}, $\varepsilon^{\mu\nu\sigma}$.)

The identities

$\eta^{abcd}\eta_{efgh} = -4!\,\delta_e{}^{[a}\delta_f{}^{b}\delta_g{}^{c}\delta_h{}^{d]},$	$\eta^{\alpha\beta\gamma}\eta_{\mu\nu\sigma} = 3!\,\delta^{[\alpha}{}_{\mu}\delta^{\beta}{}_{\nu}\delta^{\gamma]}{}_{\sigma},$

and the further identities obtained from these by contracting pairs of indices, are frequently useful in proving formulae given in the sequel. (The — sign in the general-relativity case appears because the determinant g is negative.)

2.2. *Time derivatives.* – We write the effective time derivative of a tensor T, measured by an observer moving with the standard velocity, as $\dot{T}$. Thus

(2.3) $\quad \dot{T}^{a\ldots b}{}_{c\ldots d} := T^{a\ldots b}{}_{c\ldots d;e}\,u^e$ is the covariant derivative along the particle world lines.	(2.3′) $\quad \dot{T}^{\alpha\ldots\beta}{}_{\gamma\ldots\delta} := \partial T^{\alpha\ldots\beta}{}_{\gamma\ldots\delta}/\partial t + T^{\alpha\ldots\beta}{}_{\gamma\ldots\delta,\varepsilon}\,v^\varepsilon,$ is the convective derivative for motion with the fluid.

2·3. *The acceleration vector.* – We can represent the combined effects of gravitational and inertial forces on the fluid by the vectors

(2.4) $$\dot{u}^a := u^a{}_{;b} u^b$$

the « acceleration vector », which is spacelike as (2.1) implies $\dot{u}^a u_a = 0$.

(2.4′) $$a_\nu := (v_\nu)^{\bullet} + \Phi_{,\nu}$$

where Φ is the gravitational potential (force $= -\operatorname{grad}\Phi$).

In co-moving co-ordinates, we find

$$\dot{u}^a \overset{*}{=} \Gamma^a{}_{00}$$

$$a_\nu \overset{*}{=} \frac{\partial \Phi}{\partial x^\nu}\,.$$

Note that even in Newtonian theory, we are unable to separate the gravitational and inertial parts of a_ν invariantly if the density of matter does not go to zero at infinity; see the discussions of HECKMANN and SCHÜCKING [13], BONDI [1], TRAUTMANN [14], and references cited there.

2·4. *Relative motion of neighbouring particles.*

2·4.1. The relative position vector. Consider the world-lines of neighbouring particles 0 and G, labelled by co-moving co-ordinates y^ν and $y^\nu + \delta y^\nu$ respectively. The vector X^a with components given in co-moving co-ordinates by $X^\nu \overset{*}{=} \delta y^\nu$ (and in general relativity, $X^0 \overset{*}{=} 0$) joins these same two world-lines at all times; we therefore call it a *connecting vector.*

Using general co-ordinates in general relativity, this vector has components $X^a = (\partial x^a/\partial y^\nu)\,\delta y^\nu$. If we calculate $\dot{X}^a$, we find

$$\dot{X}^a = X^a{}_{;b} u^b = \{\partial(\partial x^a/\partial y^\nu\,\delta y^\nu)/\partial x^b + \Gamma^a{}_{bc}(\partial x^c/\partial y^\nu\,\delta y^\nu)\}\frac{\partial x^b}{\partial s} =$$

$$= \{\partial(\partial x^a/\partial s)/\partial x^c + \Gamma^a{}_{cb}\,\partial x^b/\partial s\}\frac{\partial x^c}{\partial y^\nu}\,\delta y^\nu = u^a{}_{;b} X^b\,,$$

since $\partial^2 x^a/\partial y^\nu \partial s = \partial^2 x^a/\partial s\,\partial y^\nu$ and $\Gamma^a{}_{bc} = \Gamma^a{}_{cb}$.

(Geometrically, this is the statement and that the Lie derivative of $\boldsymbol{X}$ with respect to $\boldsymbol{u}$ vanishes.) A similar calculation shows that in Newtonian theory, $\dot{X}_\nu = v_{\nu,\mu} X^\mu$. Thus a connecting vector obeys the differential equation

(2.5) $$\dot{X}_a = u_{a;b} X^b\,.$$

(2.5′) $$\dot{X}_\nu = v_{\nu,\mu} X^\mu\,.$$

In general, a connecting vector will not (in general relativity) lie in the rest space of an observer moving with 4-velocity u^a. We obtain the *relative position vector* $X^a_\perp$ of G with respect to O, by projecting the connecting vector X_a into the rest-frame of O:

$$X_{\perp a} := h_a{}^b X_b . \qquad\qquad X_{\perp\nu} := h_\nu{}^\mu X_\mu = X_\nu .$$

2·4.2. Relative velocity. The *relative velocity vector* V^a of G with respect to 0 is defined to be the spatial part of the observed rate-of-change of the relative position vector:

(2.6) $\quad V^a := h^a{}_b (X_\perp{}^b)^\bullet ,$

so

$$V^a = h^a{}_b (X^c h_c{}^b)_{;d}\, u^d .$$

(2.6′) $\quad V^\mu := h^\mu{}_\nu (X_\perp{}^\nu)^\bullet ,$

so

$$V^\mu = (X^\mu)^\bullet .$$

Since X^a obeys the eq. (2.5), it follows (*) that

(2.7a) $\quad V^a = v^a{}_b X_\perp{}^b ,$

where

(2.7b) $\quad v_{ab} := h_a{}^c h_b{}^d u_{c;d} ;$

(2.7a′) $\quad V^\mu = v^\mu{}_\nu X_\perp{}^\nu ,$

where

(2.7b′) $\quad v_{\mu\nu} := v_{\mu,\nu} ;$

in each case, the relative velocity vector of a neighbouring particle is related to the relative position vector of that particle by a linear transformation, and the tensor determining that transformation is the spatial gradient of the velocity vector.

To examine this relation further, we split v_{ab} into its symmetric and skew-symmetric parts:

(2.8) $\quad v_{ab} = \theta_{ab} + \omega_{ab} ,$

where

$$\theta_{ab} = \theta_{(ab)} , \qquad \omega_{ab} = \omega_{[ab]} ,$$

so

$$\theta_{ab} u^b = 0 = \omega_{ab} u^b ,$$

(2.8′) $\quad v_{\mu\nu} = \theta_{\mu\nu} + \omega_{\mu\nu} ,$

where

$$\theta_{\mu\nu} = \theta_{(\mu\nu)} , \qquad \omega_{\mu\nu} = \omega_{[\mu\nu]} ,$$

(*) The essential properties of the covariant derivative are that it i) is linear, ii) obeys Leibniz' rule, and iii) preserves the metric, irrespective of whether indices are up or down. For example,

$$(\tfrac{1}{2} T_{ab} + f g_{ab})_{;c} = \tfrac{1}{2} T_{ab;c} + f_{;c}\, g_{ab} ;$$

$(S^{\nu\mu} X_\mu)_{,\sigma} = S^{\nu\mu}{}_{,\sigma} X_\mu + S^{\nu\mu} X_{\mu,\sigma}$. We occasionally need the expressions involving the Christoffel symbols $\Gamma^a{}_{bc}$, for example in deriving (2.5), (2.5′) and in proving the relations

$$\eta^{abcd}{}_{;e} = 0 = \eta^{\mu\nu\sigma}{}_{,\varkappa} , \qquad k^a{}_{;a} = \frac{1}{\sqrt{|g|}} \frac{\partial}{\partial x_a} (\sqrt{|g|}\, k^a) .$$

and the symmetric part into its trace and its trace-free part:

(2.9) $\theta_{ab} = \sigma_{ab} + \frac{1}{3}\theta h_{ab}$,

where

$\sigma^a{}_a = 0$, so $\sigma_{ab} = \sigma_{(ab)}$,

$\sigma_{ab} u^b = 0$ and $\theta = u^a{}_{;a}$.

(2.9′) $\theta_{\mu\nu} = \sigma_{\mu\nu} + \frac{1}{3}\theta h_{\mu\nu}$,

where

$\sigma^\mu{}_\mu = 0$, so $\sigma_{\mu\nu} = \sigma_{(\mu\nu)}$,

and $\theta = v^\nu{}_{,\nu}$.

The quantities we have defined determine the first derivatives of the velocity vector completely:

(2.10) $u_{a;b} = \omega_{ab} + \sigma_{ab} + \frac{1}{3}\theta h_{ab} - \dot{u}_a u_b$.

(2.10a) $v_{\mu,\nu} = \omega_{\mu\nu} + \sigma_{\mu\nu} + \frac{1}{3}\theta h_{\mu\nu}$,

(2.10b) $\partial v_\nu/\partial t = a_\nu - v_{\nu,\mu} v^\mu - \Phi_{,\nu}$.

We now split the relative position vector $X_{\perp_a}$ of G relative to 0 into a *relative distance* δl and a *direction* n^a; then $X_{\perp_a} = n^a \delta l$ where $n_a n^a = 1$, so $(\delta l)^2 = h^{ab} X_a X_b$ (and $n^a u_a = 0$ in general relativity).

Equating the right-hand sides of (2.6) and (2.7), we now find that *the rate-of-change of relative distance* is

$$\frac{(\delta l)^\bullet}{(\delta l)} = \theta_{ab} n^a n^b = \sigma_{ab} n^a n^b + \frac{1}{3}\theta , \tag{2.11}$$

in both cases, and that the *rate-of-change of direction* is

$$h_a{}^b (n_b)^\bullet = (\omega_a{}^b + \sigma_a{}^b - (\sigma_{cd} n^c n^d) h_a{}^b) n_b . \tag{2.12}$$

These two equations are equivalent to (2.7).

Applying these equations to a cosmological model (2.11) is a generalized Hubble law allowing for possible anisotropic expansion; it is valid for distances large enough to ensure random velocities are small compared with velocities associated with the average motion of matter, but small enough for the Hubble relation to be linear and also for the change in distance of the galaxies to be relatively small during the time of light travel between the galaxies and the observer. Thus we might expect its range of validity to be roughly from 50 to 500 Mpc. Relation (2.12), which we might expect to be valid for roughly the same length scales, gives the rate-of-change of direction of neighbouring fluid elements with respect to

a Fermi-propagated basis of orthonormal vectors:

$$h_a{}^b(e_b)^{\bullet} = 0 \tag{2.13}$$

for each basis vector e_b,

$$e^a e_a = 1\,, \qquad e^a u_a = 0\,.$$

a nonrotating basis of orthonormal vectors:

$$(e_\nu)^{\bullet} = 0 \tag{2.13'}$$

for each basis vector e_ν,

$$e_\nu e^\nu = 1\,.$$

Thus this relation gives the rate-of-change of position in the sky of neighbouring clusters of galaxies with respect to axes at rest in a local inertial rest frame (a rest frame determined by gyroscopes or other modern refinements of Newton's bucket experiment). [Detailed discussions showing that (2.13) corresponds to a dynamically nonrotating set of axes may be found in SYNGE [15] and TRAUTMANN [16].]

2'5. *The kinematic quantities.* – It is probably easiest to understand the implications of (2.11), (2.12) by considering how a sphere of fluid particles changes during the elapse of a small increment in proper time, choosing 0 at the centre of the sphere. The results are indicated in Fig. 1.

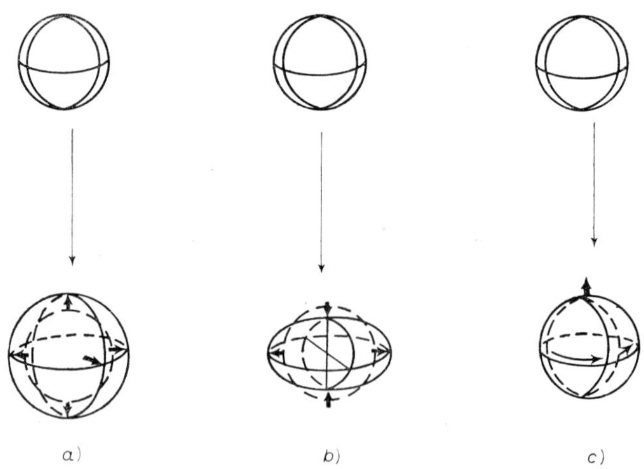

Fig. 1. – During a small time interval, *a*) the action of θ alone transforms a fluid sphere to a similar sphere of different volume but with the same orientation. *b*) The action of σ_{ab} alone distorts the sphere, leaving its volume constant and the directions of the principal axes of shear unchanged. *c*) The action of ω_a alone is to give a rigid rotation leaving one direction (the axis of rotation) fixed. As time progresses the directions of the principal axes of shear and of the axis of rotation will, in general, change.

2'5.1. Volume expansion. From (2.11) we see that θ_{ab}, the *expansion tensor*, determines the rate-of-change of distance of neighbouring particles (in our application, clusters of galaxies) in the fluid. The isotropic part of the expansion is determined by θ, the *volume expansion*. We may define a representative length l by the equation

$$\frac{l^{\cdot}}{l}=\frac{1}{3}\theta; \tag{2.14}$$

l represents completely the volume behaviour of the fluid, and corresponds to the radius function $R(t)$ in the Robertson-Walker (homogeneous and isotropic) world models. From it we can at any time t define the « Hubble constant » $H:=l^{\cdot}/l=\frac{1}{3}\theta$, and the « Deceleration parameter » $q:=-(l^{\cdot\cdot}/l)(1/H^2)$. If we plot the curve $l(t)$ which represents the direction-averaged rate-of-change of distance of clusters of galaxies as a function of proper time, H is the slope of this curve (units: $[\text{time}]^{-1}$) and q represents the curvature of this curve (in dimensionless units; q is positive if the expansion is slowing down).

2'5.2. Shear. The *shear tensor* σ_{ab} determines the distortion arising in the fluid flow, leaving the volume invariant. The directions of the principal axes of shear (*i.e.* of any eigenvector of the tensor σ_{ab}) are unchanged by the distortion, but all other directions are changed. The magnitude of σ_{ab} is the *shear* σ, defined by $\sigma^2=\frac{1}{2}\sigma^{ab}\sigma_{ab}$; $\sigma^2\geqslant 0$, and $\sigma=0\Leftrightarrow\sigma_{ab}=0$.

2'5.3. Vorticity. The *vorticity tensor* ω_{ab} determines a rigid rotation of clusters of galaxies with respect to a local inertial rest frame. We may also represent vorticity by the vorticity vector ω_a, where

(2.15) $\omega^a:=\frac{1}{2}\eta^{abcd}u_b\omega_{cd}\Leftrightarrow$

$\Leftrightarrow\omega_{ab}=\eta_{abcd}\omega^c u^d\,,$

so $\omega^a u_a=0=\omega^{ab}u_b,\ \omega^a\omega_{ab}=0\,.$

(2.15′) $\omega^\nu:=\frac{1}{2}\eta^{\nu\mu\sigma}\omega_{\mu\sigma}\Leftrightarrow$

$\Leftrightarrow\omega_{\mu\nu}=\eta_{\mu\nu\sigma}\omega^\sigma\,,$

so $\omega_{\mu\nu}\omega^\nu=0\,.$

The direction of this vector is the axis of the rotation of the matter, since if we choose n_a in the direction of ω_a when vorticity alone is nonzero, we find $h_a{}^b(n_b)^{\cdot}=0$; this direction is left invariant by the rotation. Its magnitude is the *vorticity* ω, where $\omega:=(\omega^a\omega_a)^{\frac{1}{2}}=(\frac{1}{2}\omega^{ab}\omega_{ab})^{\frac{1}{2}}$; as ω_{ab} is spacelike, $\omega\geqslant 0$ and $\omega=0\Leftrightarrow\omega_a=0\Leftrightarrow\omega_{ab}=0$.

We may note that general-relativistic cosmological models are no more and no less « anti-Machian » than Newtonian cosmological models, in the sense that in both cases a local dynamical rest-frame is nonrotating with respect to a direction average over fairly distant galaxies, if and only if $\omega=0$.

2˙5.4. General motion. In a general fluid flow, both ω and σ will be nonzero. It is then not immediately obvious from (2.12), (except in the special case in which the vorticity vector is a shear eigenvector), if there exists any direction left invariant by the rotation. However we can show that such a direction exists by the following general argument.

On a unit 2-sphere which represents the sky, mark the positions of neighbouring fluid particles at some instant. Then mark on the sphere at the position of each particle, an arrow representing the observed rate-of-change of position of that particle. The set of all such arrows forms a vector field on the 2-sphere, and any vector field on a 2-sphere must have a fixed point; in this case, the fixed point corresponding to a direction e_a left invariant by the fluid motion. Now (2.12) shows that $-e_a$ is also left invariant by the motion. Hence if we can use a fluid approximation to represent the motion of clusters of galaxies, there exists at least one direction in the sky such that the positions in the sky of nearby clusters in that direction and in the opposite direction are instantaneously fixed in a local inertial rest frame. (As the universe evolves this direction itself will, in general, change.)

2˙5.5. Kinematic quantities in the universe. Since ω_a, σ_{ab} and θ determine the relative motion of galaxies in a cosmological model, we should like to determine the values of these quantities in the observed universe. In principle we can compare observations with the theoretical expressions (2.11) and (2.12), (the left-hand side of (2.11) is observable in terms of red-shift and brightness measurements, cf. Subsect. **6**˙2.1), to evaluate θ, ω_a and σ_{ab} at the present time t_0. The value $\theta_0 = 3H_0 = 3\times(1.3\cdot 10^{10}\,\text{y})^{-1}$ is probably correct to within a factor 2 (cf. the lectures in this volume by SCIAMA and BURBIDGE), but we can only obtain rather poor limits on ω_0 and σ_0 from direct observations. The condition that the systematic motion of galaxies is *away* from us in all directions (there are no directions in which we see a systematic blue-shift effect in galactic spectra) imposes the restriction $\sigma_0 < \frac{1}{3}\theta_0$. More detailed examination of the direct evidence gives us the limits

$$\omega_0 \leqslant \tfrac{1}{3}\theta_0\,, \qquad \sigma_0 \leqslant \tfrac{1}{4}\theta_0 \tag{2.16}$$

(KRISTIAN and SACHS [17]). We can obtain much better limits on ω_0 and σ_0 by *indirect* arguments (see Subsect. **7**˙2).

2˙5.6. Other applications. While we have been considering these quantities primarily in a cosmological context, they can of course be defined in other situations where the fluid approximation is valid in astrophysics. Thus in a static (nonrotating) star model, we would have $\theta = \sigma = \omega = 0$. In the study of our own galaxy, where stars may be taken to constitute the fluid particles, observations indicate that $\theta \simeq 0$ near us, while Oort's constants A and B suffice (as the flow is approximately two-dimensional) to determine the shear and vorticity respectively (FREEMAN [18]; VOLTJER [19]).

Further discussion of these concepts in Newtonian theory may be found, for example, in SERRIN [20] and BATCHELOR [21]. A systematic development in the general-relativistic case is given in Ehlers' review article [22].

3. – Conservation of energy and momentum.

3·1. *The average velocity.* – We will take the average velocity vector to be determined by the condition that it is the « baricentric » velocity. An observer moving with an arbitrary 4-velocity can define from the rest-masses m_* and 3-velocities $\boldsymbol{V}_*$ of particles he observes in a (3-) volume element dV, the mass-current density vectors

$$(3.1)\quad J^a = \left(\frac{1}{\mathrm{d}V}\sum_{\substack{\text{particles}\\ \text{in d}V}} m_* \boldsymbol{V}_*, \frac{1}{\mathrm{d}V}\sum_{\substack{\text{particles}\\ \text{in d}V}} m_*\right).$$

$$(3.1')\quad \boldsymbol{J} = \frac{1}{\mathrm{d}V}\sum_{\substack{\text{particles}\\ \text{in d}V}} m_* \boldsymbol{V}_* .$$

Then the average velocity vector is given by

defining it as the (space-time) direction u^a of J^a:

$$J^a = \varrho u^a , \qquad u^a u_a = -1 .$$

Then ϱ is the particle rest-mass density measured by an observer travelling with 4-velocity u^a.

first defining the density ϱ of the matter:

$$\varrho = \frac{1}{\mathrm{d}V}\sum_{\substack{\text{particles}\\ \text{in d}V}} m_* .$$

Then the velocity v^ν is defined by

$$J^\nu = \varrho v^\nu .$$

Conservation of particle rest-mass now implies the conservation equations

$$(3.2)\quad (\varrho u^a)_{;a} = 0 \Leftrightarrow \dot{\varrho} + \varrho\theta = 0 ,$$

$$(3.2')\quad \frac{\partial\varrho}{\partial t} + (\varrho v^\nu)_{,\nu} = 0 \Leftrightarrow \Leftrightarrow \dot{\varrho} + \varrho\theta = 0 ,$$

which can be integrated, using (2.14), to give

$$(3.3)\qquad \varrho l^3 =: M , \qquad \dot{M} = 0 .$$

This is the equation of conservation of matter along the world-lines, implying that the total rest mass $\int_V \varrho\,\mathrm{d}V$ in any small comoving volume element V is constant.

There are two situations in which this discussion has to be modified in the relativistic case. The first is the case of particles of zero rest mass (photons, neutrinos, etc.) when the definition of u^a proceeds as before (*) if we define J^a as the number-current density instead of the mass-current density (3.1); then ϱ is the particle number density, which is again conserved in general. The second situation is that in which the fluid is far from thermal equilibrium (**), in particular when nuclear reactions or pair production are taking place. The ϱ will not be conserved, so it will be appropriate to define u^a (and a new corresponding conserved quantity ϱ) from a baryon or charge current density, instead of the mass-current density used here.

A more thorough discussion of these alternatives, and of equations (3.1)-(3.3), may be found in Ehlers' lectures.

3'2. *Conservation of energy and momentum.* – In Newtonian theory, the momentum conservation equation is the Navier-Stokes equation

$$(V_\nu)^{\cdot} = -\Phi_{,\nu} - \frac{1}{\varrho}(p_{,\nu} + \pi_\nu{}^\mu{}_{,\mu}) ,$$

where p is the isotropic pressure of matter and $\pi_{\mu\nu}$ is the (trace-free) anisotropic pressure; we can rewrite this as

$$\varrho a_\nu + (p_{,\nu} + \pi_\nu{}^\mu{}_{,\mu}) = 0 . \tag{3.4}$$

The thermal balance eq. ((3.13′) below) is deduced separately.

In general relativity, the energy-momentum tensor of matter can be split up uniquely with respect to any timelike vector u^a (and so in particular, with respect to the average velocity vector u^a) to give

$$T_{ab} = \mu u_a u_b + (q_a u_b + u_a q_b) + p h_{ab} + \pi_{ab} , \tag{3.5}$$

where $q_a u^a = 0$, $\pi^a{}_a = 0$, $\pi_{ab} u^b = 0$. Then μ is the total (relativistic) energy density of matter measured by u^a; the specific internal energy density ε of the fluid is defined by $\mu = \varrho(1+\varepsilon)$. q_a is the energy flux relative to u^a (which will represent processes such as diffusion and heat conduction), p is the isotropic pressure, and π_{ab} is the anisotropic matter pressure (due to processes such as viscosity).

(*) Providing the particles do not all move in precisely the same direction.

(**) If the material is too far from thermal equilibrium a fluid description may be inappropriate (cf. STEWART [23]).

The relativistic equations of conservation of energy and momentum are

$$T^{ab}{}_{;b} = 0 \,. \tag{3.6}$$

Using the decomposition (3.5) of T_{ab} and the decomposition of $u_{a;b}$ given in Sect. 2, we can rewrite eqs. (3.6) as the equations

$$\dot{\mu} + (\mu + p)\theta + \pi^{ab}\sigma_{ab} + q^a{}_{;a} + q^a\,\dot{u}_a = 0 \,, \tag{3.7a}$$

$$(\mu + p)\,\dot{u}_a + h_a{}^c(p_{;c} + \pi_c{}^b{}_{;b} + \dot{q}_c) + (\omega_a{}^b + \sigma_a{}^b + \tfrac{4}{3}\theta h_a{}^b)\,q_b = 0 \,. \tag{3.7b}$$

3˙3. *Equations of state.* – We only get physics into the picture when we specify further the properties of T_{ab}. We may do this by giving a prescription for defining ϱ and T_{ab} (in Newtonian theory, ϱ, p, $\pi_{\mu\nu}$ and the heat-flow vector q_ν) from a particle distribution function which obeys suitable equations (c.f. Ehlers' lectures); in the fluid approximation however, we do so by giving equations restricting ϱ, p, μ, q_a and π_{ab}.

3˙3.1. General restrictions. A general restriction one would normally put on the matter is that its energy density be positive. The restrictions of this kind we shall require the fluid to obey are

$$\mu + p > 0 \,, \tag{3.8a}$$

$$\mu + 3p > 0 \,. \tag{3.8b}$$

$$\varrho > 0 \,. \tag{3.8'}$$

(These restrictions will be used in Sect. 3˙4 and 5˙1.1). One would not expect (3.8′) to be violated under any circumstances; (3.8*a*) and (3.8*b*) can only be violated, assuming μ is positive, if the pressure takes extremely large negative values. Thus if μ is 1 g/cm³, (3.8*a*), (3.8*b*) can only be violated if $p \leqslant -10^{15}$ atm (GEROCH [24]).

Further qualitative restrictions one would usually impose are:

(3.9*a*): that the fluid should be stable against local mechanical instability; and, in the relativistic case,

(3.9*b*): that the speed of sound should be less than the speed of light.

3˙3.2. Phenomenological equations. To obtain detailed equations of state, we compare one-component fluids in General relativity and in Newtonian theory, assuming the bulk viscosity is negligible. Defining

$$v = \frac{1}{\varrho}, \tag{3.10}$$

so v is the specific volume of the fluid (*), we assume there is an equation of state of the form

$$\varepsilon = \varepsilon(p, v) \tag{3.11}$$

where ε is the specific internal energy density of the fluid. Then we can define the temperature $T(p, v)$ and the specific entropy $S(p, v)$ by

$$\mathrm{d}\varepsilon + p\,\mathrm{d}v = T\,\mathrm{d}S\,. \tag{3.12}$$

The equation of conservation of thermal energy is then

$$\varrho T\dot{S} = -(\pi_{ab}\sigma^{ab} + q^a{}_{;a} + \dot{u}_a q^a)\,, \tag{3.13}$$

$$\varrho T\dot{S} = -(\pi_{\mu\nu}\sigma^{\mu\nu} + q^\nu{}_{,\nu})\,, \tag{3.13$'$}$$

[(3.13) is just (3.7*a*) rewritten in terms of the new variables], and we find that the phenomenological equations

$$\pi_{ab} = -\lambda\sigma_{ab}\,, \tag{3.14a}$$

$$q_a = -\varkappa h_a{}^b(T_{;b} + T\dot{u}_b) \tag{3.14b}$$

$$\pi_{\mu\nu} = -\lambda\sigma_{\mu\nu}\,, \tag{3.14a'}$$

$$q_\nu = -\varkappa T_{,\nu}\,, \tag{3.14b'}$$

are necessary if the rate of generation of entropy is never negative; further the heat conduction coefficient $\varkappa(p, v)$ and the viscosity coefficient $\lambda(p, v)$ must obey the restrictions

$$\varkappa \geqslant 0\,, \qquad \lambda \geqslant 0\,.$$

[A bulk viscosity coefficient would give a contribution $-\eta\theta$ to the pressure, where η ($\geqslant 0$) is the coefficient of bulk viscosity.]

A more detailed discussion of the relativistic thermodynamics leading to the identifications (3.14) may be found in Ehlers' review paper [22] and references cited there. We may note that the difference between the energy conservation eqs. (3.13), (3.13′) and the momentum conservation eqs. (3.4), (3.7*b*) are *special*-relativistic in origin, arising partly from the inertia assigned by special relativity to all forms of energy, and partly from the $3+1$ splitting of space-time given by h_{ab}.

(*) Volume per unit mass, per particle, or per baryon if J_a is defined as the mass, number or baryon-current density vector respectively.

3·4. *Perfect fluids.* – A *perfect fluid* is characterized by negligible heat conduction and viscosity:

$$\pi_{ab} = q_a = 0 \Leftrightarrow \varkappa = \lambda = 0 .$$

In this case

$$T_{ab} = \mu u_a u_b + p h_{ab} ,$$

and u_a is uniquely defined as the timelike eigenvector of the Ricci tensor.

The momentum-conservation eqs. (3.4), (3.7*b*) now show that the acceleration of the fluid is determined by the spatial pressure gradient:

(3.15) $$\dot{u}^a = -\frac{h_a{}^b p_{;b}}{\mu + p} .$$ | (3.15′) $$a_\nu = -\frac{p_{;\nu}}{\varrho} .$$

Conditions (3.8*a*), (3.8′) ensure that these equations are determinate, and that the acceleration is always away from a high-pressure region towards a neighbouring low-pressure region. In the relativistic theory, the inertial-mass density of the fluid is $\mu + p$; a given spatial pressure gradient is therefore relatively less efficient in producing an acceleration in a general-relativistic fluid than in a corresponding Newtonian fluid, the inertial-mass density being enhanced by the contribution $p + \varepsilon\varrho$.

We may estimate that

(3.16*a*) $$\pi_{ab}|_0 \simeq 0 \simeq q_a|_0 , \qquad p|_0 \simeq 0$$

in any reasonable cosmological model, since the random velocities of galaxies are small at the present time (*). Then it is a plausible inference from the Copernican principle, but does *not* necessarily follow, that

(3.16*b*) $$\dot{u}_a|_0 \simeq 0;$$ | (3.16*b*′) $$a_\nu|_0 \simeq 0;$$

i.e. that the acceleration of the velocity field representing the average motion of galaxies is negligible at the present time. We shall assume this is true.

For a perfect fluid, eq. (3.13) is:

$$\dot{S} = 0 \Leftrightarrow \text{entropy is constant along the flow lines of the fluid.}$$

Then (3.11), (3.12) show that there is only one independent thermodynamic

(*) Random galactic velocities of order 1000 km/s imply $p/\varrho c^2 \simeq 10^{-5}$. If there is a large flux of neutrinos or gravitational waves, these estimates could be seriously incorrect.

variable along each world line; by (3.10), this can be chosen to be ϱ. As the change of ϱ along a world-line is determined by (3.2), or equivalently by (3.3) we see that

(3.17) the change of thermodynamic state along the flow lines of a perfect fluid is determined (for a given equation of state) by the average length l alone.

In the general-relativity case, the original form (3.7*a*) of (3.13) is

$$\dot{\mu} + (\mu + p)\theta = 0 \; . \tag{3.18}$$

Restriction (3.8*a*) shows that compressing the fluid increases its energy density.

We may often assume an equation of state of the form $p = p(\mu)$. Given any such equation of state, restrictions (3.9*a*), (3.9*b*) are satisfied if $1 \geqslant \partial p/\partial\mu \geqslant 0$ (Curtis [25]; see also Harrison, Thorne, Wakano and Wheeler [26]), and (3.18) gives μ as a function of l along the world lines of the fluid.

3'5. *The matter and radiation content of the universe.* – To obtain an idea of the possible thermal histories of the universe, we represent the matter content of the universe by two main components.

3'5.1. Matter content. The *matter content* of the universe (galaxies and a possible intergalactic gas) can plausibly be represented by a perfect fluid (or mixture of perfect fluids with the same 4-velocity) with equations of state

$$p = \alpha\varrho^{\gamma} \; , \qquad T = \beta p/\varrho \; , \tag{3.19}$$

determining the pressure p and temperature T, where α, β, γ are constants. Then (3.2) shows that

$$p = \frac{\text{const}}{l^{3\gamma}} \; , \qquad T = \frac{\text{const}}{l^{3(\gamma-1)}} \; ,$$

govern the change of temperature and pressure along the world lines of the fluid. For a monatomic gas, which is probably a good approximation at least at the present time, $\gamma = \frac{5}{3}$ and so

$$T_{\text{matter}} \propto \frac{1}{l^2} \tag{3.20}$$

gives the change of temperature as the volume changes.

For the relativistic case, the total energy density is the sum of the rest-mass energy and the internal energy:

$$\mu = \varrho + \frac{p}{\gamma - 1}$$

(TOOPER [27]), eqs. (3.2) and (3.18) being automatically consistent for this form of μ. The restrictions (3.9) are satisfied if $2 \geqslant \gamma > 1$. The change of μ along the world lines is clearly given by

$$\mu = \frac{(\varrho_0 l_0^3)}{l^3} + \left(\frac{P_0}{\gamma - 1}\right)\frac{l_0^{3\gamma}}{l^{3\gamma}}.$$

For recent times, we find that, to a good approximation (see (3.16) above) $p \simeq 0$ so that the energy-momentum tensor of matter in recent times is represented fairly well (for dynamical purposes) by regarding the matter as a *pressure-free fluid* (« *dust* »), with

$$p = 0\,, \qquad \mu = \varrho\,, \qquad \dot{u}_a = 0$$

and

$$\mu_{\text{matter}} \propto \frac{1}{l^3}. \tag{3.21}$$

(This is the approximation first introduced by EINSTEIN [3].)

3·5.2. Radiation content. The *radiation content* of the universe consists of many components, but the thermodynamically and dynamically dominant component appears to be the microwave radiation. We will accept this radiation as being black-body radiation at a temperature of about 3° K (this is its most probable interpretation; see Sciama's lectures for a discussion). We may represent this radiation as a perfect fluid with pressure $p = \frac{1}{3}\mu$ (*) and with the same 4-velocity as the matter. This is so at early stages of the universe's history because we would then expect (see next Section) the radiation to be in collisional equilibrium with the matter in the universe (the matter and radiation together can then be regarded as a 1-component fluid with $p \simeq \frac{1}{3}\mu$). At later stages the radiation propagates freely but may still be regarded as a perfect fluid moving with the matter 4-velocity because it is very nearly an isotropic radiation field with respect to that 4-velocity (cf. discussion in Sect. **7**).

(*) The stress-tensor of any fluid consisting of particles with zero rest mass is $T_{ab} = \sum_{\text{particles}} p_* k_{*a} k_{*b}$ where k_{*a} is null for each particle, so $T^a{}_a = 0$; but from (3.5), this is the condition $p = \frac{1}{3}\mu$.

With the condition $p=\frac{1}{3}\mu$, (3.18) shows that

$$\mu_{\text{radiation}} \propto \frac{1}{l^4}. \tag{3.22}$$

Defining the radiation temperature as usual by the condition $\mu = aT^4$ (a a constant), we see that the radiation temperature obeys the law

$$T_{\text{radiation}} \propto \frac{1}{l}. \tag{3.23}$$

We note that the radiation obeys the perfect gas equations of state (3.19) with $\gamma = \frac{4}{3}$.

We have no information as to the isotropy of neutrinos of cosmological origin. These would be freely propagating at the present time. If they are now moving isotropically, they also obey (3.23). If they move anisotropically, then a fluid approximation is no longer adequate for the discussion (cf. MISNER [28]).

3·6. *The thermal history of the universe.* – So far, we have implicitly assumed there is no effective interaction between the two fluids (the matter and the radiation) in our cosmological model (or between different matter components). This is a good approximation at the present time, but will be untrue at certain earlier times. However we can obtain a good idea of the possible thermal histories of the universe from the information already at our disposal; despite the interactions, (3.17) is still approximately true (*).

Suppose that l decreases, as we go backwards in time, continuously to zero. Then the temperature of matter and radiation both rise indefinitely, according to (3.20) and (3.23). However when the temperature is above about 3000 °K ($l/l_0 \leqslant 1/1000$) the matter is ionized; Thomson scattering of the radiation and free electrons puts the electrons and radiation into close thermal contact, so the matter becomes an opaque plasma. The matter (the ions and electrons are in close contact because of Coulomb forces) and radiation must both have the same temperature for smaller values of l. Since the radiation (present density about 10^{-33} g/cm^3) has a much greater thermal capacity than the matter (present density between 10^{-29} and 10^{-31} g/cm^3), they both then obey the radiation temperature law (3.23).

As the temperature increases, successive interaction processes become important: nuclear reactions are important from 10^8 to 10^9 °K; electron-positron pair production and annihilation are important at $\sim 6\cdot 10^9$ °K, with cor-

---

(*) Essentially because we still have an equation of state approximately of the form $p = p(\mu)$ for the mixture.

responding cooling and heating of the radiation gas; neutrino reactions become so rapid when $T \geqslant 2\cdot10^{10}$ °K, that the plasma is then opaque to neutrinos. At these temperatures, the baryons, electron-positron pairs, radiation and neutrinos are in thermal equilibrium and form an ultra-relativistic perfect gas obeying (3.23). At about 10^{12} °K, large numbers of π- and μ-meson pairs are created and annihilated; known physics is a rather poor guide as to what happens at these and higher temperatures.

With this information, we can draw a diagram (Fig. 2) showing the possible thermal histories of the universe, by giving T as a function of the average radius l. Corresponding to each function $l(t)$, we obtain a thermal history from Fig. 2; for example, if (as we go back in time) l goes to a minimum and then increases again, T goes to a maximum and decreases. We can only find the

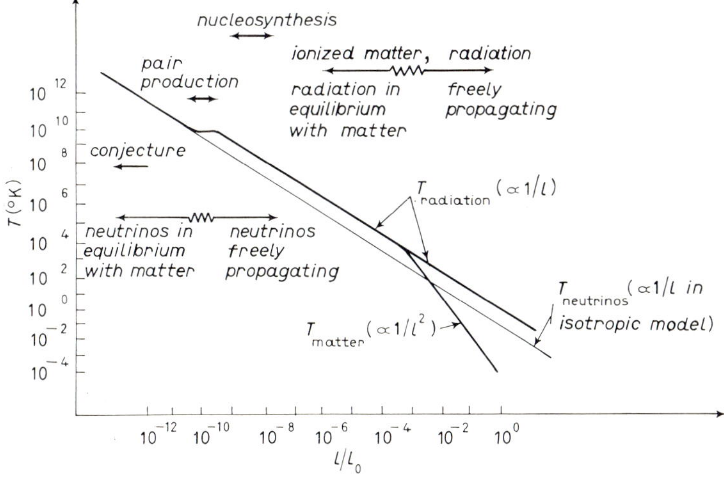

Fig. 2. – Diagram of temperature T as a function of average length l in a cosmological model. The curve for neutrinos is only valid for isotropic world models; the rest of the Figure will differ somewhat in anisotropic models, but remains essentially the same.

actual function $l(t)$ in any particular cosmological model, and so the time scales available for the various interactions and the actual thermal history in that model, from the field equations. The details of Fig. 2 depend on these time scales; if they were radically different from the time scales in the spherically symmetric models, the times available for equilibrium to be attained at each stage of expansion would be different, and various aspects of the diagram (such as the temperatures at which photons and neutrinos decouple from the matter) might change significantly.

The assumptions we have made in obtaining this diagram will be valid at nearly all times, and for any reasonable degree of inhomogeneity and anisotropy.

However (as has been pointed out by MISNER [28, 29]) there are at least two stages when irreversible processes certainly cannot be ignored: namely, when the neutrinos decouple and when the photons decouple. At these stages the matter behaves as a viscous fluid, and this might play an important role in the process of galaxy formation. The accompanying viscous heating arising (c.f. eq. (3.13)) from anisotropic fluid motion, would affect Fig. 2 at these times. (We note that the role of bulk viscosity has not yet been clarified. It may be important at some stages of the expansion of the universe, cf. a forthcoming paper by ANDERSON and STEWART.)

The detailed form of this diagram depends critically on astrophysical processes in the universe: in particular some arguments (see Sciama's lectures) suggest that an intergalactic gas might be reheated after its temperature had cooled well below the radiation temperature, at about the time that galaxy formation occurred in the universe. In this case the universe would again be filled by a ionised plasma (at a temperature of about 10^5 °K) at recent times.

The thermal history of the universe (derived for the isotropic case) is discussed in many articles, see for example DICKE, PEEBLES, ROLL and WILKINSON [30], PEEBLES [31], ALPHER, GAMOW and HERMANN [32], WAGONER, FOWLER and HOYLE [33] and HARRISON [34]. The role of irreversible processes in an anisotropic universe is discussed by MISNER [28, 29].

4. – The field equations.

4·1. *The curvature tensor.* – In general relativity, the curvature of space-time is represented by the Riemann curvature tensor R_{abcd}, which has the symmetry properties

$$R_{[ab][cd]} = R_{abcd} = R_{cdab}\,, \qquad R_{a[bcd]} = 0\,. \tag{4.1}$$

This tensor (which has 20 independent components) can be algebraically separated into the Ricci tensor R_{ab}, defined by

$$R_{ab} := R^c{}_{acb}\,, \tag{4.2}$$

and the Weyl tensor C_{abcd} (the « conformal curvature tensor ») defined by

$$C^{ab}{}_{cd} := R^{ab}{}_{cd} - 2g^{[a}{}_{[c}\, R^{b]}{}_{d]} + \frac{R}{3}\, g^{[a}{}_{[c}\, g^{b]}{}_{d]}\,, \tag{4.3}$$

where the Ricci scalar R is $R := R^a{}_a = R^{ab}{}_{ab}$. It follows from (4.3) that the Weyl tensor has all the Riemann tensor symmetries (4.1) and the additional property

$$C^{ab}{}_{ad} = 0 \, .$$

Thus we may think of R_{ab} (with 10 independent components) as the trace of R_{abcd}, and of C_{abcd} (also with 10 independent components) as its trace-free part (*).

The quantity we may regard as corresponding to R_{abcd} in Newtonian theory is the second derivative $\Phi_{,\mu,\nu}$ of the gravitational potential Φ. This can be separated algebraically into its trace $\Phi^{\cdot\nu}{}_{,\nu}$ and its trace-free part $E_{\mu\nu}$:

$$E_{\mu\nu} := \Phi_{,\mu,\nu} - \tfrac{1}{3} h_{\mu\nu} \Phi^{\cdot\sigma}{}_{,\sigma} \, . \tag{4.4}$$

4.1.1. The role of the field equations. The field equations, including a cosmological constant Λ, are (on choosing units suitably),

the Einstein equations:

$$(R_{ab} - \tfrac{1}{2} R g_{ab}) + \Lambda g_{ab} = T_{ab} \Leftrightarrow \Leftrightarrow R_{ab} = (T_{ab} - \tfrac{1}{2} T g_{ab}) + \Lambda g_{ab} \, . \tag{4.5}$$

the Poisson equation:

$$\Phi^{\cdot\nu}{}_{,\nu} + \Lambda = \varrho/2 \Leftrightarrow \Leftrightarrow \Phi^{\cdot\nu}{}_{,\nu} = \varrho/2 - \Lambda \, . \tag{4.5'}$$

In each case, the field equations determine the « trace » part of the gravitational field algebraically at each point of space-time from the matter content at that point. The « trace-free » part is related to the matter content by differential equations (see (4.21), (4.22)) and is determined by these equations in conjunction with suitable boundary conditions, initial conditions, kinematic conditions or symmetry requirements; we shall generically call such restrictions « boundary conditions ».

In Newtonian theory, we have to drop the boundary condition

$$\Phi \to 0 \text{ at } \infty \tag{4.6}$$

if we are to get any reasonable cosmological models at all (see BONDI [1] and references given there; this follows from (4.5') and is closely related to the difficulty (Subsect. 2.3) in separating a_ν invariantly into gravitational and inertial parts). In fact to get Newtonian analogues *a*) to many general-relativistic cosmological models, and *b*) to the general-relativistic expression for gravitational radiation reaction, we must also drop the condition

$$E_{\mu\nu} \to 0 \text{ at } \infty \, , \tag{4.7}$$

(*) For further details of this decomposition, see JORDAN, EHLERS and KUNDT [35].

which might seem to be a suitable generalization of (4.6). (For a) see HECKMANN and SCHÜCKING [36] and Sect. **5** below; for b) see the lectures by THORNE in this volume.)

4·2. *The Ricci identities for the velocity vector.* – In general relativity, arbitrary vector fields obey Ricci's identity (which we may regard as defining the curvature tensor). Applying this identity to the vector field u^a,

$$u_{a;d;c} - u_{a;c;d} = R_{cbcd}\, u^b \,. \tag{4.8}$$

The corresponding Newtonian identities are

$$\partial(v_{\nu,\mu})/\partial t = (\partial v_\nu/\partial t)_{,\mu}\,, \tag{4.9a}$$

$$v_{\mu,\nu,\sigma} = v_{\mu,\sigma,\nu}\,. \tag{4.9b}$$

(In orthogonal curvilinear co-ordinates these are $\partial^2 v_\nu/\partial t\,\partial x^\mu = \partial^2 v_\nu/\partial x^\mu\,\partial t$, $\partial^2 v_\mu/\partial x^\nu\,\partial x^\sigma = \partial^2 v_\mu/\partial x^\sigma\,\partial x^\nu$).

Multiplying (4.8) by u^d in the general-relativity case, we get

$$(u_{a;c})^\bullet - \dot u_{a;c} + u_{a;d}\,u^d{}_{;c} + R_{cbcd}\,u^b u^d = 0\,.$$

Projecting and remembering (2.7b), this equation is

$$h_a{}^c h_b{}^d (v_{cd})^\bullet - \dot u_a\,\dot u_b - h_a{}^c h_b{}^d\,\dot u_{c;d} + v_{ad}\,v^d{}_b + R_{acbd}\,u^c u^d = 0\,. \tag{4.10}$$

In Newtonian theory operating on (2.10b) by $_{,\nu}$ and using (4.9a), we find

$$\partial(v_{\mu,\nu})/\partial t + (v_{\mu,\nu})_{,\sigma} v^\sigma + v_{\mu,\sigma} v^\sigma{}_{,\nu} + \Phi_{,\mu,\nu} = a_{\mu,\nu}\,;$$

this can be rewritten

$$(v_{\mu\nu})^\bullet - a_{\mu,\nu} + v_{\mu\sigma} v^\sigma{}_\nu + \Phi_{,\mu,\nu} = 0\,. \tag{4.11}$$

Equations (4.10), (4.11) are propagation equations for $v_{\mu\nu}$ along the flow lines of the fluid; the similarity of these equations is essentially the similarity of the « geodesic deviation equations » in Newtonian theory and in general relativity (see PIRANI [37], pp. 260-269, for a clear account of this correspondence).

4·2.1. Raychaudhuri's equation. Contracting eqs. (4.10), (4.11) we obtain propagation equations for θ.

$g^{ab}\times$(4.10), (3.5) and (4.5) implies

$$\theta^{\bullet} + \tfrac{1}{3}\theta^2 - \dot{u}^a{}_{;a} + 2(\sigma^2 - \omega^2) + \tfrac{1}{2}(\mu + 3p) - \Lambda = 0\,, \tag{4.12}$$

$h^{\mu\nu}\times$(4.11) and (4.5′) implies

$$\theta^{\bullet} + \tfrac{1}{3}\theta^2 - a^{\nu}{}_{,\nu} + 2(\sigma^2 - \omega^2) + \tfrac{1}{2}\varrho - \Lambda = 0\,, \tag{4.12′}$$

which is Raychaudhuri's equation (RAYCHAUDHURI [38, 39]). By (2.14), $\theta^{\bullet} + \frac{1}{3}\theta^2 = 3l^{\bullet\bullet}/l$, so we can rewrite this equation in the form

$$3l^{\bullet\bullet}/l = 2(\omega^2 - \sigma^2) + \dot{u}^a{}_{;a} - \tfrac{1}{2}(\mu + 3p) + \Lambda\,. \tag{4.13}$$

$$3l^{\bullet\bullet}/l = 2(\omega^2 - \sigma^2) + a^{\nu}{}_{,\nu} - \tfrac{1}{2}\varrho + \Lambda\,. \tag{4.13′}$$

This shows how the second derivative of the curve $l(t)$ is determined directly at each space-time point by the matter density at that point, with the Λ-term acting as a constant repulsive force; rotation tends to hold the matter apart (as we might expect, representing a « centrifugal » effect); a pure distortion tends to pull the world-lines together; and acceleration affects the average distance of the world lines through its divergence.

The main difference between the Newtonian and general-relativistic cases lies in the fact that while the active gravitational mass density is ϱ in Newtonian mechanics, it is $\mu + 3p = \varrho + \varepsilon\varrho + 3p$ in general relativity. It is this additional pressure and internal energy contribution to the gravitational force which is the major cause of the problem of gravitational collapse in general relativity. Thus if we consider a static star model filled with a perfect fluid and take $\Lambda = 0$, (4.12) becomes

$$\dot{u}^a{}_{;a} = \tfrac{1}{2}(\mu + 3p)\,, \qquad\qquad a^{\nu}{}_{,\nu} = \tfrac{1}{2}\varrho\,,$$

where the acceleration is determined from the pressure gradient by (3.15). The extra terms in the relativistic case show that the pressure which tries to balance the star through acceleration tends to defeat itself, since it contributes directly to the gravitational field which tends to cause the star to collapse. A contributing factor is the relative inefficiency of a given spatial pressure gradient in causing acceleration, in the relativistic case (see Subsect. 3·4).

We can evaluate Raychaudhuri's equation in a cosmological model at the present time t_0. Remembering the definitions of q and H (Subsect. 2·5.1) we find from (2.16) and (3.16) (we make the plausible inference $\dot{u}^a{}_{;a}|_0 \simeq 0$; this is *not* a direct consequence of (3.16)) that (*)

$$q_0 \simeq \frac{\mu_0}{6H_0^2} - \frac{\Lambda}{3H_0^2}, \tag{4.13a}$$

with possible maximum errors of $-\frac{2}{3}$, $+\frac{3}{8}$ arising from the limits (2.16) on shear and vorticity (but the actual errors due to these terms are probably negligible, see Sect. **7**).

If we believe that $\Lambda = 0$, the (4.13*a*) becomes

$$q_0 \simeq \frac{\mu_0}{6H_0^2}, \tag{4.13b}$$

showing directly the deceleration caused by the matter content of the universe. It is rather difficult to estimate q_0 observationally (cf. Burbidge's lecture). We have indications that $q_0 > 0$; more specifically, that $q_0 = 1 \pm \frac{1}{2}$. However we can only be reasonably certain (**) of fairly broad observational limits, say

$$|q_0| < 5\,. \tag{4.14a}$$

On the other hand, we can find a lower limit for μ_0 on the basis of the observed matter in the universe; this lowest estimate is $\mu_0/6H_0^2 \simeq 10^{-2}$. There could exist intergalactic gas which is so far unobserved, giving values up to $\mu_0/6H_0^2 \simeq 1$ (cf. lectures by SCIAMA in this volume). Thus we can obtain plausible limits

$$10^{-2} \leqslant \frac{\mu_0}{6H_0^2} \leqslant 1\,, \tag{4.14b}$$

on the density μ_0. However it is possible that the universe is filled with a large density of matter or energy, (such as neutrinos, (RUDERMAN [40]), gravitational waves, (FIELD, REES and SCIAMA [41]), condensed stars (ZEL'DOVICH [8]), young galaxies (PARTRIDGE and PEEBLES [42]), or rocks), which is very difficult to detect. The best definite limit we can place on the effective density of such forms of energy comes directly from (4.14*a*) and (4.13*b*), which place a limit $\mu_0 \leqslant 10^{-28}$ g/cm³ on the smoothed-out energy density of such « unobservable » forms of matter.

An alternative viewpoint is to use (4.13*a*) to give limits on Λ in terms of

---

(*) The density μ is often represented by the parameter $\sigma := \mu/6H^2$ (do not confuse with the shear!).

(**) q_0 as defined is obtained by taking a suitable *directional average*, as it is defined from l.

an equivalent mass density; we find

$$-250\mu_0 \leqslant \Lambda \leqslant 300\mu_0 \tag{4.14c}$$

if we believe (4.14a), (4.14b).

4.2.2. Further propagation equations. So far, we have extracted propagation equations for θ from (4.10), (4.11). We can also obtain propagation equations for the shear and vorticity from these relations.

The skew parts of (4.10), (4.11) are equivalent to

$$h^a{}_b(l^2\omega^b)^{\bullet} = \sigma^a{}_b(l^2\omega^b) + \frac{l^2}{2}\eta^{abcd}u_b\,\dot{u}_{c;d}\,. \tag{4.15}$$

$$(l^2\omega^\mu)^{\bullet} = \sigma^\mu{}_\nu(l^2\omega^\nu) + \frac{l^2}{2}\eta^{\mu\nu\sigma}a_{\nu,\sigma}\,. \tag{4.15'}$$

Substituting into these equations for $\dot{u}_a$ from (3.15), we obtain the usual vorticity conservation laws when we have a perfect fluid with equation of state $p = p(\mu)$ (see EHLERS [22], GODEL [43], SYNGE [44]).

The symmetric, trace-free parts of (4.10), (4.11) are (using (4.12))

$$h_a{}^f h_b{}^g(\sigma_{fg})^{\bullet} - h_a{}^f h_b{}^g\,\dot{u}_{(f;g)} - \dot{u}_a\dot{u}_b + \omega_a\omega_b + \sigma_{af}\sigma^f{}_b + \tfrac{2}{3}\theta\sigma_{ab} + h_{ab}\left(-\tfrac{1}{3}\omega^2 - \tfrac{2}{3}\sigma^2 + \tfrac{1}{3}\dot{u}^c{}_{;c}\right) - \tfrac{1}{2}\pi_{ab} + E_{ab} = 0\,, \tag{4.16}$$

where

$$E_{ac} := C_{abcd}u^b u^d;$$

the Weyl tensor symmetries imply

$$E_{ab} = E_{(ab)}, \quad E^a{}_a = 0\,, \quad E_{ab}u^b = 0\,.$$

$$\dot{\sigma}_{\mu\nu} - a_{(\mu,\nu)} + \omega_\mu\omega_\nu + \sigma_{\mu\varkappa}\sigma^\varkappa{}_\nu + \tfrac{2}{3}\theta\sigma_{\mu\nu} + h_{\mu\nu}\left(-\tfrac{1}{3}\omega^2 - \tfrac{2}{3}\sigma^2 + \tfrac{1}{3}a^\nu{}_{,\nu}\right) + E_{\mu\nu} = 0\,, \tag{4.16'}$$

where

$$E_{\mu\nu} := \Phi_{,\mu,\nu} - \tfrac{1}{3}h_{\mu\nu}\Phi^{,\sigma}{}_{,\sigma}\,,$$

so

$$E_{\mu\nu} = E_{(\mu\nu)}\,, \quad E^\mu{}_\mu = 0\,.$$

Thus we see that while the expansion is affected directly by the matter at each point, the gravitational field E_{ab} (the « tidal force ») affects the fluid flow by inducing shear in the flow lines; this shear then determines the vorticity propagation and also enters into the expansion equation (tending to cause the world lines to converge). Note the terms in ω_a in (4.16), representing the distorting effect of « centrifugal forces ».

4.2.3. Constraint equations. Three further sets of equations can be obtained from (4.8) and (4.9b).

Multiply (4.8) by $g^{ac}h^{cd} \Rightarrow$

(4.17) $h^e{}_b(\omega^{bc}{}_{;c} - \sigma^{bc}{}_{;c} + \frac{2}{3}\theta^{;b}) + (\omega^e{}_b + \sigma^e{}_b)\,\dot{u}^b = q^e$.

Multiply (4.9b) by $h^{\nu\sigma} \Rightarrow$

(4.17′) $\omega^{\nu\mu}{}_{,\mu} - \sigma^{\nu\mu}{}_{,\mu} + \frac{2}{3}\theta^{,\nu} = 0$.

The vorticity vector satisfies a constraint equation:

by (4.8),

$$R^a{}_{[bcd]} = 0 \Rightarrow u_{[a;b;c]} = 0 ;$$

$\times \eta^{abcd} u_d$ shows

(4.18) $\omega^a{}_{;a} = 2\omega^b \dot{u}_b$.

(4.9b) implies

$$v_{[\mu,\nu,\sigma]} = 0 ;$$

$\times \eta^{\mu\nu\sigma}$ shows

(4.18′) $\omega^\nu{}_{,\nu} = 0$.

Finally,

(4.8) $\times \eta_{sfdc} u^f$ and symmetrization shows

(4.19) $H_{ad} = 2\dot{u}_{(a}\omega_{d)} - h_a{}^t h_d{}^s (\omega_{(t}{}^{b;c} + \sigma_{(t}{}^{b;c})\,\eta_{s)fbc} u^f$,

where

$$H_{ac} := \tfrac{1}{2}\eta_{ab}{}^{gh} C_{ghcd} u^b u^d .$$

The Weyl tensor symmetries show

$$H_{ab} = H_{ab}, \quad H^a{}_a = 0 , \quad H_{ab} u^b = 0 .$$

(4.9b) $\times \eta^{\varkappa\mu\sigma}$ and symmetrization shows

(4.19′) $(\omega_{(\varkappa}{}^{\nu,\tau} + \sigma_{(\varkappa}{}^{\nu,\tau})\eta_{\mu)\nu\tau} = 0$.

(H_{ab} has no analogue in Newtonian theory).

In the general-relativity case, eqs. (4.12) and (4.15)-(4.19) are equivalent to the Ricci identities (4.8). Through these equations we can completely determine the curvature tensor of space-time, since the Ricci tensor is determined directly by the matter content and the Weyl tensor is determined by the symmetric, trace-free 3-tensors E_{ab}, H_{ab} (it is given by

$$C_{abcd} = (\eta_{abpq}\eta_{cdrs} + g_{abpq}g_{cdrs})\,u^p u^r E^{qs} - (\eta_{abpq}g_{cdrs} + g_{abpq}\eta_{cdrs})\,u^p u^r H^{qs} ,$$

where

$$g_{abcd} := g_{ac}g_{bd} - g_{ad}g_{bc}) .$$

In the Newtonian case, eqs. (4.12′) and (4.15′), (4.16′) are equivalent to (4.9*a*) and eqs. (4.17′)-(4.19′) are equivalent to (4.9*b*).

4·3. *The Bianchi identities.* – In general-relativity, the curvature tensor has to obey the *Bianchi identities*

$$R_{ab[cd;e]}=0\Leftrightarrow C^{abcd}{}_{;d}=R^{c[a;b]}-\tfrac{1}{6}g^{c[a}R^{;b]}; \tag{4.20}$$

these two forms are equivalent only because space-time is 4-dimensional (Kundt and Trümper [45]). These identities imply the contracted identities

$$R^{ab}{}_{;b}=\tfrac{1}{2}R^{;a}\Leftrightarrow G^{ab}{}_{;b}=0$$

(which are the eq. (3.7)) and 16 further identities. Substituting into (4.20) for C_{abcd} in terms of E_{ab}, H_{ab} these equations take a form rather similar to Maxwell's equations. If the space-time is filled with a perfect fluid, they are (*)

$$:[\text{« div}\,E\,\text{»}]:h^t{}_aE^{as}{}_{;d}h_s{}^d-\eta^{tbpq}u_b\sigma_p{}^dH_{qd}+3H^t{}_s\omega^s=\tfrac{1}{3}h^t{}_b\mu^{;b}; \tag{4.21a}$$

$$:[\text{«}\,\dot H\,\text{»}]:h^{ma}h^{tc}\dot H_{ac}-h_a{}^{(m}\eta^{t)rsd}u_rE^a{}_{s;d}+2E_q{}^{(t}\eta^{m)bpq}u_b\dot u_p+h^{mt}(\sigma^{ab}H_{ab})+ \\ +\theta H^{mt}-3H_s{}^{(m}\sigma^{t)s}-H_s{}^{(m}\omega^{t)s}=0; \tag{4.21b}$$

$$:[\text{« div}\,H\,\text{»}]:h^t{}_aH^{as}{}_{;d}h_s{}^d+\eta^{tbpq}u_b\sigma_p{}^dE_{qd}-3E^t{}_s\omega^s=(\mu+p)\omega^t; \tag{4.21c}$$

$$:[\text{«}\,\dot E\,\text{»}]:h_a{}^mh_c{}^t\dot E^{ac}+h_a{}^{(m}\eta^{t)rsd}u_rH^a{}_{s;d}-2H_q{}^{(t}\eta^{m)bpq}u_b\dot u_p+h^{mt}\sigma^{ab}E_{ab}+ \\ +\theta E^{mt}-3E_s{}^{(m}\sigma^{t)s}-E_s{}^{(m}\omega^{t)s}=-\tfrac{1}{2}(\mu+p)\sigma^{tm}. \tag{4.21d}$$

Corresponding to the identities (4.21*a*), (4.21*b*), in Newtonian theory the tensor $E^{\mu\nu}$ satisfies the identities

$$E^{\mu\nu}{}_{,\nu}=\tfrac{1}{3}\varrho^{,\mu}, \tag{4.22a}$$

$$E^{(\mu}{}_{\sigma,\tau}\eta^{\nu)\sigma\tau}=0, \tag{4.22b}$$

which follow from (4.9*a*). Further the shear tensor and expansion θ satisfy the 8 equations

$$\sigma_{[\nu}{}^{[\mu}{}_{,\tau]}{}^{,\varkappa]}+\tfrac{2}{3}h^{[\mu}{}_{[\nu}\theta^{,\varkappa]}{}_{,\tau]}=0, \tag{4.22c}$$

(*) This form of the Bianchi identities is due to M. Trümper. No difficulty arises in obtaining these equations for a general fluid; they are rather complex, and we give them in an Appendix. They hold for any energy-momentum tensor; the fluid approximation limits then only through prescribing the equations of state.

which follow from (4.9*b*) and are in fact completely analogous to (4.21*c*) and (4.21*d*). One could see this by substituting into (4.21*c*) and (4.21*d*) from (4.19), obtaining second-order identities satisfied by the kinematic quantities; however it is easier to proceed by noting that (4.22*c*) is the condition $\omega^{\mu}{}_{,\nu\sigma}\eta^{\nu\sigma\tau}=0$ in Newtonian theory (TRÜMPER [46]), while eqs. (4.21*c*) and (4.21*d*) are together equivalent to the corresponding general-relativity condition

$$h^a{}_s\,\omega^s{}_{;cd}\,\eta^{cdef}u_e = h^a{}_s R^s{}_{bcd}\,\omega^b\eta^{cdef}u_e\,.$$

4·4. *The field equations.*

4·4.1. The case of zero rotation. In this special case, we can obtain a useful alternative form of some of the general-relativity equations. Standard theorems show that

$\omega=0\Leftrightarrow u_{[a}u_{b;c]}=0$	$\omega=0\Leftrightarrow v_{[\mu,\nu]}=0$
$\Leftrightarrow\exists$ locally functions	$\Leftrightarrow\exists$ locally a function
$f,g:u_a=fg_{;a}\,.$	$g:v_\mu=g_{,\mu}\,.$

Thus in each case, $\omega=0$ is the condition that there locally exist 3-surfaces in space-time (the surfaces $\{g=\text{const}\}$) orthogonal to the velocity vector field. However in general-relativity, u_a is the 4-velocity vector field; so the condition $\omega=0$ is precisely the condition that the instantaneous rest spaces defined at each point by h_{ab} should mesh together to form a set of 3-surfaces in space-time. These surfaces, which are surfaces of simultaneity for all the fluid observers, define a cosmic-time co-ordinate (the function g) determined by the fluid flow. (This time co-ordinate can be locally normalized to measure *proper* time along each world line only if $\dot{u}=0$.)

We can now use the Gauss-Codacci formulae (SCHOUTEN [4], JORDAN, EHLERS and KUNDT [35]) to relate the curvature tensor of space-time to the curvature of the 3-spaces orthogonal to u^a. If the Ricci tensor of these 3-spaces is $R^*{}_{ab}$, we find

$$(4.23)\qquad R^*{}_{ab}=h_a{}^f h_b{}^g\big(-l^{-3}(l^3\sigma_{fg})^{\boldsymbol{\cdot}}+\dot{u}_{(f;g)}\big)+\dot{u}_a\dot{u}_b+\pi_{ab}+ \\ +\tfrac{1}{3}h_{ab}\big(-\tfrac{2}{3}\theta^2+2\sigma^2+2\mu+2\Lambda-\dot{u}^c{}_{;c}\big)\,,$$

which implies that the Ricci scalar R^* of the 3-spaces is

$$(4.24)\qquad R^*=-\tfrac{2}{3}\theta^2+2\sigma^2+2\mu+2\Lambda\,;$$

these equations show how the curvature of the 3-spaces $\{g = \text{const}\}$ is determined by the matter content of the space-time through the field equations. They are essentially general-relativistic in character; R^*_{ab} has no Newtonian analogue. We can re-express (4.23) in the language of classical differential geometry (cf. EHLERS [22]): if, in a 3-space $\{g = \text{const}\}$, the Gaussian curvature of the 2-surface formed by all geodesics through the point x orthogonal to the direction e_a ($e^a e_a = 1$, $e^a u_a = 0$) at that point is $K(x, e^a)$, then

$$(4.25)\qquad K(x, e^a) = \{l^{-3}(l^3 \sigma_{fg})^{\cdot} - \dot{u}_{f;g} - \dot{u}_f \dot{u}_g - \pi_{fg}\} e^f e^g + \\ + \tfrac{1}{3}(\sigma^2 - \tfrac{1}{3}\theta^2 + \dot{u}^c{}_{;c} + \Lambda + \mu) .$$

If we separate R^*_{ab} into a trace-free part (which is essentially equivalent to E_{ab}) and its trace, then these quantities are related to each other by the Bianchi identities for the 3-spaces,

$$(4.26)\qquad h^d{}_a R^{*ab}{}_c h_b{}^c = \tfrac{1}{2} R^{*;c} h_c{}^d .$$

These identities are equivalent to the three 4-space Bianchi identities (4.21*a*), because of (4.23).

4.4.2. The field equations in general. We may regard eqs. (4.12), (4.16) and (4.17) as 9 of the 10 general-relativity field equations. The remaining field equation is one of the four first integrals which exist, when all the other conditions are satisfied, as a consequence of the contracted Bianchi identities (3.7) (*). This equation may be understood as giving a geometrical constraint, but no extra restrictions on the kinematic quantities, if all the other relations are satisfied. We may see this most easily in the case $\omega = 0$. Then the trace-free part of (4.23) is equivalent to (4.16), and the trace (eq. (4.24)) is the tenth field equation, relating R^* to the kinematic quantities and the density of matter. We only have to fulfil this equation at one point of space-time; then it is fulfilled in an open neighbourhood of that point (it is fulfilled in a 3-surface $\{g = \text{const}\}$ because of eqs. (4.21*a*) (which are equivalent to (4.26) if (4.23) is fulfilled); and it is fulfilled on timelike lines through this 3-surface because (3.7*a*) ensures that it is a first integral of the field equation). In effect, the other equations determine a solution of the field equations up to a

---

(*) Let $Z_{ab} := (R_{ab} - \frac{1}{2} R g_{ab}) + \Lambda g_{ab} - T_{ab}$; then the field equations are satisfied when $Z_{ab} = 0$. Now $Z^{ab}{}_{;b} = 0 \Leftrightarrow \partial Z_a{}^4/\partial x^4 = - Z_a{}^\nu{}_{;\nu} - \Gamma^4{}_{4d} Z_a{}^d + \Gamma_{a4}{}^d Z_d{}^4$ shows that if $Z_\mu{}^\nu = 0$ at all times and $Z_a{}^4 = 0$ on an initial surface $\{x^4 = \text{const}\}$, then $Z_a{}^4 = 0$ on an open neighbourhood of the surface; so the equations $Z_a{}^4 = 0$ are four first integrals of the other field equations. For further discussion see, for example, ANDERSON [47], CHOQUET-BRUHAT [48].

constant; eq. (4.24) determines this constant. The situation will be similar in the case $\omega \neq 0$.

In Newtonian theory, the only field equation is (4.12′), the remaining 8 equations which correspond to general-relativity field equations being simply kinematic identities.

We will not give a systematic discussion of the full set of equations here: rather, after briefly discussing the differences between the general-relativity and Newtonian equations, we shall illustrate their use by applying them to simple cosmological models.

4.4.3. Comparison of Newtonian and general-relativity equations. A comparison shows the first-order eqs. (4.12)-(4.19) satisfied by the kinematic quantities in general-relativity and in Newtonian theory are extremely similar, despite there being only one field equation in Newtonian theory and ten in general-relativity. Apart from the extra terms in the momentum equation mentioned in Sect. **3**, there are some extra terms in the general-relativity equations which are essentially special-relativistic in origin; these are the term $\dot{u}_a \dot{u}_b$ in (4.16) and the terms in $\dot{u}_a$ in (4.17)-(4.19), resulting from the $3+1$ splitting of space-time given by h_{ab}. There are also extra terms which arise for essentially general-relativistic reasons. The general-relativity equivalents of some of the Newtonian equations contain terms representing the greater geometric freedom resulting from the possibility that space-time can be curved; these are the terms p in (4.12), π_{ab} in (4.16), q_a in (4.17) and H_{ab} in (4.19). However we have substituted for many of these terms (p, π_{ab} and q_a) through the field equations, and these are determined by the other kinematic quantities through equations of state (cf. Sect. **3**). The Newtonian and general-relativity equations are most similar in the case of dust when these terms and $\dot{u}_a$ vanish.

The second-order identities (4.21), (4.22), although in 1-1 correspondence, look rather different; the extra terms in the general-relativity equations are so numerous that they dominate the equations. A further important difference arises from the different meanings E_{ab} has in the two theories: $\dot{E}_{ab}$ is locally prescribed by the general-relativity equations, but is left unprescribed in Newtonian theory. Thus, given suitable equations of state, the time development of the system off an initial surface is completely determined in the general-relativity case. For example if we have a perfect fluid with equation of state $p = p(\mu)$, $\dot{\mu}$ is given by (3.18), $p^{\cdot}$ follows from the equation of state, and the rate of change of $\dot{u}_a$ along the world-lines follows from that of p. The rates of change of θ, σ_{ab}, ω_a are determined by (4.12), (4.16), (4.15); and the time derivatives of E_{ab}, H_{ab} are determined by (4.21*b*), (4.21*d*). However in the Newtonian case, the time development of the system is not determined until some suitable restriction has been put on $E_{\mu\nu}$; for example we can choose some particular world line and then prescribe $E_{\mu\nu}$ as an arbitrary function along that world line. Thus having dropped the boundary condition (4.7) we can use the resulting freedom of choice of $E_{\mu\nu}$ (this freedom is essentially a consequence of the infinite speed of propagation of gravitational effects) to influence the fluid flow in a rather arbitrary manner (cf. HECKMANN and SHÜCKING [36]). In particular, we can use this freedom in such a way as to produce Newtonian analogues of relativistic cosmological models. When we do so, we find that the general-relativity integrability conditions are more restrictive than the Newtonian conditions; as we shall see in the next Section, we seem to be able to find Newtonian analogues for each general-relativity solution, but the converse is not true.

5. – Applications of the field equations.

We shall assume in this Section, unless otherwise stated, that the matter and radiation content of the universe can be represented as a perfect fluid with equation of state $p = p(\mu)$. This will be a good approximation at most times (see Sect. **3**).

5'1. *The Friedmann (or Robertson-Walker) models.* – These are those general-relativistic models which are locally istropic about every point of space-time. For example, exact isotropy of the number counts and radiation distribution would show $h_a{}^b \mu_{;b} = 0 \Leftrightarrow \dot{\mu} u_a = -\mu_{,a}$; then the equation of state implies $h_a{}^b p_{;b} = 0$. The surfaces $\{\mu = \text{const}\}$ (*) are therefore orthogonal to the fluid vector u_a in this case, which implies $\omega_a = 0$; and the momentum conservation equation (3.15) imply $\dot{u}_a = 0$. An isotropic Hubble law implies $\sigma_{ab} = 0$ (and in fact also implies $\dot{u}_a = 0$; see Sect. **6**). With these conditions, (4.17) shows $h_a{}^b \theta_{;b} = 0$. Thus these models can be characterized by the conditions

$$\omega_a = \sigma_{ab} = 0 = \dot{u}_a \,, \tag{5.1a}$$

which imply the further conditions

$$\mu = \mu(t) \,, \qquad p = p(t) \,, \qquad \theta = \theta(t) \,, \tag{5.1b}$$

where the surfaces $\{t = \text{const}\}$ are the surfaces orthogonal to the fluid flow vector. We may choose the co-ordinate t to measure proper time along each world line, and the scale factor l to be a function of t alone; then $l(t)$ is pre-

(*) $\theta \neq 0$ shows $\dot{\mu} \neq 0$ by (3.8*a*), (3.18), so these *are* well-defined surfaces when this condition holds.

cisely the « radius function » $R(t)$ commonly used in describing the Robertson-Walker space-times.

The Newtonian analogues of these models may be defined by (5.1a) which again implies (5.1b). The remaining nontrivial equations are

(5.2a) $\dot{\mu} + 3(\mu + p) l^{\bullet}/l = 0$,	(5.2a') $\dot{\varrho} + 3\varrho l^{\bullet}/l = 0$,
(5.2b) $3l^{\bullet\bullet}/l + \frac{1}{2}(\mu + 3p) - \Lambda = 0$,	(5.2b') $3l^{\bullet\bullet}/l + \frac{1}{2}\varrho - \Lambda = 0$,
(5.2c) $E_{ab} = H_{ab} = 0$.	(5.2c') $E_{\mu\nu} = 0$.

In the general-relativity case, eq. (4.25) reduces to

$$K(x, e^a) = \tfrac{1}{3}(\mu + \Lambda - \tfrac{1}{3}\theta^2) = K(t) \,, \tag{5.2d}$$

which shows that the 3-surfaces orthogonal to u^a are isotropic at each point, and so are 3-spaces of constant curvature $K(t)$. (We may note that (5.2c) is the statement that space-time is conformally flat (the Weyl tensor vanishes). There is in fact a converse to this (due to TRÜMPER):

$$\{E_{ab} = H_{ab} = 0\} \Rightarrow (5.1) \text{ holds} \,,$$

as the Bianchi identities (4.21) show that when $E_{ab} = H_{ab} = 0$, then $h_a{}^b \mu_{,b} = \omega_a = \sigma_{ab} = 0$. Since $p = p(\mu)$ (*),

$$\{h_a{}^b \mu_{,b} = 0\} \Rightarrow \{h_a{}^b p_{,b} = 0\} \Rightarrow \dot{u}_a = 0 \,.)$$

Equation (5.2a) shows that

(5.3) $ll^{\bullet}(\mu + 3p) = -(\mu l^2)^{\bullet}$.	(5.3$'$) $\varrho l l^{\bullet} = -(\varrho l^2)^{\bullet}$.

Thus when $l^{\bullet} \neq 0$, we can integrate Raychaudhuri's equation (5.2b) to obtain

(5.4a) $3l^{\bullet 2} - (\mu l^3)/l - \Lambda l^2 = 10E$, $E = \text{const}$	(5.4a') $3l^{\bullet 2} - (\varrho l^3)/l - \Lambda l^2 = 10E$, $E = \text{const}$

which is the Friedmann equation; it has the form of an energy equation (the term μl^3 is constant if we have a dust-filled universe, and ϱl^3 is constant in the

(*) If we drop this restriction, we also obtain a family of inhomogeneous spaces (SHEPLEY and TAUB [49]).

Newtonian case). We may note that if there are several noninteracting components of matter, (5.2) implies (5.3) independently for each such component; so we may add together the densities μ_i of such components in (5.4), or use these equations with μ representing the total energy density.

In the general-relativity case, (5.2d) is equivalent to (5.4a) if $K(t)$ is related to E by

$$K(t) = -\frac{10E}{3l^2} = \frac{k}{l^2}, \tag{5.4b}$$

where for convenience we have introduced the constant $k = -10E/3$; on multiplying $l(t)$ by a suitable number, we can normalize k to one of the values $+1$, 0 or -1.

Evaluating (5.2d) at the present time shows that

$$K(t_0) = \frac{k}{l_0^2} = H_0^2\left(\frac{\mu_0}{3H_0^2} + \frac{\Lambda}{3H_0^2} - 1\right); \tag{5.5a}$$

this equation can be combined with (4.13a) (which is now an exact relation if $p_0 = 0$) to show

$$K(t_0) = \frac{k}{l_0^2} = H_0^2(2q_0 - 1) + \Lambda\,. \tag{5.5b}$$

If the pressure vanishes at all times, we can rewrite the Friedmann equation in the form

$$H^2 := \left(\frac{l^\bullet}{l}\right)^2 = \left(\frac{l_0}{l}\right)^3 (2q_0 H_0^2 + \tfrac{3}{2}\Lambda) + \frac{\Lambda}{3} - \left(\frac{l_0}{l}\right)^2 K(t_0)\,. \tag{5.4c}$$

Given suitable equations of state, we can find $\mu(l)$ (see Sect. **3**) and then Friedmann's eq. (5.4) gives the rate of expansion as a function of average length l. We can further integrate the equation $\theta(l) + 3l^\bullet/l$ analytically or numerically to obtain $l(t)$. These equations have been studied extensively in the literature (see, for example, ROBERTSON [50, 51], BONDI [1], STABELL and REFSDAL [52], REFSDAL, STABELL and DE LANGE [53], TAUBER [54], RINDLER [5]) so we shall simply give two comments on their solutions.

5·1.1. *Expansion from a singularity*. Raychaudhuri's equation (5.2b) shows that if $\Lambda \leqslant 0$ and the energy condition (3.8) holds then, irrespective of the equation of state of matter, $l^{\bullet\bullet} < 0$. This implies $q_0 > 0$. In fact, even if $\Lambda > 0$, the condition $q_0 > 0$ is sufficient to imply $l^{\bullet\bullet} < 0$ at all earlier times (Λ is constant so (3.8) and (3.18) guarantee that $\mu/2$ remains larger than Λ). Therefore either $\Lambda \leqslant 0$ or $q_0 > 0$ imply there was a singularity in the universe when $l \to 0$ a finite time t_0 ago, where (*) (cf. Fig. 3)

$$t_0 < \frac{1}{H_0} \simeq 1.3 \cdot 10^{10}\ \text{y}\ . \tag{5.6}$$

[In the general-relativity case, the larger the pressure exerted by matter and radiation, the smaller t_0 is.] The conservation equations show that the temperature and density of matter become infinitely large near the singularity.

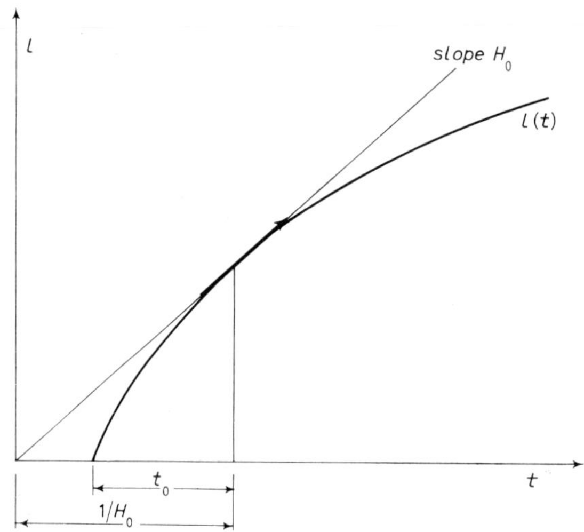

Fig. 3. – If $q_0 > 0$ and the energy conditions $\mu + p > 0$, $\mu + 3p > 0$ are always fulfilled, then the age t_0 of an isotropic universe is less than $1/H_0$.

Raychaudhuri's equation shows similarly that if $\Lambda \leqslant 0$ or $q_0 > 0$, a Robertson-Walker universe evolves very rapidly through its high-temperature phases, most of its lifetime up to the present occurring when $T_{\text{rad}} < 300\ ^\circ\text{K}$. For example, if decoupling takes place a time t_d after the singularity, then $l_d/l_0 \simeq 1/1000$ so $t_d < 10^{-3} H_0^{-1} \simeq 10^7$ y.

5·1.2. *Qualitative properties of the solutions.* If $\Lambda = 0$, there are (again irrespective of the equation of state as long as $\mu > 0$, $p \geqslant 0$) three possible kinds of solution. These are illustrated in Fig. 4 *a*). The solution either (if $k > 0$) collapses back to a second singularity, or (if $k = 0$) has just sufficient energy to escape such a collapse (so $l \to \infty$ as $t \to \infty$), or (if $k < 0$)

(*) This inequality can be sharpened and used to show, on comparing t_0 with the age of the galaxy, that $q_0 < 5.0$; see RINDLER [55].

easily escapes. The second case is the generalized Einstein-de Sitter case, for which $\mu = \frac{1}{3}\theta^2$ at all times. We may use (5.5a) to see that the values $\mu_0 > 3H_0^2$, $\mu_0 = 3H_0^2$ and $\mu_0 < 3H_0^2$ correspond to these three cases respectively; and (5.5b) shows that, up to a correction term involving p_0, they correspond to $q_0 > \frac{1}{2}$,

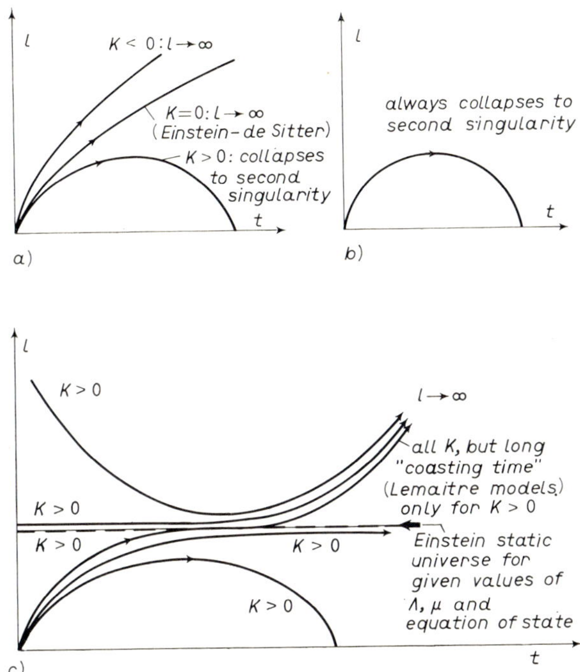

Fig. 4. – The possible characteristics of the function $l(t)$ in Robertson-Walker universes in which $\mu + p > 0$, $\mu + 3p > 0$ at all times. a) when $\Lambda = 0$; b) when $\Lambda < 0$; c) when $\Lambda > 0$. The time reverses of these solution are also solutions.

$q_0 = \frac{1}{2}$ and $q_0 < \frac{1}{2}$. The solutions for which $l \to \infty$ as $t \to \infty$ are asymptotic for large t to solutions with $p = 0$, $\mu = 0$ and $l = H_0 t$ (the *Milne* universe) if $k = -1$ and to the *Einstein-de Sitter* universe ($p = 0$, $l \propto t^{\frac{2}{3}}$) if $k = 0$.

If $\Lambda < 0$ the solution must collapse back to a second singularity.

If $\Lambda > 0$ there exists an unstable solution with $\theta = 0$, *i.e.* $l = \text{const}$; this in the *Einstein static solution* (EINSTEIN [3]). In this case,

$$(5.7) \quad \begin{cases} \Lambda = \frac{1}{2}(\mu + 3p) > 0\,, \\ K = \frac{1}{2}\mu + p = \text{const} > 0\,. \end{cases} \qquad (5.7') \quad \begin{cases} \varrho = \text{const}\,, \\ \Lambda = \frac{1}{2}\varrho > 0\,. \end{cases}$$

Thus as well as the possibilities arising when $\Lambda = 0$, we may further construct solutions asymptotic, or nearly asymptotic, to the Einstein static universe, obtaining the possibilities shown in Fig. 4 *c*). Those solutions which expand forever are asymptotic for large l to solutions with $p = 0$, $\mu = 0$ and $l = \exp[H_0 t]$ where $H_0 = \frac{1}{3}\theta = \sqrt{\Lambda}/3$, a constant. The exact general-relativity solution of this kind has $k = 0$, and is the *de Sitter* universe; it is the same space-time which, with different field equations, constitutes the steady-state universe.

We see that solutions in which there is no singularity, and so in which there is a maximum temperature during the evolution of the universe, can only occur for large positive Λ.

5˙1.3. Co-ordinates. It can be shown from (5.1), (5.2*d*) that there exist co-ordinates such that

the metric takes the form

$$ds^2 = -\,dt^2 + R^2(t)\,d\sigma^2$$

where $d\sigma^2$ is the metric of a 3-space of constant curvature $k = +1$, 0 or -1 (see, for example, HECKMANN and SCHÜCKING [6], ANDERSON [47]),

the co-ordinates of any given fluid particle are

$$x^\nu = R(t)\,c^\nu\,, \qquad c^\nu = \text{const}\,,$$

and the gravitational potential is

$$\Phi = \frac{1}{6}\left(\frac{\varrho(t)}{2} - \Lambda\right) h_{\mu\nu} x^\mu x^\nu$$

(see, for example, BONDI [1]),

where we have written $R(t) \equiv l(t)$ to conform with the notation commonly used in these models. In the general-relativity case, the co-ordinates are co-moving co-ordinates; they can be chosen so that

$$d\sigma^2 = dr^2 + f^2(r)(d\theta^2 + \sin^2\theta\, d\varphi^2)\,,$$

where

$$f(r) = \begin{cases} \sin r \\ r \\ \sinh r \end{cases} \qquad \text{if } k = \begin{cases} +1 \\ 0 \\ -1 \end{cases};$$

(these are the co-ordinates we shall use in Sect. 6).

5˙2. *Gödel's solution.* – This solution is characterised by

$$\theta = \sigma_{ab} = \dot{u}_a = 0\,, \qquad \omega \neq 0\,, \qquad \omega_{a;b} = 0 \qquad (\Rightarrow \omega = \text{const})\,.$$

The field equations and 1st-order identities take the form

(5.8a) $2\omega^2 + \Lambda = \frac{1}{2}(\mu + 3p)$,	(5.8a′) $2\omega^2 + \Lambda = \frac{1}{2}\varrho$,
(5.8b) $E_{ab} = \frac{1}{3}h_{ab}\omega^2 - \omega_a\omega_b$,	(5.8b′) $E_{\mu\nu} = \frac{1}{3}h_{\mu\nu}\omega^2 - \omega_\mu\omega_\nu$,
(5.8c) $H_{ab} = 0$,	

which imply

$h_d{}^a h_e{}^b E_{ab;c} = 0$, $\mu = \text{const}$, $p = \text{const}$.	$E_{\mu\nu,\sigma} = 0$, $\varrho = \text{const}$.

The second-order identities are

(5.9a) $-3E^t{}_s\omega^s = (\mu + p)\omega^t$,	trivially satisfied.
(5.9b) $-E_s{}^{(m}\omega^{t)s} = 0$.	

In the general-relativity case, substituting from (5.8) we find that (5.9b) is identically satisfied, but (5.9a) gives a new condition:

(5.10a) $$\mu + p = 2\omega^2 .$$

Combining this with (5.8a) shows that

(5.10b) $$\tfrac{1}{2}(\mu - p) = -\Lambda .$$

We note that $\Lambda < 0$ for most reasonable equations of state; this is the opposite sign to that in the Einstein static universe.

Given a suitable equation of state $p = p(\mu)$, in general-relativity only one of μ, ω, Λ will determine the other two through eqs. (5.10). (In particular, if $p = 0$ we find $\mu = 2\omega^2 = -2\Lambda$.) In Newtonian theory, we have only to satisfy the restriction (5.8a′); two of ϱ, ω, Λ are arbitrary, and Λ can be positive, zero or negative. Thus the family of general-relativity solutions is more restricted than the family of Newtonian solutions.

Further details of these solutions may be found in papers by

GÖDEL [56].	HECKMANN and SCHÜCKING [13].

5·3. *Further solutions.* – Gödel's solution is not a realistic model universe, as the matter in this solution does not expand. We wish to find solutions with $\theta > 0$ which are more complex than the Robertson-Walker solutions. The problem in integrating Raychandhuri's equation to obtain equations like the Friedmann equation is that we require some restriction on the Weyl tensor enabling us to find σ^2 as a function of l. We might try

5'3.1. Solutions with $\sigma = \dot{u} = 0,\ \omega\theta \neq 0$. In this case the vorticity eq. (4.15) shows $(l^2\omega^b)^{\bullet} = 0$ which implies $\omega^2 = \Omega^2/l^4,\ \dot{\Omega} = 0$. Then we can integrate Raychaudhuri's equation to obtain the generalized Friedmann equation

$$3(l^{\bullet})^2 + \frac{2\Omega^2}{l^2} - \frac{(\mu l^3)}{l} - \Lambda l^2 = 10E\,, \qquad E^{\bullet} = 0\,.$$

However it can be shown, using the integrability condition, that

there exist no such general-relativity solutions if $\partial p/\partial\mu \neq 0$ (*) or if $p = 0$ (ELLIS [57]).	there exist many such Newtonian solutions (HECKMANN and SCHÜCKING [13]) which are necessarily homogeneous (TRÜMPER [46]).

Another simple case is.

5'3.2. Solutions with $\omega = \dot{u} = 0,\ \sigma\theta \neq 0,\ K(x, e^a)$ **isotropic.** In this case the trace-free part of (4.23) (or (4.25)) shows $(l^3\sigma_{ab})^{\bullet} = 0$ which implies $\sigma^2 = \Sigma^2/l^6,\ \dot{\Sigma} = 0$. We can integrate Raychaudhuri's equation to obtain the generalized Friedmann equation.

$$3(l^{\bullet})^2 - \frac{\Sigma^2}{l^4} - \frac{(\mu l^3)}{l} - \Lambda l^2 = 10E, \qquad \dot{E} = 0\,.$$

The qualitative behaviour of $l(t)$ is the same as in the Robertson-Walker case, but the expansion time scales associated with these anisotropic solutions are less than those in isotropic solutions.

There are many such solutions in general relativity (see for example ELLIS and MCCALLUM [58], ELLIS [57]) and in Newtonian theory. For example, the popular Bianchi I space-times with a metric of the form

$$\mathrm{d}s^2 = -\,\mathrm{d}t^2 + X^2(t)\,\mathrm{d}x^2 + Y^2(t)\,\mathrm{d}y^2 + Z^2(t)\,\mathrm{d}z^2$$

in co-moving co-ordinates (see, for example, HECKMANN and SCHÜCKING [7], THORNE [59], JACOBS [60, 61]), belong to this family. These spatially homogeneous models have space sections of constant curvature $K = 0$; they are

---

(*) Then the surfaces $p = \text{const}$ are hypersurfaces orthogonal to the fluid flow vector.

among the simplest homogeneous, anisotropic general-relativity cosmological models. However $E_{ab} \neq 0$ in them so their (spatially homogeneous) Newtonian analogues also have $E_{\mu\nu} \neq 0$.

5˙3.3. Perturbations. An important application of the set of equations we have obtained is in the study in a very clear manner of *perturbations of isotropic world models* (HAWKING [62]; but note that there are some misprints in this paper, and that the expressions given for density perturbations with $p = \frac{1}{3}\mu$ are only valid for large wavelengths).

All perturbation studies agree that unless there is a very large positive Λ-term (as in extreme Lemaitre models), there has not been time for galaxies to form from thermal fluctuations in a Robertson-Walker universe (see the seminars by SALPETER and REES).

5˙3.4. Viscous effects. Finally, following MISNER [28], we may consider the effect of *viscosity* on the evolution of cosmological models. Consider the spaces of Sect. 5˙3.2, but now with a viscosity term present. The trace-free part of (4.23) is (because of (3.14))

$$(l^3 \sigma_{fg})^{\bullet} = l^3 \pi_{fg} = -l^3 \lambda \sigma_{fg} .$$

If we assume λ is approximately constant during the time of interest, we can integrate to obtain

$$\sigma^2 = \frac{\Sigma^2 \exp[-2\lambda t]}{l^6} ,$$

showing that viscosity causes an exponential decay in the shear. This is a strictly general-relativistic effect (the term π_{ab} occurs in (4.16) but not in (4.16′)). Further, terms $-\frac{1}{2}\lambda E_{ab}$ and $-\frac{1}{2}\lambda H_{ab}$ now appear on the right-hand sides of (4.21*d*), (4.21*b*) respectively, showing that viscosity will tend to make the free gravitational field (represented by the Weyl tensor) die away (HAWKING [62]).

This and similar calculations lead us to hope (MISNER [28, 29]; cf. STEWART [23]) that almost any universe model will turn out, after sufficiently realistic physical processes have been considered, to evolve into a state very like a Robertson-Walker universe.

5˙3.5. Further solutions. Much of the work in this and the last chapter is based on Ehler's very clear review article [22]. Further applications of the Bianchi identities to study the dynamics of fluids in general relativity may be found in KUNDT and TRÜMPER [45], SZEKERES [63], SHEPLEY and TAUB [49]. Solutions with $H_{ab} = 0$ are discussed by TRÜMPER [64]. Discussion and applications of the Newtonian equations in the cosmological context may be found in HECKMANN and SCHÜCKING [13, 36], HECKMANN [65], RAYCHAUDHURI [39], TRAUTMANN [14], TRÜMPER [46]. Solutions with $E_{\mu\nu}=0$ are discussed by NARLIKAR [66].

Besides those mentioned in this chapter, many other exact general-relativity solutions have been found; a classification of these space-times according to their symmetries, and further references, may be found in ELLIS [57], STEWART and ELLIS [67], ELLIS and MACCALLUM [58].

6. – Observations in cosmological models.

In this Section we discuss observations in a general curved space-time. Although the sources and observers move with a unique velocity at each space-time point in the applications to cosmological models we have in mind, many of the relations in **6·1-6·4** and **6·6** are valid for sources and observers moving with arbitrary 4-velocities at arbitrary points.

6·1. *The geometric optics approximation.* – The radiation which conveys information in a cosmological model may be represented by a geometric optics solution (*) of Maxwell's equations. The electromagnetic field F_{ab} is regarded as a test field (*i.e.* we can neglect its effect on the curvature of space-time) in a charge- and current-free space-time, and so obeys Maxwell's source-free equations

$$F_{[ab;c]}=0 \Leftrightarrow \exists \Phi_a : F_{ab}=\Phi_{b;a}-\Phi_{a;b}\,, \tag{6.1a}$$

$$F^{ab}{}_{;b}=0\,. \tag{6.1b}$$

The potential Φ_a will be chosen to obey the *gauge* condition

$$\Phi^a{}_{;a}=0\,. \tag{6.1c}$$

We assume there exist solutions of these equations of the form

$$\Phi_a=g(\varphi)A_a+\text{small tail terms}\,, \tag{6.2}$$

where *a*) $g(\varphi)$ is an arbitrary function of the phase φ, and *b*) g varies rapidly compared with the amplitude A_a in the sense that

$$g'\,k_{[a}A_{b]}\gg gA_{[a;b]}\,, \tag{6.3}$$

(*) For further discussion of this approximation, see EHLERS [68].

where $g' := \partial g/\partial \varphi$ and we have defined the propagation vector k_a by

$$k_a := \varphi_{;a} \,. \tag{6.4}$$

(a) is the condition that arbitrary information can be propagated by the signal (cf. TRAUTMANN [69]), and b) is the condition that the signal represents a high-frequency wave with a relatively slowly varying amplitude.) Substituting (6.2) into (6.1b), ignoring the tail terms, and equating to zero separately the coefficients of g, $g' = \partial g/\partial\varphi$ and $g'' = \partial^2 g/\partial\varphi^2$ (which we may do as g is arbitrary), we find

$$k^a k_a = 0 \,, \tag{6.5a}$$

$$A_{a;b} k^b = -\tfrac{1}{2} A_a k^b{}_{;b} \,, \tag{6.5b}$$

$$(A^a)^{;b}{}_{;b} + R^a{}_b A^b = 0 \,. \tag{6.5c}$$

The third equation will play no further part in the present discussion; its essential effect is the show that we cannot in general omit the tail terms if (6.2) is to be an exact solution of (6.1).

From (6.1a) and (6.3) we find that the electromagnetic field has the approximate form

$$F_{ab} \simeq g'(k_a A_b - A_a k_b) \tag{6.6}$$

and the electromagnetic stress tensor S_{ab}, defined by

$$S_{ab} := F_{ac} F_b{}^c - \tfrac{1}{4} g_{ab} F^{cd} F_{cd} \,,$$

has the form (*)

$$S_{ab} \simeq A^2 (g')^2 k_a k_b \,, \tag{6.7}$$

where we have defined $A^2 := A^a A_a$.

An observer with 4-velocity u^a finds the radiation flux across a surface perpendicular to k^a to be the same as the instantaneous energy density of the radiation, both being equal to

$$\mathfrak{f} = S_{ab} u^a u^b = A^2 (g')^2 (k_a u^a)^2 \,. \tag{6.8}$$

Equation (6.5a) implies $k^a k_{a;b} = 0$. However (6.4) shows $k_{a;b} = k_{b;a}$, so

(*) This is the stress-tensor of a particle (the photon) moving with 4-velocity k^a. The gauge condition (6.1c) implies $k^a A_a = 0$ and so ensures A_a is spacelike.

we find

$$k_{a;b}\,k^b = 0\ . \tag{6.9}$$

Thus the light rays (the curves whose tangent vector field is k^a) are *null geodesics.* It follows that light rays are bent by an anisotropic gravitational field. Since this deflection need not be the same for every ray in a small bundle of light rays, such a tube may be differentially bent. Thus a curved space-time will in general distort optical images (SACHS [70]).

6·2. *Red-shifts.* – The rate-of-change of $g(\varphi)$ measured by an observer moving with 4-velocity u^a is $g_{;a}u^a = g'(k_a u^a)$. If observers with 4-velocities u_1^a, u_2^a measure the rate of change of the same signal $g(\varphi)$, these rates of change are in the ratio $(k_a u^a)_1/(k_b u^b)_2$. We can think of this as a *time-dilatation* effect: if a proper time interval $\mathrm{d}t$ is observed to elapse between particular signals (such as pulses emitted at unit time intervals by one of the observers), then $\mathrm{d}t_2/\mathrm{d}t_1 = (k_a u^a)_1/(k_b u^b)_2$. In particular, the observed frequencies ν of light or radio waves are related by

$$\frac{\nu_1}{\nu_2} = \frac{(k_a u^a)_1}{(k_b\, u^b)_2}\ .$$

The *red-shift* z of a source as measured by an observer is defined in terms of wavelengths by

$$z := \frac{\lambda_{\text{observed}} - \lambda_{\text{emitted}}}{\lambda_{\text{emitted}}} =: \frac{\Delta\lambda}{\lambda_{\text{emitted}}}\ . \tag{6.10a}$$

We therefore find that

$$1 + z = \frac{\lambda_{\text{observed}}}{\lambda_{\text{emitted}}} = \frac{\nu_{\text{emitted}}}{\nu_{\text{observed}}}\ ,$$

so

$$1 + z = \frac{(u^a\, k_a)_{\text{emitter}}}{(u^b\, k_b)_{\text{observer}}} \tag{6.10b}$$

determines the red-shift from the 4-velocity vectors $u^a|_{\text{observer}}$, $u^a|_{\text{emitter}}$ and from the tangent vector k^a to the null geodesic. This relation is true no matter what the separation of emitter and observer, and holds independent of an interpretation of the red-shift as a « Doppler » or « gravitational » red-shift. as an example, if both observer and emitter are at the « same » point (*) and

(*) In the cosmological context, at distances up to, say, 10 Mpc.

the emitter moves radially away from the observer, we can write

$$u_a = \cosh\beta u_a + \sinh\beta e_a\,, \qquad e_a e^a = 1\,, \qquad e^a u_a = 0\,, \tag{6.11a}$$

as the 4-velocity of the emitter, where u^a is the observer's 4-velocity, e_a is the direction of motion of the emitter in the observer's rest frame, and $V = \operatorname{tgh}\beta$ is the velocity of relative motion. Then a null ray representing a signal from the emitter to the observer is $k^a = k(u^a - e^a)$; from (6.10*b*) we immediately find

$$1 + z = \exp[-\beta] = \sqrt{\frac{1+V}{1-V}}\,, \tag{6.11b}$$

the standard result for the red-shift due to radial motion of the source relative to the observer when both are at the same space-time point.

For later use, we introduce the following decomposition of k^a. Consider an observer with 4-velocity u^a and let v be an affine parameter along the null geodesic with tangent vector k^a; so $k^a = \mathrm{d}x^a/\mathrm{d}v$. If n^a is the unit vector in the direction of the projection of k^a into the rest space of the observer, then

$$k^a = (-u_b k^b)(u^a + n^a)\,, \qquad n^a n_a = 1\,, \qquad n^a u_a = 0\,. \tag{6.12}$$

Thus a small increment $\mathrm{d}v$ in the affine parameter will be considered by the observer to correspond to a time difference $\mathrm{d}t$ and a spatial displacement $\mathrm{d}l$, with

$$|\mathrm{d}t| = |\mathrm{d}l| = (-k^a u_a)\,\mathrm{d}v\,. \tag{6.13}$$

6.2.1. The linear red-shift relation in a cosmological model. In a given cosmological model, the emitter and observer coincide with particular galaxies moving with the unique fluid velocity u^a. The change in $(u^a k_a)$ occurring in a parameter distance $\mathrm{d}v$ along the null geodesic is

$$\mathrm{d}(u^a k_a) = (u^a k_a)_{;b} k^b\,\mathrm{d}v = (u_{a;b} k^a k^b)\,\mathrm{d}v + u_a(k^a{}_{;b} k^b)\,\mathrm{d}v\,.$$

The second term vanishes by (6.9). Substituting from (2.10) and (6.12),

$$\mathrm{d}(u^a k_a) = (\theta_{ab} n^a n^b + \dot{u}_a n^a)(u^c k_c)^2\,\mathrm{d}v\,.$$

As (6.10) implies that the change $\mathrm{d}\lambda$ in any wavelength λ in the parameter distance $\mathrm{d}v$ is given by

$$\frac{\mathrm{d}\lambda}{\lambda} = -\frac{\mathrm{d}(u_a k^a)}{(u_b k^b)}\,,$$

the change of red-shift along the null geodesic is

(6.14*a*) $$\frac{d\lambda}{\lambda} = (\theta_{ab} n^a n^b + \dot{u}^a n^a)\, dl\,.$$

Using (2.11), this can be written (EHLERS [22])

(6.14*b*) $$\frac{d\lambda}{\lambda} = (dl)^{\bullet} + (\dot{u}^a n_a)\, dl\,.$$

Thus the red-shift has been split (because we have a unique 4-velocity u^a determined at each point) into a radial « Doppler » part (the first term, equivalent to (6.11*b*) for small distances when (2.11) is valid, since the latter then implies the motion is slow) and a « gravitational » part (the second term). Further, we can see how this red-shift-distance relation varies with direction in the sky; since the angular dependence of the terms due to θ, σ_{ab} and $\dot{u}_a$ are different, we can in principle determine these quantities directly from the linear red-shift-distance relation around that point, estimating the distances from the observed brightness of the sources (see Subsect. **6**'4).

6'2.2. Spherically symmetric cosmological models. To illustrate these relations, we consider a Robertson-Walker universe in the co-ordinates of Subsect. **5**'1.3. By the homogeneity and isotropy of these space-times all future-directed null geodesics are equivalent, so it suffices to consider future-directed radial null geodesics through the origin of co-ordinates. The corresponding solution of the geodesic equation is

(6.15) $$k^a = \frac{1}{R}\left(1, \frac{1}{R}, 0, 0\right) \Leftrightarrow k_a = \frac{1}{R}(-1,\ R, 0, 0)\,.$$

Since the fluid velocity vector is $u^a = (1, 0, 0, 0)$, we find

(6.16*a*) $$-u^a k_a = \frac{1}{R}\,.$$

Thus in these models,

(6.16*b*) $$1 + z = \frac{R_{\text{observer}}}{R_{\text{emitter}}}\,.$$

This result can also be obtained by direct integration of (6.14), or by simple geometric methods (cf. NARLIKAR and DAVIDSON [9], and the interesting discussion by SCHRÖDINGER [71]).

6·3. *Polarization.* – It follows from (6.5*b*) that the state of polarization of the light is completely unaffected by the curvature of space-time. More precisely, any numerical parameters describing the polarization are unchanged along the null geodesics, while any directions associated with the polarization are parallely propagated along the null geodesics.

6·4. *Luminosity.* – The quantiy $\mathfrak{f}$ in eq. (6.8) is the instantaneous flux of the radiation; the rate of change of $\mathfrak{f}$ along the null geodesics is determined by the equation

$$(A^2)_{;a}k^a = -A^2 k^a{}_{;a} \tag{6.17}$$

which follows from (6.5*b*). However what is measured in practice is not $\mathfrak{f}$, but a time-average of $\mathfrak{f}$ over a fairly large number of high-frequency oscillations. Thus the observed flux (*) is $A^2(G(\varphi))^2(k^a u_a)^2$ where $G(\varphi)$ is a suitable average of $g'(\varphi)$, and is a slowly varying function of φ. As $G(\varphi)$ is constant along the null geodesics, it can be absorbed into A without affecting (6.17). The measured flux F can therefore be written in the form

$$F = A^2(k_a u^a)^2 \,. \tag{6.18}$$

6·4.1. The area law. Consider a bundle of null geodesics diverging from a radiation source, with cross-sectional area $\mathrm{d}S$ perpendicular to the propagation vector $\boldsymbol{k}$ at a point with affine parameter value v. We quote two geometrical results describing the geometry of such a bundle of null geodesics (see SACHS [70] or PIRANI [37] for proofs):

a) the measurement $\mathrm{d}S$ is independent of the 4-velocity of the observer measuring $\mathrm{d}S$;

b) the change of $\mathrm{d}S$ along the null geodesics is determined by

$$\frac{\mathrm{d}}{\mathrm{d}v}(\mathrm{d}S) = (\mathrm{d}S)_{;a}k^a = \mathrm{d}S\, k^a{}_{;a} \,. \tag{6.19}$$

Combining (6.19) with (6.17) shows that $(A^2\mathrm{d}S)$ is constant along the geodesics. Thus

$$A^2\mathrm{d}S|_1 = A^2\mathrm{d}S|_2 \tag{6.20}$$

describes the change of A along the bundle of null geodesics. Combining

---

(*) Rate at which radiation crosses unit area per unit time.

(6.18) and (6.20) shows that

$$F \propto \left[\frac{(k^a u_a)_{\text{observer}}}{(k^b u_b)_{\text{emitter}}}\right]^2 \frac{1}{\mathrm{d}S},$$

(the factor $(k_b u^b)_{\text{emitter}}$ being constant along the geodesic), so from (6.10) we find

$$F = \frac{\text{constant}}{(1+z)^2 \mathrm{d}S}, \tag{6.21}$$

gives the measured flux at any point along the bundle of null geodesics. The $(1+z)$ factors may be understood as arising from i) the loss of energy suffered by each photon due to the red-shift, and ii) the lower measured rate of arrival of photons due to the time dilatation. Apart from these factors, the flux is proportional to $(\mathrm{d}S)^{-1}$; this expresses the conservation of photons along the bundle of null geodesics. When the energy conditions (3.8) are fulfilled, the space-time curvature tends to cause the bundle of null geodesics to con-

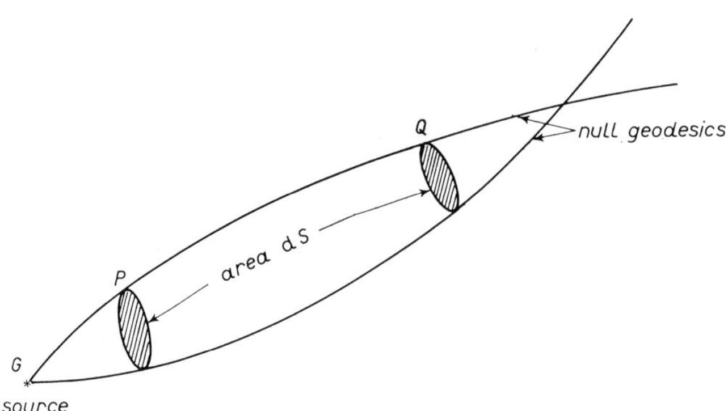

Fig. 5. – The refocussing of a bundle of null geodesics emanating from a source G; the cross-sectional areas of the bundle at P and Q are the same. As a result, the source will appear to be anomalously bright, and to have an anomalously large angular diameter, to an observer at Q.

verge (*) (SACHS [70], PENROSE [72], BERTOTTI [73]). If there is sufficient matter present to reconverge the null geodesics, so that the cross-sectional area $\mathrm{d}S$ is the same at two points (P and Q in Fig. 5), then the factor A^2 will be the same at these two points. Thus the source will seem anomalously bright to an observer at Q; if he and an observer at P both adjust their velocities so as to see the same red-shift, they both measure the same flux of radiation

(*) This is true even if space-time is empty.

from the source (in practice the red-shift factors will often nullify this effect). Near a point where the null geodesics are refocussed, this gravitational lens effect tends to produce very high fluxes (cf. Salpeter's seminar).

6·4.2. Relation to the source luminosity. The constant in eq. (6.21) has still to be related to the source characteristics. The luminosity of the source is defined as the total rate of emission of radiant energy by the source. In principle, the luminosity of the source at some instant t_0 would

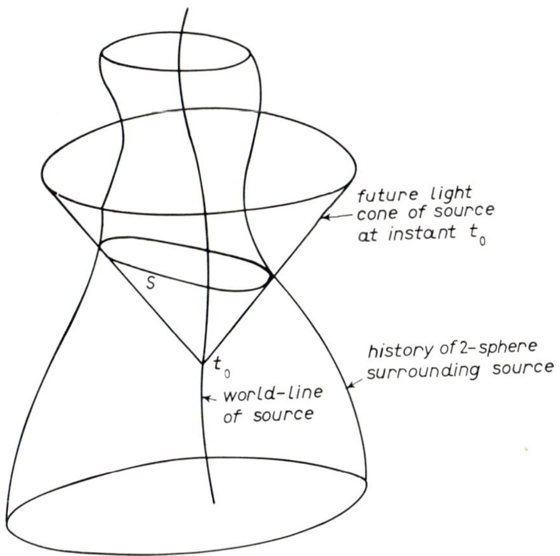

Fig. 6. – A space-time diagram of the intersection S of the future light cone of the source at an instant t_0, with the history of a 2-sphere surrounding the source. The total light emitted by the source at instant t_0 can be measured on S.

have to be measured by enclosing it in a 2-sphere and measuring the rate at which radiation emitted at time t_0 crosses each surface elements $\mathrm{d}S$ (*) of the sphere (in the space-time diagram of this situation, Fig. 6, we measure the radiation on the 2-sphere S). Then we form the integral

$$L := \int_S (1+z)^2 F \,\mathrm{d}S\,. \tag{6.22}$$

By (6.21), this is a constant, independent of the choice of the 2-sphere and of its motion; this constant is the source luminosity L.

(*) Remember that $\mathrm{d}S$ is defined as the area projected perpendicular to k^a.

In practice, we observe the flux from the source along some bundle of geodesics which subtends a small angle $d\Omega_G$ at the source and has cross-sectional area dS_G at the observer. (Subscript G denotes the bundle of geodesics diverging from the galaxy.) Consider a unit sphere lying in the locally Euclidean space-time near the source, and centered on the source (this implies that its 4-velocity is the same as that of the source, so $(1+z)=1$ on this sphere). Let the value of F on this sphere be denoted by F_G. Then, assuming the source radiates spherically symmetrically,

$$L = \int_{\text{2-sphere}} F_G \, dS = 4\pi F_G \tag{6.23}$$

relates the source luminosity to F_G. Now (6.21) applied to the bundle of null rays shows that

$$\left((1+z)^2 F \, dS_G\right) = \text{const} = F_G \, d\Omega_G \,, \tag{6.24}$$

where we have evaluated the constant on the unit 2-sphere. Therefore if we define the *galaxy area distance* r_G of the source from the galaxy by

$$dS_G = r_G^{\,2} \, d\Omega_G \,, \tag{6.25}$$

eqs. (6.23), (6.24) show that

$$F = \frac{L}{4\pi} \frac{1}{(r_G)^2 (1+z)^2} \,, \tag{6.26a}$$

is the observed flux from the source in terms of the galaxy luminosity L, red-shift z and area distance r_G. The effects of the curvature of space-time are contained in the factor $(1/r_G)^2$, as definition (6.25) determines r_G as a function of the affine parameter (and so, in a cosmological model, of the red-shift). In the context of astronomical measurement, this flux is called the *observed luminosity* of the source; the *apparent magnitude* m of the source is defined by

$$m = -2.5 \log_{10} F + \text{const}$$

so we can rewrite (6.26a) as

$$m = -2.5 \log_{10} L + 5 \log_{10} r_G (1+z) + \text{const} \,. \tag{6.26b}$$

6.4.3. Distance definitions. Since we cannot measure the solid angle $d\Omega_G$, the galaxy area distance r_G is not a measurable quantity. However we can define an analogous quantity, the *observer area distance* r_0, which is, in principle, observable. Let $\mathrm{d}\Omega_0$ be the solid angle subtended by a bundle of null geodesics diverging from the observer, and let $\mathrm{d}S_0$ be the cross-sectional area of this bundle at some point. Then the observer area distance r_0 of this point from the observer is defined by

$$\mathrm{d}S_0 = r_0{}^2 \mathrm{d}\Omega_0 \,. \tag{6.27}$$

Thus we can find r_0 if we can measure the solid angle subtended by some object (e.g. a galaxy of given type, or H_{II} regions in a galaxy) whose cross-sectional area can be found from astrophysical considerations. When the distortion effect is not large (if it is large, we should be able to detect it), we can determine the area distance to reasonable accuracy from the observed angular diameter α of some object whose linear dimension d perpendicular to the line of sight can be estimated; then $d \simeq r_0 \alpha$.

If we consider a given galaxy and observer, there are defined two seemingly independent area distances (r_0 and r_G) between them. An important geometrical result, discovered by ETHERINGTON in 1933 and rediscovered by SACHS and PENROSE in 1966, is that these area distances are essentially equivalent. More precisely, we have the

Reciprocity Theorem. The area distances r_0 and r_G are related by

$$(r_G)^2 = (r_0)^2 (1+z)^2 \,. \tag{6.28a}$$

Proof. Let a bundle of null geodesics diverging from G with solid angle $\mathrm{d}\Omega_G$ have tangent vector k^a, and let a bundle of null geodesics converging to O with solid angle $\mathrm{d}\Omega_0$ have tangent vector k'^a, where OG is a null geodesic common to both bundles (see Fig. 7); k^a and k'^a are to coincide on OG. Let v, v' be affine parameters and p^a, p'^a be connecting vectors for k^a, k'^a respectively. Then $k^a = \partial x^a / \partial v$, $k^a{}_{;b} k^b = 0$, $\mathrm{D}p^a/\mathrm{D}v := p^a{}_{;b} k^b = k^a{}_{;b} p^b$ hold, together with the corresponding primed equations (the proof of the last equation is essentially identical to the proof of eq. (2.5)). These equations imply the *geodesic deviation equation* $\mathrm{D}^2 p^a / \mathrm{D}v^2 = -R^a{}_{bcd} k^b p^c k^d$, and the corresponding primed equation. Since $R^a{}_{bcd} k^b k^d = R^a{}_{bcd} k'^b k'^d$ and $v = v'$ along OG, these equations and the Riemann tensor symmetries (4.1) show

$$p'^a \frac{\mathrm{D}^2 p_a}{\mathrm{D}v^2} - p^a \frac{\mathrm{D}^2 p'_a}{\mathrm{D}v^2} = 0 \text{ along } OG \,.$$

Therefore

$$p'^a \frac{\mathrm{D}p_a}{\mathrm{D}v} - p^a \frac{\mathrm{D}p'_a}{\mathrm{D}v} = \text{constant along } OG.$$

Evaluating this constant at O (where $p'^a = 0$) and at G (where $p^a = 0$), we find

$$p^a \bigg|_O \frac{\mathrm{d}p'_a}{\mathrm{d}v}\bigg|_O = -p'_a\bigg|_G \frac{\mathrm{d}p^a}{\mathrm{d}v}\bigg|_G . \tag{6.29}$$

To completely determine the set of connecting vectors we consider, we specify that p^a will be orthogonal to the observers 4-velocity at O and p'^a will be orthogonal to the galaxies 4-velocity at G; then the set of connecting vectors p^a, p'^a satisfying these conditions is 2-dimensional at each point on OG. We write the magnitudes of $\boldsymbol{p}, \boldsymbol{p}'$ as p, p'.

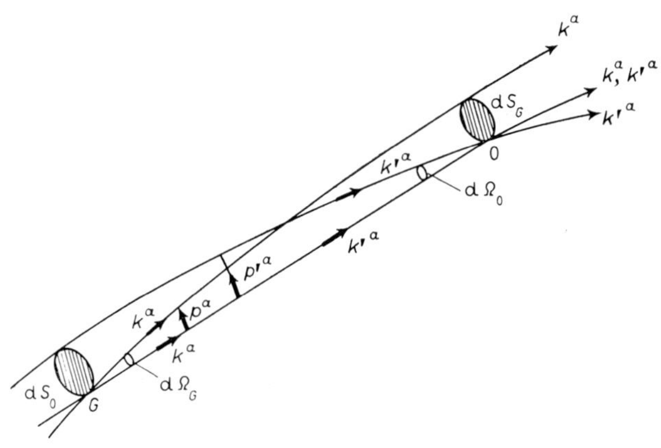

Fig. 7. – Bundles of null geodesics diverging from a source G and converging to an observer O, with a null geodesic OG common to both bundles. If O moves so as to see no redshift in the source spectrum, then equal areas $\mathrm{d}S_O$, $\mathrm{d}S_G$ subtend equal solid angles $\mathrm{d}\Omega_O$, $\mathrm{d}\Omega_G$.

We wish to use relation (6.29) to relate the areas $\mathrm{d}S_O$, $\mathrm{d}S_G$ of the bundles of geodesics at G, O respectively. To do so, we choose from the connecting vectors $\boldsymbol{p}$ a pair $\boldsymbol{p}_1, \boldsymbol{p}_2$ such that

$$\frac{\mathrm{d}p_1{}^a}{\mathrm{d}v}\bigg|_G \frac{\mathrm{d}p_{2a}}{\mathrm{d}v}\bigg|_G = 0 . \tag{6.30a}$$

In general, these vectors will not be orthogonal at O; let the angle between them at O be $\pi/2 + \psi$. If we rotate the pair $\boldsymbol{p}_1, \boldsymbol{p}_2$ at G through an angle $\pi/2$, this maps $\boldsymbol{p}_1$ to $\boldsymbol{p}_2$ and $\boldsymbol{p}_2$ to $-\boldsymbol{p}_1$, and so changes the angle at O to $\pi/2 - \psi$. Thus if we rotate the pair through all angles up to $\pi/2$ at G, the angle between them at O will certainly take all values between $\pi/2 + \psi$ and $\pi/2 - \psi$. We can therefore find a particular pair of connecting vectors $\boldsymbol{p}_1, \boldsymbol{p}_2$ which satisfy

(6.30*a*) and also

$$p_1{}^a|_O\, p_{2a}|_O = 0\,. \tag{6.30b}$$

We now choose from the connecting vectors $\boldsymbol{p}'$ a pair $\boldsymbol{p}'_1, \boldsymbol{p}'_2$ determined by the condition

$$p'_{1a}\bigg|_G \frac{\mathrm{d}p_{2a}}{\mathrm{d}v}\bigg|_G = 0\,, \qquad p'_{2a}\bigg|_G \frac{\mathrm{d}p_1{}^a}{\mathrm{d}v}\bigg|_G = 0\,.$$

Then (6.29) shows

$$p_2{}^a\bigg|_O \frac{\mathrm{d}p_1{}'_a}{\mathrm{d}v}\bigg|_O = 0\,, \qquad p_{1a}\bigg|_O \frac{\mathrm{d}p_2{}'^a}{\mathrm{d}v}\bigg|_O = 0\,,$$

i.e. we have (because of (6.30*a*), (6.30*b*)) the conditions

$$p_1{}'^a|_G\, p_2{}'_a|_G = 0\,, \qquad \frac{\mathrm{d}p_1{}'^a}{\mathrm{d}v}\bigg|_O \frac{\mathrm{d}p_2{}'_a}{\mathrm{d}v}\bigg|_O = 0\,. \tag{6.30c}$$

With this choice of connecting vectors,

$$\mathrm{d}S_G = p_1|_O\, p_2|_O\,, \qquad \mathrm{d}S_O = p'_1|_G\, p'_2|_G\,, \qquad \mathrm{d}\Omega_O = \frac{\mathrm{d}p'_1}{\mathrm{d}l}\bigg|_O \frac{\mathrm{d}p'_2}{\mathrm{d}l}\bigg|_O\,,$$

$$\mathrm{d}\Omega_G = \frac{\mathrm{d}p_1}{\mathrm{d}l}\bigg|_G \frac{\mathrm{d}p_2}{\mathrm{d}l}\bigg|_G\,.$$

However (6.29) shows

$$(p_1 p_2)\bigg|_O \left(\frac{\mathrm{d}p'_1}{\mathrm{d}v}\frac{\mathrm{d}p'_2}{\mathrm{d}v}\right)\bigg|_O = (p'_1 p'_2)\bigg|_G \left(\frac{\mathrm{d}p_1}{\mathrm{d}v}\frac{\mathrm{d}p_2}{\mathrm{d}v}\right)\bigg|_G\,.$$

Using (6.13) this relation is

$$\mathrm{d}S_G\, \mathrm{d}\Omega_O (k^a u_a)^2|_O = \mathrm{d}S_O\, \mathrm{d}\Omega_G (k_a u^a)^2|_G\,,$$

which is relation (6.28) in view of (6.25), (6.27) and (6.10).

This result is simply a consequence of the geodesic deviation equation. It tells us that if $z = 0$, then equal surface elements $\mathrm{d}S_O$, $\mathrm{d}S_G$ subtend equal solid angles $\mathrm{d}\Omega_O$, $\mathrm{d}\Omega_G$, irrespective of the curvature of space-time (The factor $(1+z)^2$ is simply the special relativistic correction to solid-angle measurements.) Note that if there is a gravitational lens effect leading to anomalously large source brightness, this is accompanied by an anomalously large source solid angle. In fact, if we have the situation (see Fig. 5) that light is refocussed, the angular diameter of a given object decreases to a minimum and then starts increasing again as that object is moved further down the past light cone of the observer.

The reciprocity theorem allows us to extend (6.26*a*) to

(6.26*c*)
$$F = \frac{F_G}{(r_G)^2(1+z)^2} = \frac{F_G}{(r_O)^2(1+z)^4}\,,$$

so we can rewrite (6.26*b*) as

(6.26*d*)
$$m = -2.5\log_{10} L + 5\log_{10} r_O(1+z)^2 + \text{const}\,.$$

Another way of stating the result of the theorem is to define a *corrected luminosity distance* r by

$$r^2 = \frac{F_G}{F(1+z)^4}\,.$$

(see KRISTIAN and SACHS [17]); then the theorem is

(6.28*b*)
$$r = r_O\,.$$

Since we can in principle find the corrected luminosity distance r for any source with known intrinsic luminosity by measuring the flux from the source, we can (in principle) verify eq. (6.28*b*) experimentally. However in practice (see Subsect. 6.6.2) we are usually unable to measure r, r_O independently.

In the literature, r, r_O and D (defined by $D^2 := F_G/F$) have all been called luminosity distances; this has caused some confusion about what the correct red-shift factors in various formulae are (*). A further distance defined by the null geodesics is the *parallax distance* r_p, defined by $r_p = 2/k^a{}_{;a}|_G$, which is equivalent to the usual definition of parallax distance when there is no distortion (JORDAN, EHLERS and SACHS [74]); however this distance is clearly not measurable in the cosmological context.

These distances are all (in principle) directly measurable quantities, which reduce to the usual special relativistic distance *a*) for slowly moving nearby sources, and *b*) in a static situation in a flat space-time. However in the situation in which there is refocussing and so a minimum angular diameter for any given object, some or all of these distances will be double-valued. Even the cosmological red-shift, which is often a single-valued directly observable di-

---

(*) When we refer to a « luminosity distance » it will be the uncorrected distance D, and a « corrected luminosity distance » will be $r = r_O$.

stance indicator, may be double-valued in certain circumstances (for example, in a Robertson-Walker universe ($\Lambda > 0$, $K > 0$) which contracts from an infinite to a finite radius and then re-expands to an infinite radius).

6'4.4. Spherically symmetric cosmological models. As an example, consider a Robertson-Walker universe with the metric given in Subsect. 5'1.3, and the source taken at the centre of the spatial co-ordinates; we use (6.15) to give the components of k^a. Let the time at the observev be t_0 and the time at the source be t_G.

A small affine parameter change $\mathrm{d}v$ along the null geodesics corresponds to co-ordinate changes $\mathrm{d}t$, $\mathrm{d}r$ where

$$\mathrm{d}v = R\,\mathrm{d}t = R^2\,\mathrm{d}r\,, \tag{6.31}$$

(cf. eq. (6.13)). The divergence of k^a is

$$k^a{}_{;a} = \frac{1}{\sqrt{-g}}\frac{\partial}{\partial x^a}(\sqrt{-g}\,k^a) = \frac{2}{R^2}\left(R^{\bullet} + \frac{f'(r)}{f(r)}\right).$$

Since (6.19) implies $[\log \mathrm{d}S]_G^O = \int_G^O k^a{}_{;a}\,\mathrm{d}v$, we can use (6.31) together with this equation and the definition of $\mathrm{d}\Omega_G$ to find

$$\log(\mathrm{d}S_G/\mathrm{d}\Omega_G) = \int_G^O \frac{2R^{\bullet}}{R^2}\cdot R\,\mathrm{d}t + \int_G^O \frac{2f'(r)}{R^2 f(r)}\cdot R^2\,\mathrm{d}r = \log\left(R^2 f^2(r)\right)\Big|_G\,, \tag{6.32}$$

since

$$\lim_{r\to 0}\left(\frac{\mathrm{d}S}{R^2 f^2(r)}\right)\Bigg|_G = \mathrm{d}\Omega_G\,.$$

Here r is the radial co-ordinate difference between the source and the observer; to express the result in a manner independent of the spatial co-ordinates used, we note that

$$r = \int_G^O \mathrm{d}r = \int_{t_G}^{t_0} \frac{\mathrm{d}t}{R(t)}\,,$$

the equivalence holding because the light moves on a null geodesic (see (6.31)). Thus if we define

$$u := \int_{t_G}^{t_0} \frac{\mathrm{d}t}{R(t)}\,, \tag{6.33}$$

then (6.32) is equivalent to

$$r_G^2 = R^2|_O f^2(u) , \tag{6.34a}$$

where

$$f(u) = \begin{cases} \sin u \\ u \\ \sinh u \end{cases} \text{if} \quad k = \begin{cases} +1 \\ 0 \\ -1 \end{cases} .$$

Similarly we can show that

$$r_O^2 = R^2|_G f^2(u) = (1+z)^{-2} R^2|_O f^2(u) , \tag{6.34b}$$

which is consistent with (6.16) (*). Whenever $R^{\bullet} \neq 0$ we can express u in the form

$$u = \int_{R_G}^{R_O} \frac{\mathrm{d}R}{RR^{\bullet}} = \int_0^z \frac{\mathrm{d}z}{(1+z)R^{\bullet}} ,$$

and then substitute from the Friedmann equation to obtain $u(z)$ and so $r_O(z)$. For example when pressure-free matter is the dominant energy component in the universe (as it is at recent times) we can use the Friedmann equation in the form (5.4*c*) and numerically integrate to obtain $r_O(z)$. (See, for example, REFSDAL, STABELL and DE LANGE [53].) In the particular case $\Lambda = 0$ we can integrate analytically to find (using (5.5*b*) to evaluate R_0 when $k \neq 0$) that

$$r_0 = \frac{1}{H_0 q_0^2 (1+z)^2} \left\{ q_0 z + (q_0 - 1)\left((1 + 2q_0 z)^{\frac{1}{2}} - 1\right) \right\} , \tag{6.35a}$$

when $q_0 \neq 0$, and

$$r_0 = \frac{1}{2H_0} \left(1 - \frac{1}{(1+z)^2} \right) , \tag{6.35b}$$

when $q_0 = 0$. (cf. MATTIG [75], SANDAGE [76]). Combining these with (6.26*d*) gives the usual *m*-*z* relations.

We see that when $q_0 \neq 0$, there is a maximum value of r_0 for some value of z, and thereafter r_0 decreases as z increases; *i.e.* we *do* have refocussing of

(*) We can also derive these equations direct from the Robertson-Walker geometry, (cf. DAVIDSON and NARLIKAR [9]).

null geodesics in these models. This is still true if we include a pressure term in the Friedmann equation, so it is true in all nonempty Robertson-Walker universes which expand from a singularity (we can always ignore a Λ-term at early enough times). However the (uncorrected) luminosity distance $D = r_0(1+z)^2$ always decreases as z increases; *i.e.* although the area factor in (6.26*c*) tends to increase F after a certain point, the cosmological red-shift factors suffice to ensure that the observed flux from identical sources always decreases as distance increases.

Equations (6.35*a*), (6.35*b*) imply a simple expression for z as a function of D; in fact

$$1+z = q_0(1+DH_0)-(q_0-1)\sqrt{1+2DH_0}\,. \tag{6.35c}$$

6'5. *Number counts in a cosmological model.* – Consider a small affine parameter displacement $\mathrm{d}v$ at a point A on a bundle of past null geodesics subtending a solid angle $\mathrm{d}\Omega_0$ at 0; this corresponds to a distance $\mathrm{d}l = (-k^a u_a)\,\mathrm{d}v$

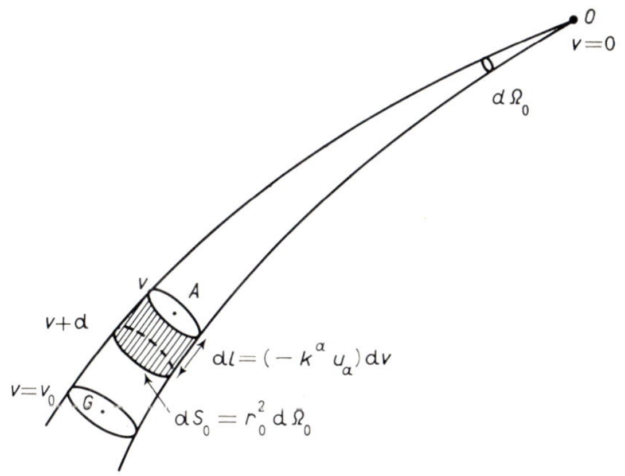

Fig. 8. – A section $(v, v+dv)$ of a bundle of null geodesics which subtends a solid angle $\mathrm{d}\Omega_0$ at 0.

in the rest-frame of a galaxy at A. The cross-section area of the bundle is $\mathrm{d}S_0 = r_0{}^2\mathrm{d}\Omega_0$, so if the number-density of radiation sources is n per unit proper volume, then the number of sources in this section of the bundle (see Fig. 8) is

$$\mathrm{d}N = r_0{}^2\mathrm{d}\Omega_0(n(-k^a u_a))_A\,\mathrm{d}v\,. \tag{6.36}$$

Integrating this expression with respect to v gives the number $N(v)$ of sources seen by O in the solid angle $\mathrm{d}\Omega_0$ which lie at affine parameter distances less than v. We can combine $N(v)$ with the functions $r_0(v)$, $z(v)$ to obtain $N(F)$, the number of sources with intrinsic luminosity seen in $d\Omega_0$ with observed flux greater than F, or $N(z)$, the number of sources seen in $d\Omega_0$ with red-shift less than z.

6·5.1. Spherically symmetric models. In a Robertson-Walker universe, if the number of sources is conserved then $n = n_0(R_0^3/R^3)$, and (6.36) becomes

$$dN = \frac{n_0 R_0^3}{R^3} . d\Omega_0 . R^2 f^2(u) \frac{1}{\dot{R}} . R^2 du ,$$

on using (6.16*a*), (6.31), (6.34*b*). Thus

$$N(u) = 4\pi n_0 R_0^3 \int_0^u f^2(u')\, du' = 4\pi n_0 R_0^3 \begin{cases} \frac{1}{2}(u - \sin u \cos u) , & \text{if } k = +1 , \\ u^3/3 & \text{if } k = 0 , \\ \frac{1}{2}(\sinh u \cosh u - u) , & \text{if } k = -1 , \end{cases}$$

is the number of sources in all directions at distances up to that characterized by u. Using (6.34) and (5.5*b*) we can write this as a function of $r_0(1+z)$, (*i.e.* of r_G):

$$N = \begin{cases} \dfrac{2\pi n_0}{(H_0^2(2q_0 - 1) + \Lambda)^{\frac{3}{2}}} (\arcsin f \mp f\sqrt{1-f^2}) , & \text{if } k = +1 , \\ \dfrac{4\pi}{3} n_0((1+z) r_0)^3 , & \text{if } k = 0 , \\ \dfrac{2\pi n_0}{(H_0^2(1 - 2q_0) - \Lambda)^{\frac{3}{2}}} (f\sqrt{1+f^2} - \arg\sinh f), & \text{if } k = -1 , \end{cases} \tag{6.37a}$$

where

$$f = \left(k(H_0^2(2q_0 - 1) + \Lambda)\right)^{\frac{1}{2}} \cdot r_0(1+z) , \qquad \text{when } k \neq 0 . \tag{6.37b}$$

(Care has to be taken in choosing the correct sign $\mp$ when $k = +1$ (ROEDER and MCVITTIE [77]); note also that N is monotonic with affine distance although r_0 is not.) Combining this expression with $r_0(z)$, we obtain $N(z)$; alternatively we can eliminate z to obtain N as a function of one of the area distances. In particular if $\Lambda = 0$ in a dust-filled universe, (6.35*c*) shows we can obtain the simple expresssion

$$r_0(1+z) = \frac{D}{q_0(1 + DH_0) - (q_0 - 1)\sqrt{1 + 2DH_0}} ,$$

which, combined with (6.37) expresses $N(D)$ in terms of simple analytic functions. (cf. MATTING [78], SANDAGE [76].)

6'6. *Observed specific intensity.* – The theoretical relations we have derived so far need to be modified in three ways to correspond more closely to what is actually observed.

6'6.1. Specific flux. So far, we have considered a total (bolometric) luminosity L and flux F. However we actually observe in very restricted wavelength ranges (in radio observations, nearly at one frequency: in optical observations usually in the U, B, V bands). To allow for this, we represent the source spectrum by a function $\mathscr{I}(\nu)$, where $L\mathscr{I}(\nu)\,\mathrm{d}\nu$ is the rate at which radiation is emitted by the source at frequencies between ν and $\nu + \mathrm{d}\nu$; $\mathscr{I}(\nu)$ is normalized by the condition $\int_0^\infty \mathscr{I}(\nu)\,\mathrm{d}\nu = 1$. The frequency ν_o measured by some observer is related to the frequency ν_G of the same radiation in the rest-frame of the emitting galaxy by $\nu_o = \nu_G/(1+z)$ (cf. Subsect. 6'2), which implies $\mathrm{d}\nu_o = \mathrm{d}\nu_G/(1+z)$. Equation (6.26) can therefore be written

$$F = \frac{L}{4\pi}\frac{\int_0^\infty \mathscr{I}(\nu_G)\,\mathrm{d}\nu_G}{r_o^2(1+z)^4} = \frac{L}{4\pi}\frac{\int_0^\infty \mathscr{I}\big(\nu_o(1+z)\big)\,\mathrm{d}\nu_o}{r_o^2(1+z)^3}\,.$$

Thus the flux measured in the frequency range ν, $\nu + \mathrm{d}\nu$ by the observer is

$$F_\nu\,\mathrm{d}\nu = \frac{L}{4\pi}\frac{\mathscr{I}\big(\nu(1+z)\big)\,\mathrm{d}\nu}{r_o^2(1+z)^3} = \frac{L}{4\pi}\frac{\mathscr{I}\big(\nu(1+z)\big)\,\mathrm{d}\nu}{r_G^2(1+z)}\,. \tag{6.38a}$$

We call F_ν the *specific flux* (in the case of radio sources, it is called the *flux density*) of the radiation. We can often assume $\mathscr{I}(\nu)$ has the form $\mathscr{I}(\nu) = \mathrm{const}\,\nu^{-\alpha}$, where α is a constant (the « spectral index »); (this is a good approximation for many radio sources when $0.7 \leqslant \alpha \leqslant 0.9$ and for many optical sources at wavelengths longer than 5000 Å when $\alpha \simeq 2$). Then we find

$$F_\nu = \frac{F_G\mathscr{I}(\nu)}{r_o^2(1+z)^{3+\alpha}} = \frac{F_G\mathscr{I}(\nu)}{r_G^2(1+z)^{1+\alpha}}\,. \tag{6.38b}$$

An alternative way of allowing for the effect of the source spectrum is to introduce a correction term, the « K-correction » (SANDAGE [79]) which represents the difference between F_ν (given by (6.38*a*)) and F (given by (6.26)).

6'6.2. Intensity. So far we have implicitly assumed the sources observed are point sources. In practice we usually observe extended sources, and instead of measuring the flux from the source, our direct measurements tell us the flux per unit solid angle from the source (HOYLE [80]), that is, the *intensity* of radiation from the source. Considering a source (or part of a source) of area A, we find from (6.26*c*) (6.27) that the intensity is

$$I := \frac{F}{\mathrm{d}\Omega_0} = \frac{I_G}{(1+z)^4}, \tag{6.39}$$

where the factor $I_G := F_G/A$ depends simply on the source characteristics. For a given source, the intensity is independent of the area distance r_0 and depends only on the red-shift z; if $z = 0$, I has the value I_G which is the *surface brightness* of the source. Thus, for example, the density of marking of an image on a photographic plate is independent of the space-time curvature, depending simply on the source surface brightness and observed red-shift; as we can find z independent of I, we can in principle find the surface brightness of the source direct from our measurements.

To find the flux F from an extended source, we have to measure I and then integrate over the image to obtain F; therefore we have in fact (except in the case of quasars or very distant galaxies), explicitly or implicitly, to estimate the solid angle subtended by the source before we can deduce F from our direct observations (see HOYLE [80] for further discussion). It is for this reason that we are in practice unable to measure the corrected luminosity distance r and area distance r_0 independently (see Subsect. **6**˙4.3). We may in fact lose useful information if we consider only the flux F (the « magnitude » of the source) rather than the intensity and solid angle information which is combined to give F.

We may note that it is expression (6.39) which is involved in Olber's paradox (BONDI [1]; see also WHITROW and YALLOP [81], HARRISON [82] and references cited there); this expression shows that we have the alternatives of assuming I_G is very low for the sources along almost all null geodesics from us (perhaps because of the source evolution which must occur in view of the finite source lifetime), or that the red-shift increases indefinitely along almost all null geodesics.

6˙6.3. Specific intensity. Combining the effects considered in Sections **6**˙6.1 and **6**˙6.2, we see that what we usually measure directly is the flux per unit solid angle in some frequency range $(\nu, \nu + \mathrm{d}\nu)$. That is, we measure the *specific intensity* I_ν of radiation from the source; (6.27), (6.38) show

$$I_\nu \mathrm{d}\nu := \frac{F_\nu \mathrm{d}\nu}{\mathrm{d}\Omega} = I_G \cdot \frac{\mathscr{I}\big(\nu(1+z)\big)\,\mathrm{d}\nu}{(1+z)^3}, \tag{6.40}$$

which for a given source depends simply on the redshift of the source ($I_G \cdot \mathscr{I}(\nu)$ is the surface brightness of the source at frequency ν). To estimate the specific flux (*e.g.* to estimate the visual magnitude m_V) from an extended source, we have to integrate the observed specific intensity, which is what is actually measured by the photographic plate or radio receiver, over the image of the source.

As an application of (6.40), we consider measurement of radiation from a source emitting black-body radiation. Defining the function $g(\nu)$ by $g(\nu) = I_G \mathscr{I}(\nu)/\nu^3$, we can rewrite eq. (6.40) as

(6.41*a*) $$I_\nu \, \mathrm{d}\nu = g(\nu(1+z))\, \nu^3 \, \mathrm{d}\nu ,$$

which gives the observed specific intensity at each frequency ν for any observer who measures the source red-shift as z. On the other hand,

$$(I_\nu \, \mathrm{d}\nu)|_G = f\left(\frac{\nu}{T_e}\right) \nu^3 \, \mathrm{d}\nu ,$$

is the specific intensity of radiation at the source (*), where $f(\nu/T_e)$ is the Planck function for black-body radiation at temperature T_e. Comparing these expressions at the source shows $g(\nu) = f(\nu/T_e)$, which implies

$$g(\nu(1+z)) = f\left(\frac{\nu(1+z)}{T_e}\right).$$

Hence we can rewrite eq. (6.41*a*) as

(6.41*b*) $$I_\nu \mathrm{d}\nu = f\left(\frac{\nu}{T}\right) \nu^3 \, \mathrm{d}\nu ,$$

where we have defined

(6.42) $$T := \frac{T_e}{1+z};$$

i.e. an observer who measures the source red-shift as z, sees the radiation as black-body radiation at temperature T where T is defined by (6.42). Thus black-body radiation propagates with an unchanged spectrum through curved space-time, the observed intensity of radiation depending only on the observer's 4-velocity. (cf. Ehler's lectures for a different derivation of this result). We may note that in the Robertson-Walker models, (6.42) enables us to find the change of black-body temperature along the null geodesics, (3.23) enables us to find it along timelike geodesics, and (6.16) shows these results are consistent.

(*) This form follows direct from Wien's law.

This calculation is equally valid if we regard it as relating to observations at one point. Suppose an observer S moving with 4-velocity u^a measures the temperature of black-body radiation, which has a propagation vector k^a, as T, while an observer S' at the same point moving with 4-velocity u'^a measures the temperature of the same radiation as T_0. Then

$$T = \frac{T_0}{1+z} = T_0 \frac{(k^a u_a)}{(k^b u'_b)},$$

where z is the red-shift S measures for radiation from a source moving with 4-velocity u'^a. If the relative velocity of S and S' has magnitude v and the angle measured by S between the direction of $\boldsymbol{k}$ and the direction of motion of S' is θ, we can directly evaluate this equation by using (6.11a) and (6.12). (Note that $e^a n_a = \cos\theta$). We find (*)

$$T = \frac{T_0}{\cosh\beta - \sinh\beta\cos\theta} = T_0 \frac{(1-v^2)^{\frac{1}{2}}}{(1-v\cos\theta)}, \tag{6.43}$$

where $v = \operatorname{tgh}\beta$. In particular, if S' observes an isotropic black-body radiation field with temperature T_0, then S observes a black-body radiation field with temperature distribution in the sky given by (6.43). Conversely, if S sees such a radiation field, then there exists an observer S' who sees an isotropic radiation field. S can find the 4-velocity of S' by noting the maximum and minimum black-body temperatures $T_{\max}$, $T_{\min}$ on his celestial sphere; then the velocity of S' relative to S is in the direction in which S observes $T_{\min}$ and has magnitude

$$v = \frac{T_{\max} - T_{\min}}{T_{\max} + T_{\min}},$$

while the isotropic radiation temperature measured by S' is $T_0 = \sqrt{T_{\max} T_{\min}}$. (Remember we are using units in which the speed of light is 1.)

6'6.4. Absorption and emission in a cosmological model. Finally, we should allow for absorption and emission of radiation along the line of sight from the source G to the observer O. Let $I_\nu(v)$ be the specific intensity of radiation travelling along a bundle of null geodesics from G to O as measured by an observer moving with the average velocity u^a at a point A (affine parameter distance v from O) in a cosmological model. We consider the change in I_ν as v increases to $v + dv$ (see Fig. 8). By eq. (6.40), we can represent the change in I_ν due to geometrical and red-shift effects alone by the differential

(*) This is just the generalization of (6.11 b) to include the transverse Doppler shift.

equation

$$\frac{\mathrm{d}I_{\nu'}}{\mathrm{d}v} = \frac{3}{1+z} I_{\nu'} \frac{\mathrm{d}z}{\mathrm{d}v},$$

where $\nu' := \nu(1+z)$ is the frequency of radiation at A which, when red-shifted to O, is observed at frequency ν. Let $S(v, \nu)\,\mathrm{d}\nu$ be the rate of emission of radiation by each source at A per unit solid angle in the frequency range ν to $\nu + \mathrm{d}\nu$, let $n_s(v)$ be the number density of sources at A, let $n_a(v)$ be the number density of particles scattering or absorbing radiation at A, and let $\sigma(v, \nu)$ be the interaction cross-section of these particles at frequency ν (we may allow for re-emission by suitable additions to either σ or S; similarly we may, if we wish, represent absorption processes by a negative contribution to S). Allowing for these processes in the volume $\mathrm{d}l\,\mathrm{d}S_0 = (-u_a k^a)\,\mathrm{d}v\,\mathrm{d}S_0$ at A, the change in I'_ν along the geodesic can be represented by the differential equation

$$\frac{\mathrm{d}I_{\nu'}}{\mathrm{d}v}(v) - \frac{3}{1+z} I_{\nu'}(v) \frac{\mathrm{d}z}{\mathrm{d}v} - n_a(v)\,\sigma(v, \nu')\, I_{\nu'}(v).(-k_a u^a)(v) = -n_s(v)\, S(v, \nu').(-k_a u^a)(v)\,,$$

(remembering that when $\mathrm{d}v$ is positive, S acts as a *negative* term and σ as a *positive* term). Integrating this equation along the geodesic from $O(v=0)$ to $G(v = v_*)$ we find the specific intensity at 0 is

$$(6.44a) \qquad I_\nu = \int_0^{v_*} \frac{n_s(v)\, S\big(v, \nu(1+z)\big)}{(1+z)^3} .\exp[-p(v, \nu)](-k_a u^a)(v)\,\mathrm{d}v + \frac{I_{\nu(1+z_*)}(v_*)}{(1+z_*)^3} .\exp[-p(v_*, \nu)]\,,$$

where the optical depth $p(v, \nu)$ between A and O for radiation observed at O at frequency ν is

$$(6.44b) \qquad p(v, \nu) := \int_0^v n_a(v')\,\sigma\big(v', \nu(1+z')\big)(-k_a u^a)(v')\,\mathrm{d}v'\,;$$

in these expressions, z and $(k^a u_a)$ are regarded as known functions of v.

This equation determines the specific intensity of radiation we observe in any direction in the sky; the second term represents radiation propagating according to eq. (6.40) from G but attenuated by absorption, while the first term represents the integrated emission from sources between O and G, again attenuated by absorption. This equation implies a very similar equation, which can also be derived directly from (6.39), determining the (integrated) intensity I; the only essential differences are that the factors $1/(1+z)^3$ are replaced by factors $1/(1+z)^4$, and the arguments containing the frequency ν are omitted (*).

6'6.5. **Spherically symmetric models.** To illustrate the use of these formulae, we will apply (6.44) to a Robertson-Walker universe, taking the matter content to be dust and ignoring the effects of radiation on the dynamics of the universe. Then, since (6.13) shows

$$(-k_a u^a)\,\mathrm{d}v = -\,\mathrm{d}t = -\frac{1}{R^\bullet}\frac{\mathrm{d}R}{\mathrm{d}z}\,\mathrm{d}z\,.$$

we find from (6.16), (5.4*c*) and (5.5*b*) that

$$(6.45)\qquad (-k_a u^a)\,\mathrm{d}v = \frac{\mathrm{d}z}{(1+z)^2\{H_0^2(1+2q_0 z)+(\Lambda/3)\big(2z-1+(1+z)^{-2}\big)\}^{\frac{1}{2}}}\,.$$

(We can easily allow for noninteracting radiation in (6.45) (cf. the comments following (5.4*a*)) if integrated emission from early times is important; if other field equations than those of general relativity are used, one can allow for this by using a suitable replacement of the Friedmann equation to determine a new form for (6.45).) For simplicity we shall further take $\Lambda=0$ and assume that the radiation sources and absorbing particles are conserved; then

$$n(z) = n(0)(R_0^3/R^3) = n(0)(1+z)^3.$$

With these assumptions, the contribution to I_ν from sources up to a red-shift z_*, ignoring absorption, is

$$(6.46)\qquad I_\nu = \frac{n_s(0)}{H_0}\int\limits_0^{z_*}\frac{S\big(z,\nu(1+z)\big)\,\mathrm{d}z}{(1+z)^2(1+2q_0 z)^{\frac{1}{2}}}\,.$$

We might further assume that the source emission can be described by a source spectrum $\mathscr{I}(\nu)$ which is independent of z, and an amplitude $\bar{S}(z)$; then

$$(6.47a)\qquad S(z,\nu) = \bar{S}(z)\,\mathscr{I}(\nu)\,.$$

If in particular the emission is that of a line spectrum at frequency ν_e we can set $\mathscr{I}(\nu)=\delta(\nu-\nu_e)$ where δ is the Dirac delta function; the integrated emis-

(*) For a derivation of an equivalent expression in the Robertson-Walker case, see DAVIDSON and NARLIKAR [9].

sion $\mathscr{G}_\nu(\nu_e)$ in this case is

$$\mathscr{G}_\nu(\nu_e) = \begin{cases} \dfrac{n_s(0)}{H_0 \nu_e} \dfrac{\bar{S}(z_\nu)}{(1+z_\nu)(1+2q_0 z_\nu)^{\frac{1}{2}}}, & \text{if } z_\nu \leqslant z_* , \\ 0 \qquad , & \text{if } z_\nu > z^* , \end{cases}$$

where $z_\nu := (\nu_e/\nu) - 1$. (Line emission from galaxies may be important in the infrared (GOULD and SCIAMA [83]), microwave (PETROSIAN, BAHCALL and SALPETER [84]), X-ray (GOULD and G. R. BURBIDGE [85]), and γ-ray (CLAYTON and SILK [86]), regions, while line emission from an intergalactic gas could be important in the radio (PENZIAS and WILSON [87]) and ultra-violet and infrared (WEYMANN [88]) regions.) Regarding a source of form (6.47a) as built up by such line emission spectra, we can re-express (6.46) in the form

$$I_\nu = \int_0^\infty \mathscr{I}(\nu_e)\, \mathscr{G}_\nu(\nu_e)\, \mathrm{d}\nu_e \,. \tag{6.47b}$$

While in general we have to integrate by machine, for simple spectra we can sometimes obtain analytic expressions for I_ν. For example if the sources have a spectral index α and their amplitude varies as $(R(t))^m$, then

$$S(t, \nu(1+z)) = S(t_0, \nu)(1+z)^{m-\alpha};$$

so in an Einstein-de Sitter universe ($q_0 = \frac{1}{2}$) we find from (6.46)

$$I_\nu = \frac{n_s(0)\, S(\nu, t_0)}{H_0(\alpha + \frac{3}{2} - m)} \left\{1 - \frac{1}{(1+z_*)^{\frac{3}{2}+\alpha-m}}\right\}, \qquad \text{when } m \neq \alpha + \tfrac{3}{2} ,$$

and in a Milne universe ($q_0 = 0$) we find

$$I_\nu = \frac{n_s(0)\, S(\nu, t_0)}{H_0(\alpha + 1 - m)} \left\{1 - \frac{1}{(1+z_*)^{1+\alpha-m}}\right\}, \qquad \text{when } m \neq \alpha + 1;$$

in the exceptional cases $q_0 = \frac{1}{2}$, $m = \alpha + \frac{3}{2}$ and $q_0 = 0$, $m = \alpha + 1$ we find

$$I_\nu = \frac{n_s(0)\, S(\nu, t_0)}{H_0} \log_e (1 + z_*) \,.$$

Thus if the source brightness increases faster than $(1+z)^{\alpha+\frac{3}{2}}$ in an Einstein-de Sitter universe, or than $(1+z)^{\alpha+1}$ in a Milne universe, there must be a cut-off in the source evolution before some value of z which can be determined by comparing these expressions with the observed values of I_ν. The source of radiation might be galaxies or other discrete sources (see, for example, REES and SETTI [89], and discussions by DAVIDSON and NARLIKAR [9] of the radio, LOW and TUCKER [90] of the infra-red, WOLFE and BURBIDGE [91] of the microwave, and GOULD [92] of the ultra-violet) or an intergalactic gas (see, for example, FIELD and HENRY [93] and WEYMANN [88]).

With the assumptions discussed at the beginning of this Section, (6.45) shows that the optical depth up to a red-shift z in a Robertson-Walker universe is

$$p(z, \nu) = \int_0^z \frac{n_a(0)}{H_0} \sigma\big(z', \nu(1+z')\big) \frac{(1+z')\,\mathrm{d}z'}{(1+2q_0 z')^{\frac{1}{2}}} . \tag{6.48}$$

Considering some particular scattering or absorption process, it is often reasonable to assume

$$\sigma(z, \nu) = \bar{\sigma}(z)\, \mathscr{A}(\nu) . \tag{6.49a}$$

In particular, we can represent line absorption at some frequency ν_a by setting $\mathscr{A}(\nu) = \delta(\nu - \nu_a)$; the resulting optical depth $A_\nu(z, \nu_a)$ is

$$A_\nu(z, \nu_a) = \begin{cases} \dfrac{n_a(0)}{H_0 \nu_a} \bar{\sigma}(z_\nu) \dfrac{(1+z)^2}{(1+2q_0 z_\nu)^{\frac{1}{2}}} , & \text{for } z_\nu < z_* , \\ 0 , & \text{for } z_\nu \geqslant z_* , \end{cases}$$

where $z_\nu := (\nu_a/\nu) - 1$. (Lack of observed 21 cm (FIELD [94]; PENZIAS and SCOTT [95]) and Ly α (GUNN and PETERSON [96]; BAHCALL and SALPETER [97]) absorption gives useful limits (see *e.g.* SCIAMA [98]) on the intergalactic medium; Ly α absorption could be important in Q.S.O. observations (REES [99]).) Regarding an absorption process with cross-section (6.49a) as built up of such line absorptions, we can re-express (6.48) in the form

$$p(z, \nu) = \int_0^\infty A_\nu(z, \nu_a)\, \mathscr{A}(\nu_a)\, \mathrm{d}\nu_a . \tag{6.49b}$$

While in general we have to integrate these equations numerically, in special cases we can integrate them analytically. In particular, when $\sigma(z, \nu) = \sigma_0 = \text{const}$, we can integrate (6.48) to show

$$p(z, \nu) = \frac{\sigma_0 n_a(0)}{H_0} \cdot \frac{1}{3q_0^2} \{(3q_0 + q_0 z - 1)(1 + 2q_0 z)^{\frac{1}{2}} - (3q_0 - 1)\} , \tag{6.50a}$$

when $q_0 \neq 0$, and

$$p(z, v) = \frac{\sigma_0 n_a(0)}{H_0} \cdot \frac{1}{2}(1+z)^2 , \tag{6.50b}$$

when $q_0 = 0$. Equation (6.50a) would be applicable to the Thomson scattering arising if an intergalactic medium were ionized (BAHCALL and SALPETER [97]; BAHCALL and MAY [100]); for a fully ionized gas in an Einstein-de Sitter universe with $n_a(0) \simeq 10^{-5}$ the optical depth would be unity at a red-shift of about 7, while in a low-density universe with $n_a(0) \simeq 10^{-7}$, the optical depth would be unity at a red-shift of about 30. We can also use this expression to calculate the probability of absorption effects arising from galaxies (WAGONER [101]; ROEDER and VERREAULT [102]) or other randomly distributed matter (BAHCALL and PEEBLES, [103]) intervening between an observer and some galaxy or quasi-stellar object in a Robertson-Walker universe.

In some situations, one would have to consider absorption and emission processes together (see, for example, studies of the spectrum of the background radiation by WEYMANN [88, 104], ZEL'DOVICH and SUNYAEV [105], and of absorption and emission by an intergalactic gas (PEEBLES [106])). Finally we note that while we have been considering (6.44) in the context of a cosmological model, one can apply it in other situations when a fluid approximation is appropriate in astrophysics; (for example, to determine the propagation of radiation in a star).

6·7. *Null cone observations in a cosmological model.* – In any given cosmological model, we can deduce the following quantities:

i) the proper motion in the sky observed for distant galaxies (see eq. (2.12) for first-order effect) and the distortion of images caused by the curved space-time, both of which can be found by integrating the geodesic equation (6.9);

ii) the red-shift observed for any distant sources in the model, given by (6.10b) (see eq. (6.14) for the first-order effect);

iii) the area distance r_0 of any source, obtained from $k^a{}_{;a}$ by integrating (6.17) and using the definition (6.27);

iv) the number N of sources of any given type observed up to some affine parameter distance v in any direction, obtained by integrating (6.36);

v) the specific intensity of radiation from any given source, and the specific intensity of background radiation in the model as a function of position in the sky (given by (6.44));

and find relations between them. For example, in a Robertson-Walker model filled with dust and with $\Lambda = 0$, there is no proper motion or distortion effect;

the area distance r_0 is related to the red-shift by (6.35) and the number density to the area distance and red-shift (and so to the luminosity distance D) by (6.37); eq. (6.44) together with (6.45) can be used to determine the observed spectrum from any source at red-shift z_* and to find the spectrum of background radiation (taking z_* to correspond to an optical depth of unity).

Observations of the black-body radiation in certain homogeneous anisotropic models have been discussed by THORNE [59], MISNER [28], REES [107] and HAWKING [108], and in perturbations of a Robertson-Walker model by SACHS and WOLFE [109], REES and SCIAMA [110], and others. The observational relations for discrete sources have been derived in certain homogeneous anisotropic models by SAUNDERS [111, 112], TOMITA [113] and MACCALLUM and ELLIS [114]; in an exact study of some inhomogeneities in Robertson-Walker models by KANTOWSKI [115]; and in perturbations of a Robertson-Walker model by BERTOTTI [73], ZIPOY [116], GUNN [117], PETROSIAN and SALPETER [118], and others. They may be obtained in the form of a power-series expansion about « here and now » in any space-time as discussed in a very interesting paper by KRISTIAN and SACHS [17] (cf. also MCCREA [119]). The divergence $k^a{}_{;a}$ is needed before r_0 can be found (*) in a general model, so it is often useful to know that $k^a{}_{;a}$ obeys an equation very similar to Raychaudhuri's equation (see for example SACHS [70], PENROSE [72]); in fact there is a close correspondence between equations obeyed by the kinematic quantities defined for a timelike congruence (see Sect. **4** above) and equations obeyed by analogous quantities (« optical scalars ») defined for a congruence of null geodesics; see JORDAN, EHLERS and SACHS [74], SACHS [70], NEWMAN and PENROSE [120]. (This correspondence is based on the fact that both timelike and null geodesics obey the geodesic deviation equation.)

In principle we can determine whether any of these space-times could be good models of the universe or not by comparing these theoretical relations directly with observation of classes of sources with known physical characteristics, and with observations of the background radiation. The essential problem in carrying out this programme is that we do not know *a priori* the properties of the particular sources we observe (cf. the discussions in this volume by E. M. BURBIDGE and G. R. BURBIDGE). Apart from statistical scatter in the properties of sources of a given type and selection effects which are exacerbated by the fact that we often have only a poor idea of what kind of object we are observing, there are two systematic effects we shall briefly mention. Firstly, the average luminosity L, spectrum $\mathscr{I}(\nu)$ and cross-section area A of a given class of sources will in general change in time, and so be functions of the red-shift z; we have to determine this evolution somehow. Secondly, we can determine r_0 either directly by measuring the angular diameters of

(*) There are other methods of obtaining r_0 which are essentially equivalent.

sources of known diameters, or indirectly by measuring the specific flux of radiation from the source and using eq. (6.38*a*). However the sources we observe in a cosmological context seldom have sharp edges: we therefore measure the angular diameter of the source up to some isophote (SANDAGE [79]) rather than measuring a metric diameter. (We can obtain angular measurements corresponding to a metric diameter by measuring the angles observed between centres of galaxies in a cluster, or between radio emission centres in a multiple radio source, but then we have considerable difficulties in estimating the dimensions of the system.) Correspondingly a measurement of the flux from the source may omit a contribution from the outer regions which are lost in the general background. Thus we can only measure the area distance r_0 of an extended source accurately if we know the surface brightness of the source as a function of frequency, time, and position on the surface of the source (cf. STOCK and SCHÜCKING [121]). If we could measure I_ν separately from r_0, this might give some guide as to the evolution taking place in the source.

7. – The observable universe.

In this Section we aim to give a brief qualitative description of the observable universe, adopting conventional interpretations of the observations. (For an alternative viewpoint, see Hoyle's Bakerian Lecture [122].)

7·1. *Causality*. – In special relativity we assume that no signal or particle can travel faster than the speed of light; it follows that an event can be causally influenced only by events within or on its past light cone, and can causally influence only events in or on its future light cone. As the past and future light cones never intersect in flat space-time, these sets of events are disjoint. In a general curved space-time, this need not be so (see KRONHEIMER and PENROSE [123] and Geroch's lectures in this volume for very general discussions). Further, imposing the field equations is not sufficient to prevent causality violations, as Gödel's solution (see Subsect. 5·2) contains closed timelike lines (GÖDEL [56]); an observer in this space-time can travel into, and influence, his own history.

We will regard solutions in which this can happen as physically unreasonable. We shall therefore assume that space-time obeys the *strong causality condition*: every point in space-time is contained in a small open neighbourhood such that every timelike (and null) curve that leaves this neighbourhood never re-enters it.

It follows from this postulate that the past light cone of any point is part of the boundary of the past of that point (HAWKING [124]); so the total region of space-time which can have a causal influence on any observer lies within (*) or on his past light cone.

7·2. *Isotropy of the microwave radiation.* – The microwave background radiation is known (see Sciama's course) to be extremely isotropic about us, so the Copernican principle (Sect. **1**) leads us to believe the same is true for any observer moving with the average velocity u^a. If this isotropy were exact, it would follow from a theorem of EHLERS, GEREN and SACHS [125] that the universe was exactly a Robertson-Walker universe. This theorem states that if freely-propagating radiation is isotropic with respect to some velocity field u^a everywhere in an open set in space-time, then the shear σ_{ab} of u^a vanishes in that open set. However we can assume that in recent times $\theta > 0$ and $\dot{u}_a = 0$ (see eq. (3.16)); it then follows from the work of EHLERS, GEREN and SACHS, or from the theorem quoted in Subsect. **5**·3.1, that $\omega_a = 0$ also. But these conditions ensure that space-time is a Robertson-Walker universe (cf. Subsect. **5**·1).

In view of this result, we shall regard the very high degree of isotropy of the radiation as evidence that the universe has been very nearly a Robertson-Walker universe since the time of last scattering of the microwave radiation.

We can in fact use the limits on the anisotropy of the radiation to place limits on the anisotropy and inhomogeneity of the universe (see THORNE [59], SACHS and WOLFE [109], REES and SCIAMA [110], HAWKING [126], WOLFE [127]) and on our motion relative to the mean velocity u^a (SCIAMA [128], STEWART and SCIAMA [129]; and cf. Subsect. **6**·6.3); in particular, the measurements may be taken to indicate limits

$$\omega_0 < 10^{-3}\theta_0\,, \qquad \sigma_0 < 10^{-3}\theta_0$$

on the present shear and vorticity of the universe.

The isotropy of radio source counts (**) (HUGHES and LONGAIR [130]) and of background radiation at other wavelengths seems to confirm this picture.

7·3. *The existence of singularities.* – If the universe were exactly a Robertson-Walker universe, with the energy conditions (3.8) satisfied, then the condition $\Lambda \leqslant 0$ would imply there had been a singularity a finite time t_0 ago in the past; if $q_0 > 0$ the same conclusion would hold irrespective of the sign of Λ (cf. Subsection **5**·1.1). Even if q_0 were not positive it would be plausible that the same

---

(*) GEROCH has pointed out that if there are « wormholes » in the universe, part of the « interior » of the light cone can reappear *outside* the light cone.

(**) Isotropy of the number counts may also be used to put limits on the vorticity, cf. GÖDEL [43].

result were true, because it is plausible to assume that the microwave radiation is black-body radiation (cf. Sciama's course) which has been thermalized by being in equilibrium with ionized matter. Decoupling of the primeval fireball of matter and radiation would take place when $R/R_0 \sim 1/1000$ (cf. Sect. **2**), but if a dense intergalactic gas had been ionized by reheating, the last effective scattering of radiation could occur as late as a red-shift of about 7 (corresponding to an optical depth unity, see Sect. **6**). However the reheated gas could not thermalize the radiation to its observed temperature, so we could assume that the existence of black-body radiation implied that at some time, $R/R_0 \leqslant 1/1000$. But at that time $\mu \geqslant 10^9 \mu_0$, so the Λ-term was totally negligible (cf. estimate (4.14*c*)). In fact even if the reheated matter were responsible for the thermalization we would find $R/R_0 \leqslant \frac{1}{8}$ at some time; but then $\mu \geqslant 512 \mu_0$, so use of estimate (4.14*c*) of Λ in Raychaudhuri's equation would show that $l^{\bullet\bullet} < 0$ at that and all earlier times.

Thus any of the conditions i) $\Lambda \leqslant 0$, ii) $q_0 > 0$, or iii) the existence of black-body radiation indicating a previous time when matter and radiation were in equilibrium, would imply that the radius function increased monotonically from a singularity a finite time previously. Two consequences that would follow are

a) there would be a minimum angular diameter observed for sources of the same metric size (this follows from the results of Subsect. **6**˙4.4), and

b) at an early state there would be a ionized plasma filling the universe (cf. Sect. **3**).

The existence of the plasma implies a cut-off to optical and radio observations; this must occur by a red-shift ~ 1000, but could possibly happen by as low a red-shift as 7, in a high-density universe. Similarly, at early times the universe would be opaque to neutrinos; in fact there would be a maximum red-shift for observations by each other kind of signal.

The existence of minimum angular diameters implies that there are 2-surfaces on which the past null geodesics generating the observer's light cone are reconverging (see Fig. 9); these 2-surfaces are « closed trapped surfaces » in the sense defined by PENROSE [131]. In fact it can be shown that the reconvergence will occur before the optical depth is unity (HAWKING and ELLIS [131*a*]).

The real universe is not exactly a Robertson-Walker universe. Since however we can assume it is very like a Robertson-Walker universe at least up to the time of last scattering, we expect the qualitative features *a*), *b*) to remain true in the real universe. Figure 9 illustrates these features; if $q_0 > 0$ we can estimate the time to the surface of last scattering as being of the order of, but less than, $1/H_0$ (cf. Subsect. **5**˙1.1).

Perhaps the most important feature is the existence of the closed trapped surface. HAWKING and PENROSE [132] have shown i) the existence of this surface, ii) the strong causality condition (Subsect. **7**·1. above), iii) the energy condition: for every timelike vector V^a at each point, $R_{ab}V^aV^b \geqslant 0$ (in the case of a perfect fluid and $\Lambda = 0$, this is equivalent to $\mu + p \geqslant 0$, $\mu + 3p \geqslant 0$, cf. (3.8)), together with a « generality condition » which would certainly be satisfied in the real universe (*), imply that there must be a singularity in the universe. We can therefore deduce, on the basis of reasonable physical assumptions, that the inhomogeneity and anisotropy of the real universe is una-

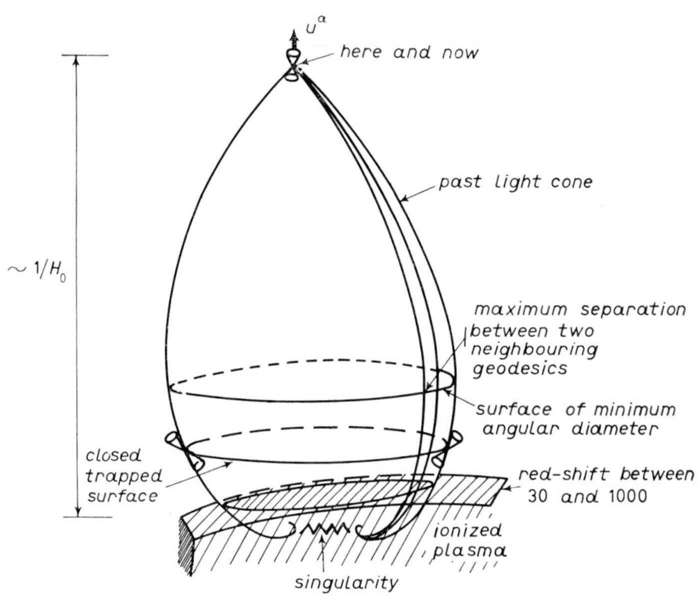

Fig. 9. – A closed, trapped surface in our past light cone. We cannot observe by optical or radio telescopes beyond the surface of ionization; this surface lies beyond the surface of minimum angular diameter. Numbers are for a low-density universe with $\Lambda = 0$.

ble to prevent the existence of a singularity in our past (**) if $\Lambda \leqslant 0$, and unlikely to prevent it even if $\Lambda > 0$. To postulate a « bounce » in an oscillating universe means abandoning either general relativity (and similar theories such as the Brans-Dicke theory, in which gravity is always attractive) or the energy condition (which we can do only by introducing large negative pressures or densities, if $\Lambda \leqslant 0$) or the causality condition (but in general, a violation of causality cannot prevent a singularity occurring (HAWKING [133])).

(*) For example, it is fulfilled if the universe is filled with a fluid with $\mu > 0$.

(**) To show it is in the *past*, we have to show that all the past timelike geodesics from the observer also reconverge in a compact region; this we can do (HAWKING and ELLIS [131*a*]). (Then we no longer need the « generality condition ».)

7·4. *The early universe.* – While we have concluded that the universe is very like a Robertson-Walker universe since the decoupling of matter and radiation, we have no direct observational evidence about earlier times, and so cannot derive the same conclusion for these times. In fact we may suspect that the universe was *not* very like a Robertson-Walker universe at early times, for the following reasons.

a) There must have been fluctuations larger than purely thermal fluctuations present in the universe at times soon after the decoupling of matter and radiation, in order that galaxy formation could have taken place (cf. the seminars by SALPETER and REES). Allowing for the damping effect of viscosity when photons decoupled and earlier when neutrinos decoupled, there must have been very large fluctuations early on in order to provide sufficiently large fluctuations later.

b) There are particle horizons (RINDLER [134]) in nonempty Robertson-Walker models which develop from a singularity (since this is true for all such models with $\mu > 0$, $p \geqslant 0$). This means that at early times in these models each particle was causally connected only to a small number of other particles; we can now observe radiation from parts of the universe which, if the early universe was very like a Robertson-Walker universe, were not causally connected to each other at the time of emission of the radiation. However both the physical state and the physical parameters of the matter in these regions appear to be very similar, despite their having been unable to interact with each other. This seems a rather implausible situation (*). By contrast, MISNER [28, 29] has suggested that the early universe could have been highly anisotropic and inhomogeneous; dissipative processes would tend to smooth out the anisotropy and inhomogeneity, thus providing a mechanism to explain the present high degree of homogeneity and isotropy. If the universe were highly anisotropic or inhomogeneous, the horizon structure could be very different from that in a Robertson-Walker universe (MISNER [135]) and so the previous problem need not arise.

c) One can argue (HAWKING and ELLIS [131*a*]) that solutions of the general-relativity field equations in which singularities have a spacelike character (as is the case in those Robertson-Walker universes in which particle horizons occur, see PENROSE [136]) are rather special; it may be that more general solutions have singularities with a timelike character (as in the Reissner-Nordström and Kerr solutions). This suggests that a realistic solution would be unlike a Robertson-Walker solution in the vicinity of its singularities (incidentally implying a different horizon structure, cf. *b*)).

(*) Unless one can characterize in detail a plausible creation process that must of necessity proceed in a uniform way.

d) Observations of a fairly uniform distribution of ^{4}He (about 25 % by weight) suggests that many parts of the universe evolved through the helium formation phase with time scales similar to those in the Robertson-Walker universes (see PEEBLES [31], THORNE [59], HAWKING and TAYLER [137], WAGONER, FOWLER and HOYLE [33]). However very low helium abundances observed in the atmospheres of some halo stars (cf. BURBIDGE and SCIAMA in this volume) might indicate inhomogeneous element production resulting from large density inhomogeneities in the early universe.

7·5. *The nature of the singularity.* – The existence of a singularity in our past implies a breakdown of ordinary physical laws. However the theory has been extrapolated to circumstances where we would expect some quantized theory of gravitation to be a more appropriate description of space-time than general relativity; it is not yet clear whether a real physical singularity would be predicted by such a theory or not (cf. the discussion by MISNER [138]). Thus although the singularity theorem does not necessarily imply the existence of a real physical singularity (an end to space-time) in the universe, it may be taken to imply that conditions in our past were so extreme that general relativity was no longer a valid theory.

Although it appears that a singularity is predicted in every sufficiently large co-moving volume in the universe (HAWKING and ELLIS [131*a*]) this does not necessarily imply that all the matter in the universe experienced indefinitely high densities in the past. In fact if the singularities are timelike (cf. *c*) above), it is conceivable that a contracting phase in the universe could have changed to an expanding phase with most of the matter in the universe passing between isolated singularities. A consequence of this would be that no Cauchy surface would exist: complete knowledge of the state of the universe on any one space-section would be insufficient to determine its complete time development.

In any case, it is clear that the existence of a « singularity » in our past implies a breakdown of our ability to predict, at least on the basis of present theory; perhaps we should regard this as the essential feature of the singularity.

7·6. *The observable universe.* – Our optical and radio observations give us information about the distribution of matter and the curvature of space-time on our past light cone, back to the time when the universe was opaque. In principle we can obtain information about conditions on our past light cone at even earlier times (for example, by neutrino telescopes). Using the field equations, we can extrapolate inwards to find the geometry of space-time inside our light cone; in practice our observations of distant sources give us relatively little information so we can only extrapolate a short way in for these sources, while we can reasonably extrapolate the history of nearby sources for a rather greater time.

We can also use astrophysical, geophysical, and geological observations to give evidence about the history of the universe near the world line of our galaxy up to very early times. (For example, both the theory of element formation combined with observations of element abundances and the theory of stellar evolution based on observations of relatively nearby stars give restrictions on the physical conditions and the thermal evolution of the early universe near our galaxy.) Thus the part of the universe about which we can obtain reasonably good evidence is the shaded region in Fig. 10 (cf. HOYLE [80]).

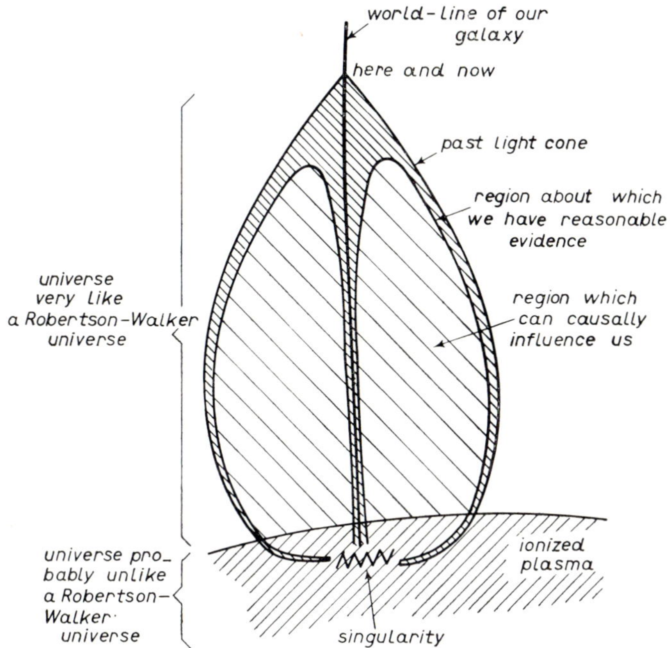

Fig. 10. – The observable universe, and the part of the universe we can observe in some detail. The part of our past since decoupling of matter and radiation is very like a Robertson-Walker universe; the hidden part is probably rather inhomogeneous in structure.

Using the Copernican principle we have concluded that the region of our past light cone since decoupling is very like a Robertson-Walker universe; we should therefore like to find a particular Robertson-Walker universe which is a good model for this space-time region by comparing the theoretical relations (Sect. **6**) with the observations. (The model would be determined by μ_0 and $H_0^2 q_0$, and the epoch at which we observe it by H_0.) We might further aim to describe in detail the differences between the actual universe and the Robertson-Walker universe by comparing observations with theoretical relations in a perturbed Robertson-Walker universe, and to find the detailed distribution of matter and radiation in this perturbed universe. Finally we may aim to determine something of the early universe by indirect arguments based on the astrophysical evidence (see, for example Subsect. **7**.3, **7**.4, **7**.5 above); in doing so we find useful powerful general theorems (such as those developed by PENROSE and HAWKING) and detailed studies of anisotropic and inhomogeneous cosmological models (cf. Sect. **5**).

Having collated this information about the universe within our past light cone, the Copernican principle enables us to assume that conditions for other observers would be much the same. However we must use this principle with caution. Although we cannot necessarily assume particle horizons exist in the universe (since the early universe, about which we have no direct observational information, may be very different from a Robertson-Walker space-time), there will in general be *effective* particle horizons limiting the galaxies we can have observed since the universe became transparent to light and radiation (cf. SATO [139]) (but in a Lemaitre model ($\Lambda >$, $K > 0$) one might be able to see round the universe since decoupling occurred (*).) Further there are space-time events we have not yet observed and which no astronomer will be able to observe within, say, the next 10 000 years (although if there were no effective particle horizons, we could in principle obtain sufficient information to predict conditions at these events).

While we may in principle obtain *some* information about the space-time outside our light-cone by measuring the gravitational « Coulomb » field due to matter outside, it is as yet unclear precisely what we might be able to determine in this way. Thus statements about distant events outside our light cone are, in a fundamental way, unverifiable statements, and may be very misleading. (Consider for example the situation of an astronomer in a high-density region in a Lemaitre universe, the high density region recollapsing to a second singularity while most of the universe expands indefinitely.) Thus if we postulate a « cosmological principle » on philosophic grounds it may be impossible to disprove it; in this case, its scientific status is rather obscure.

(*) Personally, I support the view that $\Lambda = 0$ because otherwise it is a field which acts on everything but is not acted on by anything, which seems unreasonable (cf. RINDLER [5]).

APPENDIX

For a general matter tensor the left-hand sides of eqs. (4.21) are unchanged; the right-hand sides are:

$$(4.21a) \qquad := \tfrac{1}{3}h^{t}\mu_{,b} - \tfrac{1}{2}h^{t}{}_{c}\pi^{cb}{}_{;b} - \tfrac{3}{2}\,\omega^{t}{}_{b}q^{b} + \tfrac{1}{2}\sigma^{t}{}_{b}q^{b} + \tfrac{1}{2}\pi^{t}{}_{a}\dot{u}^{a} - \tfrac{1}{3}\theta q^{t}\,,$$

$$(4.21b) \qquad := \tfrac{1}{2}\sigma^{(t}{}_{e}\eta^{m)bef}u_{b}q_{f} - \tfrac{1}{2}h_{c}{}^{(t}\eta^{m)bef}u_{b}\pi^{c}{}_{e;f} + \tfrac{1}{2}(h^{mt}\omega_{c}q^{c} - 3\omega^{(m}q^{t)})\,,$$

$$(4.21c) \qquad := (\mu + p)\,\omega^{t} + \tfrac{1}{2}\eta^{tbef}u_{b}q_{[e;f]} + \tfrac{1}{2}\eta^{tbef}u_{b}\pi_{ec}(\omega^{c}{}_{f} + \sigma^{c}{}_{f})\,,$$

$$(4.21d) \qquad := -\tfrac{1}{2}(\mu + p)\,\sigma^{tm} - \dot{u}^{(t}q^{m)} - \tfrac{1}{2}h^{t}{}_{a}h^{m}{}_{c}q^{(a;c)} -$$

$$- \tfrac{1}{2}h^{t}{}_{a}h^{m}{}_{c}\dot{\pi}^{ac} - \tfrac{1}{2}\pi^{b(m}\omega_{b}{}^{t)} - \tfrac{1}{2}\pi^{b(m}\sigma_{b}{}^{t)} - \tfrac{1}{6}\pi^{tm}\theta + \tfrac{1}{6}(q^{a}{}_{;a} + \dot{u}_{a}q^{a} + \pi^{ab}\sigma_{ab})\,h^{mt}\,.$$

REFERENCES

[1] H. BONDI: *Cosmology* (Cambridge, 1960).

[2] G. C. MCVITTIE: *Handbuch der Physik*, Vol. **53**, edited by S. FLÜGGE (Berlin, 1959), p. 445.

[3] A. EINSTEIN: *Sitzberichte der Preuss. Akad. d. Wissens.* (1917); see *The Principle of Relativity* (New York, 1923).

[4] J. SCHOUTEN: *Ricci Calculus* (Berlin, 1954).

[5] W. RINDLER: *Essential Relativity* (New York, 1969).

[6] O. HECKMANN and E. SCHÜCKING: *Handbuch der Physik*, edited by S. FLÜGGE, Vol. **53** (Berlin, 1959), p. 489.

[7] O. HECKMANN and E. SCHÜCKING: *Gravitation*, edited by L. WITTEN (New York, 1962), p. 438.

[8] YA. B. ZEL'DOVICH: *Adv. in Ast. and Ast.*, **3**, 242 (1965).

[9] W. DAVIDSON and J. NARLIKAR: *Rep. Progr. Phys.*, **29**, 539 (1966).

[10] W. RINDLER: *Phys. Today*, **20**, 23 (1967).

[11] E. SCHÜCKING: *Lectures in Applied Mathematics*, edited by J. EHLERS, Vol. **8**, (Providence, R. I., 1967), p. 217.

[12] A. R. SANDAGE: *Observatory*, **88**, 91 (1968).

[13] O. HECKMANN and E. SCHÜCKING: *Zeits. f. Astrophys.*, **38**, 95 (1955).

[14] A. TRAUTMAN: *Perspectives in Geometry and Relativity* (Hlavaty Festschrift), edited by B. HOFFMANN (1966), p. 413.

[15] J. L. SYNGE: *Relativity, the General Theory* (Amsterdam, 1960).

[16] A. TRAUTMAN: *Lectures on General Relativity* (Brandeis Summer Institute in Theoretical Physics, Vol. **1**), edited by S. DESER and K. FORD (1964), p. 7.

[17] J. KRISTIAN and R. K. SACHS: *Astrophys. Journ.*, **143**, 379 (1966).

[18] K. C. FREEMAN: duplicated lecture notes, Cambridge University (1967).

[19] L. Woltjer: *Lectures in Applied Mathematics*, edited by J. Ehlers, Vol. **9** (Providence, R. I., 1967), p. 1.
[20] J. Serrin: *Handbuch der Physik*, edited by S. Flügge, Vol. **8**/1 (Berlin, 1959), p. 125.
[21] G. K. Batchelor: *An Introduction to Fluid Dynamics* (Cambridge, 1967).
[22] J. Ehlers: *Abh. Akad. Wiss. und Lit. Mainz, Math. Nat. Kl.*, No. 11 (1961).
[23] J. M. Stewart: *Month. Not. Roy. Astr. Soc.*, **145**, 347 (1969).
[24] R. P. Geroch: Ph.D. Thesis (Princeton University, 1967).
[25] A. R. Curtis: *Proc. Roy. Soc.*, A **200**, 248 (1950).
[26] B. K. Harrison, K. S. Thorne, M. Wakano and J. A. Wheeler: *Gravitation Theory and Gravitational Collapse* (Chicago, 1965).
[27] R. F. Tooper: *Astrophys. Journ.*, **142**, 1541 (1965).
[28] C. W. Misner: *Astrophys. Journ.*, **151**, 431 (1968).
[29] C. W. Misner: *Nature*, **214**, 40 (1967).
[30] R. H. Dicke, P. J. E. Peebles, P. G. Roll and D. T. Wilkinson: *Astrophys. Journ.*, **142**, 414 (1965).
[31] P. J. E. Peebles: *Astrophys. Journ.*, **142**, 542 (1966).
[32] R. A. Alpher, G. Gamow and R. C. Herman: *Proc. Nat. Acad. Sci.*, **58**, 2179 (1967).
[33] R. Wagoner, W. Fowler and F. Hoyle: *Astrophys. Journ.*, **148**, 3 (1967).
[34] E. R. Harrison: *Physics Today*, **21**, 31 (1968).
[35] P. Jordan, J. Ehlers and W. Kundt: *Abh. Akad. Wiss. und Lit. Mainz, Mat. Nat. Kl.*, No. 2 (1960).
[36] O. Heckmann and E. Schücking: *Zeits. f. Astrophys.*, **40**, 81 (1956).
[37] F. A. E. Pirani: *Lectures in General Relativity* (Brandeis Summer Institute in Theoretical Physics, Vol. 1) edited by S. Deser and K. Ford (1964), p. 249.
[38] A. Raychaudhuri: *Phys. Rev.*, **98**, 1123 (1955).
[39] A. Raychaudhuri: *Zeits. f. Astrophys.*, **43**, 161 (1957).
[40] M. Ruderman: *Rep. Progr. Phys.*, **28**, 411 (1965).
[41] G. B. Field, M. J. Rees and D. W. Sciama: *Comments Astrophys. Space Sci.*, **1**, 187 (1969).
[42] R. B. Partridge and P. J. E. Peebles: *Astrophys. Journ.*, **147**, 868 (1967).
[43] K. Gödel: *Proc. Int. Cong. Math.* (*Camb., Mass.*), **1**, 175 (1952).
[44] J. L. Synge: *Proc. Lond. Math. Soc.*, **43**, 376 (1937).
[45] W. Kundt and M. Trümper: *Abh. Akad. Wiss. und Lit. Mainz., Mat. Nat. Kl.*, No. 12 (1962).
[46] M. Trümper: *Zeits. f. Astrophys.*, **66**, 215 (1967).
[47] J. M. Anderson: *Principles of Relativity Physics* (New York, 1967).
[48] Y. Choquet-Bruhat: *Gravitation*, edited by L. Witten (New York, 1962), p. 130.
[49] L. C. Shepley and A. Taub: *Commun. Math. Phys.*, **5**, 237 (1967).
[50] H. P. Robertson: *Rev. Mod. Phys.*, **5**, 62 (1933).
[51] H. P. Robertson: *Pub. Astr. Soc. Pac.*, 82 (1955).
[52] R. Stabell and S. Refsdel: *Month. Mot. Roy. Astr. Soc.*, **132**, 379 (1966).
[53] S. Refsdal, R. Stabell and G. de Lange: *Mem. Roy. Astr. Soc.*, **71**, part 3 (1967).
[54] G. Tauber: *Journ. Math. Phys.*, **8**, 118 (1967).
[55] W. Rindler: *Astrophys. Journ.*, **157**, L148 (1969).
[56] K. Gödel: *Rev. Mod. Phys.*, **21**, 447 (1949).
[57] G. F. R. Ellis: *Journ. Math. Phys.*, **8**, 1171 (1967).

[58] G. F. R. Ellis and M. A. H. MacCallum: *Commun. Math. Phys.*, **12**, 108 (1969).
[59] K. S. Thorne: *Astrophys. Journ.*, **148**, 51 (1967).
[60] K. C. Jacobs: *Astrophys. Journ.*, **153**, 661 (1968).
[61] K. C. Jacobs: *Astrophys. Journ.*, **155**, 379 (1969).
[62] S. W. Hawking: *Astrophys. Journ.*, **145**, 544 (1966).
[63] P. Szekeres: *Journ. Math. Phys.*, **7**, 751 (1966).
[64] M. Trümper: *Journ. Math. Phys.*, **6**, 584 (1965).
[65] O. Heckmann: *Astronom. Journ.*, **66**, 205 (1961).
[66] J. V. Narlikar: *Month. Not. Roy. Astr. Soc.*, **126**, 203 (1963).
[67] J. M. Stewart and G. F. R. Ellis: *Journ. Math. Phys.*, **9**, 1072 (1968).
[68] J. Ehlers: *Zeits. f. Naturforsch.*, **22** a, 1328 (1967).
[69] A. Trautman: *Recent Developments in General Relativity* (Infeld Festschrift) (Warsaw, 1962), p. 459.
[70] R. Sachs: *Proc. Roy. Soc.*, A **264**, 309 (1961).
[71] E. Schrödinger: *Expanding Universes* (Cambridge, 1956).
[72] R. Penrose: *Perspectives in Geometry and Relativity* (Hlavaty Festschrift), edited by B. Hoffman (Bloomington, 1966), p. 259.
[73] B. Bertotti: *Proc. Roy. Soc.*, A **294**, 195 (1966).
[74] P. Jordan, J. Ehlers and R. Sachs: *Abh. Akad. Wiss. und Lit. Mainz, Mat. Nat. Kl.*, No. 1 (1961).
[75] W. Mattig: *Astr. Nach.*, **284**, 109 (1958).
[76] A. R. Sandage: *Journ. Soc. Indust. Appl. Math.*, **10**, 781 (1962).
[77] R. G. Roeder and G. C. McVittie: *Astrophys. Journ.*, **138**, 899 (1963).
[78] W. Mattig: *Astr. Nach.*, **285**, 1 (1959).
[79] A. R. Sandage: *Astrophys. Journ.*, **133**, 355 (1961).
[80] F. Hoyle: *Rendiconti S.I F.*, XX Corso (New York, 1960), p. 141.
[81] G. J. Whitrow and B. D. Yallop: *Month. Not. Roy. Astr. Soc.*, **127**, 301 (1964).
[82] E. R. Harrison: *Month. Not. Roy. Astr. Soc.*, **131**, 1 (1965).
[83] R. J. Gould and D. W. Sciama: *Astrophys. Journ.*, **140**, 1634 (1964).
[84] V. Petrosian, J. Bahcall and E. Salpeter: *Astrophys. Journ.*, **155**, L57 (1969).
[85] R. J. Gould and G. R. Burbidge: *Astrophys. Journ.*, **138**, 969 (1963).
[86] D. Clayton and J. Silk: *Astrophys. Journ.*, **158**, L43 (1969).
[87] A. Penzias and R. Wilson: *Astrophys. Journ.*, **156**, 799 (1969).
[88] R. Weymann: *Astrophys. Journ.*, **147**, 887 (1967).
[89] M. J. Rees and G. Setti: *Nature*, **219**, 127 (1968).
[90] F. J. Low and W. H. Tucker: *Phys. Rev. Lett.*, **21**, 1538 (1968).
[91] A. Wolfe and G. R. Burbidge: *Astrophys. Journ.*, **156**, 345 (1969).
[92] R. J. Gould: *Ann. Rev. Ast. and Ast.*, **6**, 195 (1968).
[93] G. Field and R. C. Henry: *Astrophys. Journ.*, **140**, 1002 (1964).
[94] G. Field: *Astrophys. Journ.*, **129**, 525 (1959).
[95] A. Penzias and E. Scott: *Astrophys. Journ.*, **153**, L7 (1969).
[96] J. Gunn and B. A. Peterson: *Astrophys. Journ.*, **142**, 1633 (1965).
[97] J. Bahcall and E. Salpeter: *Astrophys. Journ.*, **142**, 1677 (1965).
[98] D. W. Sciama: *Rendiconti S.I.F.*, XXV Corso (New York, 1966), p. 418.
[99] M. J. Rees: *Astrophys. Lett.*, **4**, 61 (1969).
[100] J. Bahcall and R. M. May: *Astrophys. Journ.*, **152**, 37 (1968).
[101] R. Wagoner: *Astrophys. Journ.*, **149**, 465 (1967).
[102] R. G. Roeder and R. Verreault: *Astrophys. Journ.*, **155**, 1047 (1969).
[103] J. Bahcall and P. J. E. Peebles: *Astrophys. Journ.*, **156**, L7 (1969).
[104] R. Weymann: *Astrophys. Journ.*, **145**, 560 (1966).

[105] Ya. B. Zel'dovich and R. Sunyaev: *Ast. and Space Sci.*, **4**, 301 (1969).
[106] P. J. E. Peebles: *Astrophys. Journ.*, **157**, 45 (1969).
[107] M. J. Rees: *Astrophys. Journ.*, **153**, L1 (1968).
[108] S. W. Hawking: *Month. Not. Roy. Astr. Soc.*, **142**, 129 (1969).
[109] R. Sachs and A. Wolfe: *Astrophys. Journ.*, **147**, 73 (1967).
[110] M. J. Rees and D. W. Sciama: *Nature*, **217**, 511 (1968).
[111] P. T. Saunders: *Month. Not. Roy. Astr. Soc.*, **141**, 427 (1968).
[112] P. T. Saunders: *Month. Not. Roy. Astr. Soc.*, **142**, 213 (1969).
[113] K. Tomita: *Progr. Theor. Phys.*, **40**, 264 (1968).
[114] M. A. H. MacCallum and G. F. R. Ellis: *Commun. Math. Phys.*, **19**, 31 (1970).
[115] R. Kantowski: *Astrophys. Journ.*, **155**, 89 (1969).
[116] D. Zipoy: *Phys. Rev.*, **142**, 825 (1966).
[117] J. Gunn: *Astrophys. Journ.*, **150**, 737 (1967).
[118] V. Petrosian and E. Salpeter: *Astrophys. Journ.*, **151**, 411 (1968).
[119] W. H. McCrea: *Zeits. f. Astrophys.*, **18**, 98 (1939).
[120] E. Newman and R. Penrose: *Journ. Math. Phys.*, **3**, 566 (1962).
[121] J. Stock and E. Schücking: *Astronom. Journ.*, **62**, 98 (1957).
[122] F. Hoyle: *Proc. Roy. Soc.*, A **308**, 1 (1969).
[123] E. H. Kronheimer and R. Penrose: *Proc. Camb. Phil. Soc.*, **63**, 481 (1967).
[124] S. W. Hawking: *Proc. Roy. Soc.*, A **294**, 511 (1966).
[125] J. Ehlers, P. Geren and R. K. Sachs: *Journ. Math. Phys.*, **9**, 1344 (1968).
[126] S. W. Hawking: *Month. Not. Roy. Astr. Soc.*, **142**, 129 (1969).
[127] A. Wolfe: *Astrophys. Journ.*, **156**, 803 (1969).
[128] D. W. Sciama: *Phys. Rev. Lett.*, **18**, 1065 (1967).
[129] J. M. Stewart and D. W. Sciama: *Nature*, **216**, 748 (1967).
[130] R. G. Hughes and M. S. Longair: *Month. Not. Roy. Astr. Soc.*, **135**, 131 (1967).
[131] R. Penrose: *Phys. Rev. Lett.*, **14**, 57 (1965).
[131*a*] S. W. Hawking and G. F. R. Ellis: *Astrophys. Journ.*, **152**, 25 (1968).
[132] S. W. Hawking and R. Penrose: *Proc. Roy. Soc.*, **314**, 529 (1970).
[133] S. W. Hawking: *Proc. Roy. Soc.*, A **300**, 187 (1967).
[134] W. Rindler: *Month. Not. Roy. Astr. Soc.*, **116**, 662 (1956).
[135] C. W. Misner: *Phys. Rev. Lett.*, **22**, 1071 (1969).
[136] R. Penrose: *Relativity, Groups and Topology* (Les Houches Summer School, 1963), edited by C. de Witt and B. de Witt (New York, 1964), p. 563.
[137] S. W. Hawking and R. J. Tayler: *Nature*, **209**, 1278 (1966).
[138] C. W. Misner: *Phys. Rev.*, **186**, 1319 (1969).
[139] H. Sato: *Progr. Theor. Phys.*, **40**, 781 (1968).

Astrophysical Cosmology.

D. W. SCIAMA

Department of Applied Mathematics and Theoretical Physics
University of Cambridge - Cambridge

These lectures are intended to complement ELLIS's course on Relativistic Cosmology. In that course the theory was in the forefront and the observations in the background. We shall reverse this emphasis. The style of the lectures is determined by the fact that, despite much recent progress, most of the observations are uncertain and their interpretation controversial. Accordingly we shall avoid detailed analyses and elaborate mathematical discussion. The main aim of the lectures is to explain what new issues arise from the recent observations. These issues will not be settled until further observations are made. The next major step forward will come when orbiting satellites are readily available for measurements from above the atmosphere at frequencies ranging from, say, 1 MHz (low-frequency radio waves) to the highest-energy γ-rays that can be detected (perhaps $3 \cdot 10^{24}$ MHz or 10^{16} eV). Such a step is probably not far away.

Meanwhile it is convenient to divide the existing observations into three phases according to the time when they were first established in something like their present form: the classical (up to 1936), the semi-classical (1955), and the modern (1963).

1. – The classical period.

The essential discovery of this period was the expansion of the universe as manifested in the redshift in the spectra of galaxies. This redshift z is defined by

$$1 + z = \frac{\lambda_{\text{obs}}}{\lambda_{\text{rest}}} .$$

Hubble's great discovery was that the redshift of a galaxy is proportional to its distance r. The first-order Doppler formula then gives for the velocity v of recession

$$v = \frac{r}{\tau},$$

where τ, the Hubble constant, is independent of r at any one cosmic epoch (that is, may be treated as a constant for galaxies whose light-time away is very much less than τ). The determination of τ, which involves determining the distance-scale of extragalactic objects, is discussed by E. M. BURBIDGE. We shall adopt the round value of 10^{10} y, which should be correct to within a factor two.

It is also useful to write the Hubble law in the form

$$v = Hr,$$

where H is also sometimes known as the Hubble constant. For $\tau \sim 10^{10}$ y, we have $H \sim 100\ \text{km}\cdot\text{s}^{-1}/\text{Mpc}$. Thus τ gives us a time scale for the universe and H gives us the expansion rate—a galaxy adds on 100 km/s to its recession velocity for each megaparsec of its distance away.

The Hubble law is usually represented by a red shift—apparent magnitude diagram for suitably chosen galaxies with a small spread in their absolute magnitudes. The modern version of such a Hubble diagram is due to SANDAGE [1] and is discussed in detail by E. M. BURBIDGE. The main developments since HUBBLE [2] are the change in the distance scale (HUBBLE thought τ was about $2\cdot 10^9$ y, which gave rise to a time-scale difficulty when comparison was made with the ages of the Earth, Sun, and Milky Way), and the addition of radio galaxies, which have led to the discovery of galaxies with larger redshift than were known previously. Unfortunately this has not yielded a decisive determination of a second-order term in the Hubble law. Such a term would be expected on the basis of most relativistic models, and knowledge of it would be needed to decide whether the expansion of the universe will continue for ever (see below).

2. – Robertson-Walker models.

The decisive theoretical contribution of the classical period was the discovery of the Robertson-Walker (or Friedmann) models of the universe. These models are isotropic and homogeneous at each instant of time. The metric may be written in co-moving co-ordinates as

$$\mathrm{d}s^2 = \mathrm{d}t^2 - \frac{R^2(t)}{(1+\frac{1}{4}kr^2)^2}\,\mathrm{d}l^2,$$

where dl represents the usual flat 3-space metric, $k=-1, 0$ or $+1$ according to whether the $t=$ constant space sections have negative, zero, or positive curvature, and $R(t)$ is the expansion factor which determines how the distance between the material particle of the substratum changes with time.

The Hubble constant τ is then given by

$$\tau = \frac{R}{\dot{R}} .$$

Another important quantity is the deceleration parameter q defined by

$$q = -\frac{R\ddot{R}}{\dot{R}^2} ,$$

a dimensionless number that characterizes the effectiveness of gravity in slowing down the expansion of the universe. A cosmical constant λ in Einstein's field equations would also contribute to q, but in this course we shall assume that $\lambda=0$. The most important model with $\lambda \neq 0$ is the Lemaître model, which is discussed by SALPETER.

The Robertson-Walker metric can be deduced by purely kinematical considerations based only on the assumed symmetry and the hypothesis that light travels on null geodesics and particles on timelike geodesics. If we now impose Einstein's field equations we find that

$$q = \frac{4\pi}{3} G\varrho\tau^2 ,$$

where G is the Newtonian constant of gravitation and ϱ is the energy-density (the 4, 4 component of the energy-momentum tensor $T^{\mu}{}_{\nu}$).

We now consider the properties of the models according to the choice of k.

i) $k=0$. If the pressure $p=0$, we have $q=\frac{1}{2}$ at all epochs and also

$$\frac{8\pi}{3} G\varrho\tau^2 = 1 .$$

The expansion factor $R(t)$ is given by

$$R(t) \propto t^{\frac{2}{3}} .$$

The universe just expands to infinity in the sense that the velocity of expansion tends asymptotically to zero. The present age t_0 of the model since the singularity at $t=0$ is $\frac{2}{3}\tau$ or $7\cdot 10^9$ y. This brings us close to a time-scale difficulty, but in view of the uncertainty in the observed value of τ this difficulty is not yet serious.

The most important physical property of the model is the density ϱ, which for our value of τ is given by

$$\varrho = \varrho_m \sim 2 \cdot 10^{-29} \text{ g/cm}^3 .$$

We shall call this value of ϱ the «magic density», because it represents the dividing line between denser models in which self-gravitation brings the expansion to a halt and turns it into a contraction, and less dense models which expand forever. In term of hydrogen atoms the magic particle density is

$$n \sim 10^{-5}/\text{cm}^3 .$$

The question whether the universe has this density plays a major part in modern astrophysical cosmology.

This simple model is called the Einstein-de Sitter model, because EINSTEIN and DE SITTER proposed it jointly in 1932 [3]. It must not be confused with the Einstein model [4] or with the de Sitter model [5], which were proposed independently in 1917. As EDDINGTON put it, the Einstein model has matter but no motion, while the de Sitter model has motion but no matter. The Einstein-de Sitter model has both, and is the simplest of the Robertson-Walker models which could be a good representation of the universe over most of its lifetime.

It is important to note that the Einstein-de Sitter model has zero pressure. This may be a good approximation today even with the cosmic black-body radiation (see later) but this radiation would have dominated over matter in the early stages, and then the pressure cannot be neglected. In a radiation-dominated universe we have

$$R(t) \propto t^{\frac{1}{2}} \qquad t \text{ small} .$$

ii) $k > 0$. In this case

$$q > \tfrac{1}{2} , \qquad \varrho > \varrho_m , \qquad t_0 < \tfrac{2}{3} \tau .$$

This is a closed universe which oscillates from one singularity to another. The time-scale difficulty is somewhat worse than for the Einstein-de Sitter model, but still not decisively so.

iii) $k < 0$. Here we have

$$0 < q < \tfrac{1}{2} , \qquad \varrho < \varrho_m , \qquad \tau > t_0 > \tfrac{2}{3} \tau .$$

This is an ever-expanding model, with a smaller time-scale difficulty. An interesting limiting case is the Milne model which has $q = 0$, $\varrho = 0$, $t_0 = \tau$ and $R(t) \propto t$. This universe is both empty and flat (that is, the four-dimensional curvature tensor vanishes) but the time co-ordinate is chosen so that the space sections $t =$ constant have negative curvature. It is the only Robertson-Walker model that has neither a particle nor an event horizon (see below) but, of course, according to Einstein's theory it represents an unrealistic limiting case. Nevertheless if the known galaxies represent the main contents of the universe, its density is only a few per cent of the magic density and the Milne model would provide a good approximation for many purposes, except at very early times.

We should also mention the steady-state model of BONDI and GOLD [6] and HOYLE [7]. This has $k = 0$, $R \propto \exp[t/\tau]$, and $q = -1$, which are the relations for the de Sitter model. However, the continual creation of matter required by the steady-state model implies a deviation from Einstein's field equations which would permit the de Sitter model to be nonempty. We shall see in these lectures that the evidence against the steady-state model is now very strong. It is perhaps worth remarking that the steady-state model has an event horizon (that is, the world line of every particle of the substratum contains a last event that we could see, even with an infinitely powerful telescope). By contrast the models we have discussed which conform to Einstein's field equations have a particle horizon (that is, there exist galaxies which have not yet had time to communicate with us). (For an introduction to the concept of horizons in cosmology see RINDLER [8].)

We may summarize the relation between the Robertson-Walker models and observation by saying that q remains essentially undetermined, and that we do not know the sign of k.

3. – The semi-classical period.

This period began in 1955 with the first attempt to use counts of radio sources to draw cosmological conclusions (RYLE and SCHEUER [9]). The present position has been described by SCHEUER in one of his lectures and also in his review article (SCHEUER [10]). There is also a review article by RYLE [11]. Here I shall simply summarize the implications of the most comprehensive survey to date (POOLEY and RYLE [12]). The data are usually presented by plotting logarithmically the number N of sources per steradian whose measured flux density at a fixed observing frequency exceeds the quantity S, against S. The observed slope of the $\log N$-$\log S$ relation is about -1.8 except at the lowest flux-densities of the Pooley-Ryle survey where it has flattened off to about -1. The conventional interpretation of this result is that the initial slope requires for its explanation an *intrinsic evolution* in the average properties of the radio sources with cosmic epoch, the *co-ordinate* density or the power of the sources increasing in the past. Such evolution would not be compatible with the steady-state model of the universe. The final flattening of the slope is then ascribed to the effects of the redshift eventually overcoming the effects of evolution. Such a flattening is in any case required to ensure that the total integrated brightness of all radio sources does not exceed the (fairly reliable) estimates of the extragalactic background radiation (in cosmological language we must avoid an Olbers' paradox). In this connection it is of great interest that the integrated radiation from the sources included in the Pooley-Ryle survey already accounts for about half of this estimated background. Moreover a reasonable extrapolation of the $\log N$-$\log S$ relation would nearly account for the other half (*). There is not much scope therefore for introducing hypothetical new populations of extragalactic radio sources, as cosmologists often like to do. This consideration severely limits possible local hypotheses for quasars, as RYLE [11] explains in detail.

The required rate of evolution of the radio sources has been discussed by many authors (*e.g.* DAVIDSON [13], LONGAIR [14], ROWAN-ROBINSON [15]). The main difficulty with such considerations is that the observed $\log N$-$\log S$ relation contains the unknown evolution rate folded into the unknown cosmological model. The problem contains too many free parameters and no decisive result has been obtained. Moreover our understanding of the physics of the radio sources is so poor that we cannot significantly constrain possible evolutionary schemes by purely physical considerations. This is a disappointing result. Moreover the one striking conclusion from some of the earlier analyses, that there has to be a sharp cut-off in the number of radio sources at some critical redshift in the vicinity of 5, is now known to be incorrect (ROWAN-ROBINSON [15]). The only compensation for losing this striking result is that one of the most plausible explanations of the diffuse X-ray background relies on the prevalence of radio sources at such large redshifts (see p. 223).

Further evidence for the evolution of radio sources comes from a separate analysis of the data on quasars (which really belongs to the modern period). My own conversion from blind adherence to the steady-state theory came when REES pointed out to me that there were far too many quasars with large redshifts (SCIAMA and REES [16]). The best discussion of this evidence to date is due to SCHMIDT [17]. He takes the homogeneous radio sample of quasars in the *3C* revised catalogue and allows for the optical selection effects which arise in the identification of *3CR* sources with quasars. He then finds that at a redshift of 1 there are about 100 times as many quasars per unit *co-ordinate* (co-moving) volume as there are locally and that at a redshift of 1.5 there

(*) This remark refers to radio frequencies at which the 3 °K microwave background (see later) constitutes a small fraction of the total extragalactic background.

are about 500 times as many (the actual numbers depend upon the cosmological model adopted). The analysis becomes unreliable at a redshift of 2, but there is no obvious sign of a different type of behaviour out to this redshift.

This rapid rate of evolution raises in urgent form the well-known question, why is there a rapid cut-off in the number of known quasars with redshifts exceeding 2? It does not appear to be a selection effect connected with the use of an ultra-violet excess in the identification procedure (SCHMIDT [18]). It remains one of the outstanding questions of quasarology. We shall meet a possible solution when we come to discuss the intergalactic medium (p. 192).

4. – The modern period.

The modern period is associated with the discovery of quasars (1963), of the 3 °K cosmic microwave background (1965), and of the cosmic X-ray background (1963). The question whether the redshift of quasars is cosmological is discussed by G. R. BURBIDGE. In this course I shall assume that the cosmological hypothesis is correct, although we still await the really decisive evidence. The cosmological significance of the distribution of quasar red shifts has already been mentioned (p. 188). Here I would only add that they cannot be used to extend the Hubble diagram out to red shifts greater than 0.5 (the limiting redshift in Sandage's Hubble diagram) because of the large spread in their intrinsic optical and radio power.

However the quasars do have further significance for the cosmologist, because their spectra contain important information about the intergalactic medium. This is discussed in detail by SCHEUER, so we shall here describe only the main points.

5. – The intergalactic medium.

It should be said at once that there is as yet no positive evidence for the existence of any intergalactic gas. Nevertheless since galaxies probably contribute only a few per cent of the magic density it has often been proposed that the universe may yet have this density, the « missing matter » being made up of intergalactic gas (although it could also be in other forms such as gravitational waves, neutrinos, rocks or faint stars). The failure to detect the gas enables us to place stringent limits on its physical properties, especially its kinetic temperature, if it is to have the magic density. (For systematic reviews see GOULD [19], SCHEUER [10]).

The most powerful limit is imposed by the lack of observed absorption in the continuous spectra of quasars with redshifts ~ 2 due to Lyman α scattering by intergalactic atomic hydrogen (SCHEUER [20], GUNN and PETERSON [21], BURBIDGE and BURBIDGE [22]). This gives us the limit

$$n_{HI} < 10^{-11}\ \text{cm}^{-3}$$

at redshifts ~ 2. By contrast the lack of 21 cm effects in absorption or emission leads only to the limit

$$n_{HI} < 5 \cdot 10^{-6}\ \text{cm}^{-3}$$

for the intergalactic gas in our own vicinity (PENZIAS and WILSON [23]). The latter limit is free of any uncertainties concerning the whereabouts of the quasars, but is not really strong enough to rule out the magic density even if the intergalactic gas is entirely neutral. At some time in the future it may be possible to find another limit from considering absorption effects on intergalactic X-rays (GOULD and SCIAMA [24]; REES, SCIAMA and SETTI [25]), but at the moment too many uncertainties surround our understanding of the X-ray background (see p. 223). It is also possible to place limits on the abundance of intergalactic molecular hydrogen (FIELD, SOLOMON and WAMPLER [26]) and of heavy elements, especially carbon, nitrogen and oxygen (BAHCALL and SALPETER [27]; REES and SCIAMA [28]).

The Lyman-α limit is so low that even apart from the question of the magic density it seems natural to explain it by assuming that the intergalactic gas is highly ionized. If this ionization is associated with a correspondingly high kinetic temperature T then the magic density would require $T > 10^6$ °K. As we shall see when we discuss the 3 °K microwave background and the hot big bang, such a high temperature would require a heat input into the gas occurring at redshifts less than about 200. Such a heat input may be associated with the formation of galaxies, radio galaxies and quasars from which cosmic rays and ultra-violet radiation would leak into intergalactic space. This may seem an *ad hoc* resolution of the problem, but in fact it was proposed that the gas could be heated in this way to temperatures of 10^5 °K or greater before the discovery of quasars with redshifts of 2 (SCIAMA [29]). The heat input required is not in fact very great in terms of available energy. The problem is rather to decide which of several possible heating mechanisms is the dominant one.

On the other hand one cannot permit the gas to be heated to a temperature very much in excess of 10^6 °K (say 10^7 °K) because then, with the magic density, the bremsstrahlung emitted by the hot gas would lead to a kilovolt X-ray flux at the earth in excess of the observed flux (GOULD and BURBIDGE [30], FIELD and HENRY [31]). What one requires is a detailed thermal history of the gas which is compatible with both the Lyman-α and the X-ray observations. Such a history has been calculated by WEYMANN [32] (for a review see GOULD [19], cf. also BERGERON [33]). Roughly speaking we may say that the observations available to WEYMANN permitted the temperature at a redshift ~ 2 to lie between 10^6 °K and 10^7 °K. Since then X-ray observations have been made at $\frac{1}{4}$ keV (these are discussed on p. 224) which potentially would drive down the upper limit on the temperature, since a cooler gas produces predominantly softer X-rays. In fact there is at the moment a controversy about both the observations of the soft X-rays and their interpretation. However, there is general agreement that the magic density with a temperature $\sim 10^6$ °K at a redshift ~ 2 can be made compatible with the present data.

Recently SUNYAEV [34] has pointed out that the ultra-violet emission from such a gas might be expected to ionize the hydrogen at the edges of galaxies such as M 31 (Andromeda), in contradiction to observation. However it is not difficult to protect the edges of galaxies. All one needs is a small abundance of atomic hydrogen either in the local group or in intergalactic space. This latter would be possible since the intergalactic gas in our vicinity need not be so highly ionised as at a redshift of 2. (I owe this remark to Dr. M. J. REES). This question could be settled by satellite observations of relatively nearby galaxies, to see whether Lyman-α absorption is present in their continuous spectra—in this case not red-shifted into the visible but observed directly in the far ultra-violet (*).

Other possible methods of detecting a highly ionized intergalactic gas were discussed in a previous set of Varenna lectures (SCIAMA [35]), namely, from absorption of low-frequency radio waves, Thomson scattering, Faraday rotation, and dispersion. None of these has yet been successful. Since Thomson scattering is important for a number of theoretical considerations (*e.g.* HAWKING and ELLIS [36], BAHCALL and MAY [37]) we give here the formula for the optical depth τ_T out to a redshift z for a $\lambda = 0$ cosmological model with present deceleration parameter q_0 made entirely of ionized hydrogen:

$$\tau_T = \frac{0.04}{q_0}\{(1+2q_0 z)^{\frac{1}{2}}(3q_0 + q_0 z - 1) - (3q_0 - 1)\}\,. \tag{1}$$

For the Einstein-de Sitter model the relation is

$$\tau_T = 0.04\{(1+z^{\frac{3}{2}}) - 1\}\,,$$

which leads to an optical depth of unity at $z \sim 7$.

(*) *Note added in proofs.* – J. E. FELTEN and J. BERGERON: *Astrophys. Lett.*, **4**, 155 (1969), have shown that galaxies are probably protected from intergalactic ultraviolet radiation by the H_{II} regions that would surround them.

The possibility that dispersion might be detected (HADDOCK and SCIAMA [38]) has been somewhat enhanced by the discovery of pulsars in our Galaxy. The question is whether a pulsar-type phenomenon could be detected in other galaxies, or even in quasars. If quasars « quase » with a period exceeding that of galactic pulsars by the ratio of the masses involved we might be dealing with periods $\sim 10^6$ s (for a quasar mass $\sim 10^6\, M_\odot$). Now the plasma frequency ν_p for the magic density ~ 30 Hz. Radio waves of frequency $\nu > \nu_p$ would propagate through such a gas with a time delay (*)

$$\frac{1}{2}\frac{\nu_p^2}{\nu^2}\{2\tau((1+z)^{\frac{3}{2}}-1)\} .$$

For $z \sim 1$ and $\nu \sim 30$ MHz the delay would be about 10^5 s, which would in principle be observable for a pulse repetition period of 10^6 s. In practice there would be difficulties of interpretation such as dispersion occurring in the quasar itself and the smoothing effects of synchrotron self-absorption. Clearly this is a problem for the future (cf. also MORRISON [39]).

A new role for the intergalactic gas has recently been proposed by REES [40]. He suggests that its absorptive effects may explain the sharp cut-off in quasar redshifts above $z \sim 2$, mentioned earlier (p. 189). What is required is a thermal history in which $n_H > 10^{-9}$ cm^{-3} above a critical redshift z_c (~ 2.3), rapidly decreasing to $n_H < 10^{-11}$ cm^{-3} below this redshift. The resulting absorption in the spectra of quasars with $z > z_c$ would change their U B, V magnitudes so as to bring them into the region of the U-B, B-V plane occupied by ordinary red stars. Quasars with such colours would not be picked out for spectral study by the selection procedures used up to date. If such quasars exist the only hope of finding them would be to study the spectra of apparently ordinary stars near otherwise unidentified radio sources. It is important to note that if this procedure were successful it would amount inter alia to the detection of an intergalactic gas (in particular these quasars would show a Scheuer-Gunn-Peterson Lyman-α absorption trough) but not necessarily to a gas with the magic density. Success would also, of course, demonstrate that these quasars are at cosmological distances and that the steady-state model is wrong (since the ionization state of the intergalactic gas would be z-dependent) (**).

(*) In equation (1) of [38] $R(t)$ should be replaced by $R^2(t)$. I am grateful to Prof. S. WEINBERG for pointing out my error.

(**) *Note added in proofs.* – See also J. ARONS and R. MCCRAY: *Astrophys. Lett.*, **5**, 123 (1969); V. PETROSIAN: *Astrophys. Lett.*, **6**, 71 (1970); R. LYNDS and D. WILLS: *Nature*, **226**, 532 (1970) (which reports the discovery of a quasar with $z = 2.877$).

6. – The microwave background radiation.

The discovery of the excess microwave background by PENZIAS and WILSON [41] and its interpretation (DICKE, PEEBLES, ROLL and WILKINSON [42]) as the black-body radiation predicted by GAMOW [43] from his hot big-bang theory of the origin of the universe have often been described. Recent review articles are those of FIELD [44] and PARTRIDGE [45]. The latter article is particularly useful for those interested in the experimental techniques. In addition both articles give an excellent account of the theoretical issues involved.

The existence of these excellent reviews makes it unnecessary for me to give a complete account here of the subject from its very beginning to the present date. I prefer to pick out particular points of interest, concentrating especially on those questions which have arisen most recently. However, for the benefit of readers who are not interested in the fine details, but who would like to have a general picture of the present situation, I shall give a brief self-contained account of the whole problem.

The present state of the observations is shown in Fig. 1, with a comparison black-body spectrum at 2.7 °K. The points labelled CN, CH and CH^+ will be discussed later. The remaining points are direct measurements made at the surface of the earth by radio astronomers. Some of the observations have been criticized, and there is a general danger of a « bandwagon effect » hanging over the subject. In addition there is a discordant measurement (derived from rocket flights) in the bandwidth between 0.04 and 0.13 cm (SHIVANANDAN, HOUCK and HARWIT [46], HOUCK and HARWIT [47]), not marked on Fig. 1, and corresponding to a flux about 100 times greater than that expected from a black-body spectrum at 2.7 °K (*).

We thus face the question, does the spectrum of the microwave radiation have the black-body character expected from the Gamow theory? There is a second related question that we should ask at this point: can the microwave radiation (which is about 100 times more intense than that produced by *known* radio sources) be due to a hitherto unknown type of radio source? Clearly we must know the answers to both these questions before the cosmological significance of the observations can be analyzed.

As regards the first question it is evident that the vital wavelength range is in the region close to the peak of the 2.7 °K spectrum at ~ 2 mm. Unfortunately this wavelength region must be studied from above the atmosphere.

(*) *Note added in proofs.* – D. MUEHLNER and R. WEISS: *Phys. Rev. Lett.*, **24**, 742 (1970), have reported results obtained from a balloon-borne far infra-red radiometer. They find excess radiation which could be concentrated in a line between 0.08 and 0.1 cm.

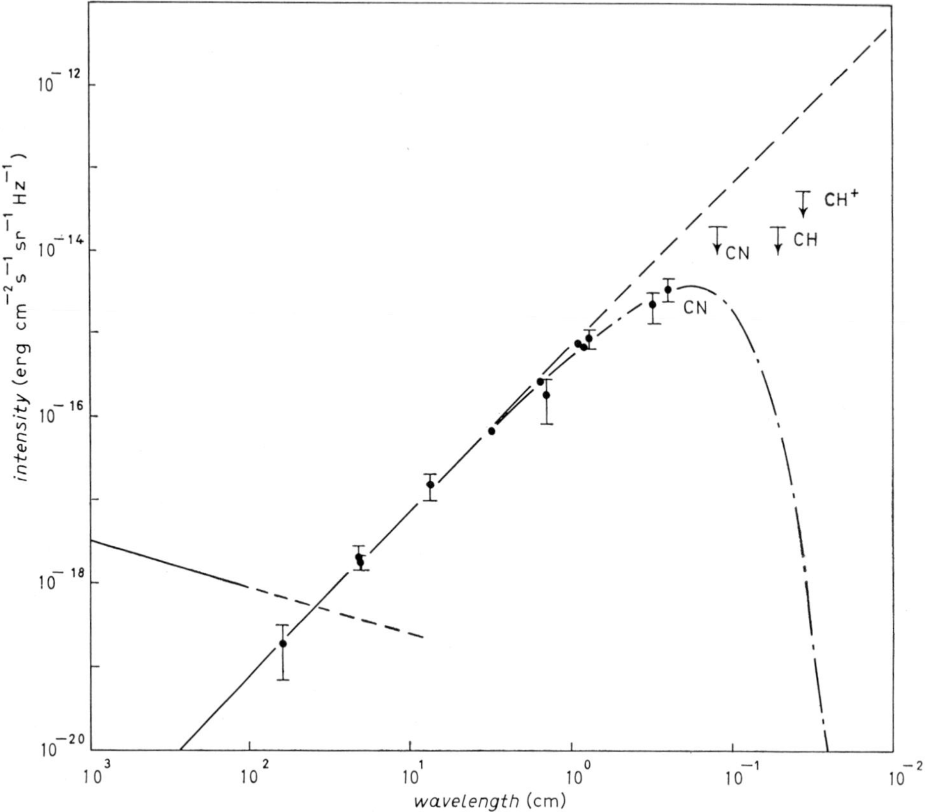

Fig. 1. – Spectrum of background radiation. Points labelled CN, CH, CH^+ are indirect determinations using interstellar molecules, while all other points are direct radio observations. The point at 74 cm wavelength has been corrected for nonthermal emission. The points at short wavelength show that the background is not due to a dilute hot source (« power law »); rather its spectrum bends over at ~ 1 mm as expected for a black-body of 2.7 °K. ——— galactic background; – – – power law; —·—·— 2.7 °K black-body.

We have just mentioned the rocket experiments in the far infra-red beyond the peak, with the enormous flux which they imply. It is most unlikely that this flux could be representative of the flux throughout our Galaxy, because as we shall see (pp. 195-196) it is almost certainly incompatible with the CH observations, unless it is confined to narrow lines, in which case it is irrelevant to the question of the continuous spectrum. Clearly the region near the peak must be thoroughly studied before we can draw firm conclusions.

The second question is whether the excess microwave background could be due to the integrated radiation from a new class of radio source with appropriate spectral properties. This possibility was first discussed in detail

(SCIAMA [48]) at a time when I was hoping to be able to save the steady-state theory, which of course has no hot big-bang to provide excess radiation and, more important, to thermalize it in a denser universe than we have to-day. The idea has been taken up again recently (HOYLE and WICKRAMASINGHE [49], NARLIKAR and WICKRAMASINGHE [50] but see SHAKESHAFT and WEBSTER [51], GOLD and PACINI [52], PARIISKI [53], WOLFE and BURBIDGE [54]). There are two constraints on model-building of this kind. The first is that the newly introduced sources must be compatible with existing source surveys which are complete down to a given flux density at the observing frequency. The second constraint is that the discrete source models must not lead to greater fluctuations with direction in the background than are observed. The first constraint led to great difficulties already in 1966, but is not mentioned in the later papers. The second constraint is referred to by WOLFE and BURBIDGE, but their analysis was concerned with the resolving of individual sources. It is, however, sufficient to check for the expected $N^{-\frac{1}{2}}$ fluctuations if N sources are simultaneously in the beam. This point has been examined by PENZIAS, SCHRAML and WILSON [55], SMITH and PARTRIDGE [56], and HAZARD and SALPETER [57]. The result of these investigations is that a discrete source model is very unlikely to satisfy all the requirements. From now on we shall assume for the purpose of these lectures that the microwave background has a black-body spectrum and comes from the hot big bang, although it should be noted that for some of our considerations it would suffice if intergalactic space contains radiation of energy density $\sigma T^4 \sim 1$ eV cm^{-3}, whatever its origin.

7. – The molecular measurements.

Soon after the observations of PENZIAS and WILSON at 7 cm, it was realized that interstellar CN molecules act as a thermometer for radiation at a wavelength of 2.6 mm. (FIELD and HITCHCOCK [58], THADDEUS and CLAUSER [59]). The reason is that the spectra of certain stars contain absorption lines at 3874.61 Å due to electronic transitions in nonrotating CN molecules, and also at 3874.00 Å and 3875.76 Å due to transitions in CN molecules with one quantum of angular momentum. The molecular spectroscopist calls these transitions $R(0)$, $R(1)$ and $P(1)$ respectively (see Fig. 2*a*). The relative intensities of the different lines enable one to infer the relative proportions of CN with zero and one quantum of angular momentum. The result can be expressed in terms of a rotational temperature T_{rot} by the Boltzmann relation

$$\frac{n_1}{n_0} = \frac{g_1}{g_0} \exp\left[-\chi/kT\right],$$

where g_0 and g_1 are the statistical weights of the two states (1 and 3 respectively in the present case) and χ is the energy difference between the two states.

A photon of energy χ could convert a CN molecule from the 0 state to the 1 state, and such a photon would have a wavelength of 2.6 mm. If the

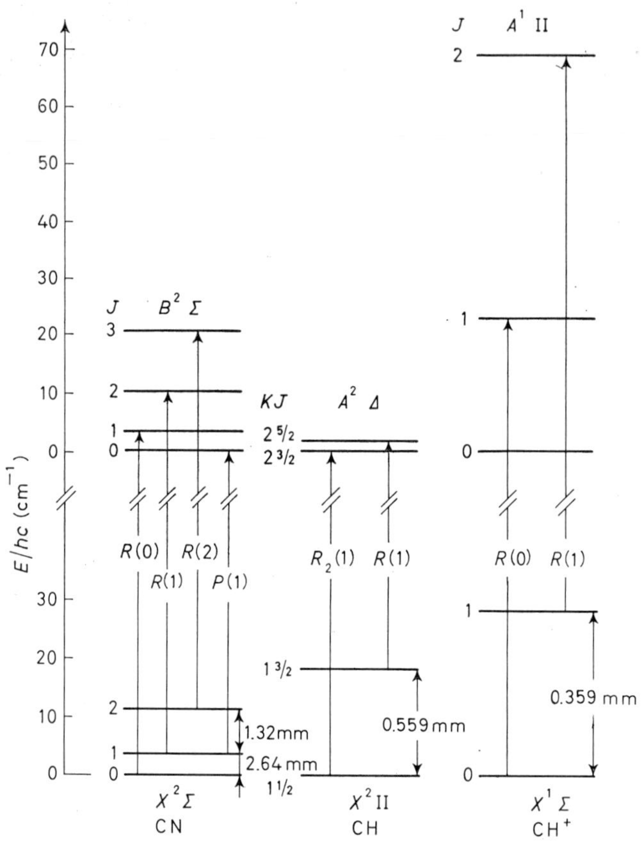

Fig. 2*a*). – Term diagrams showing rotational levels of ground electronic states which can be excited by background radiation. Both $J=0$ and $J=1$ of CN have been observed, yielding a rotational temperature. Other levels have not yet been observed, but provide upper limits on I_λ at the corresponding wavelengths.

excitation process is predominantly radiative in origin the observations then give us the effective temperature at 2.6 mm of the radiation field at the location of the CN molecules. The first such observations were made in 1941 (McKellar [60]) and gave a rotational temperature ~ 2.3 °K. At the time this was regarded as anomalously large and remained an unsolved problem until the discovery of the excess microwave background and the suggestion that this background has a black-body spectrum with $T \sim 3$ °K. It then became natural to suggest that the same black-body radiation field at 2.6 mm is exciting the CN.

It has, however, recently been realized that collisional excitation by protons and electrons may be more important than was first thought. It is therefore essential to measure T_{rot} for stars in widely separated regions of the Galaxy, for which the physical environments of the CN along the different lines of sight would be expected to differ by appreciable amounts. This has

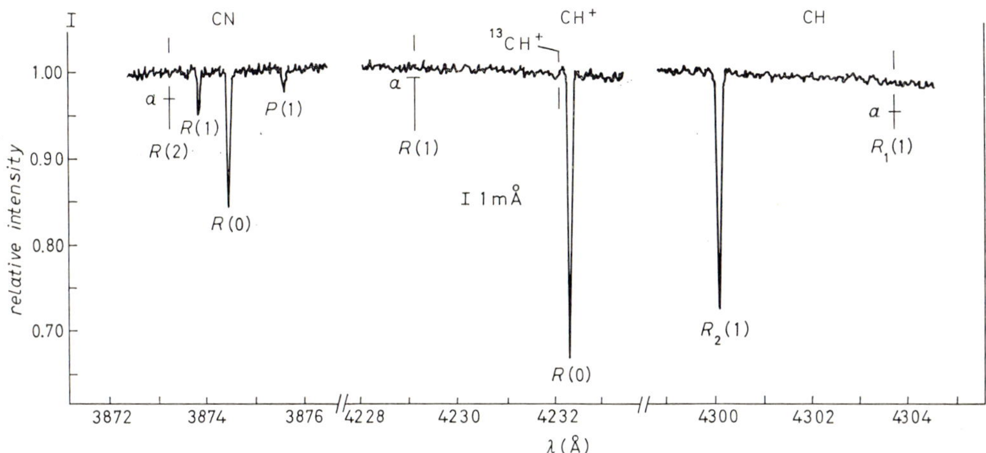

Fig. 2*b*). – Three regions of the interstellar spectrum of ζ Ophiuchi, showing observed spectral lines arising from the levels of Fig. 2 *a*). *R*(1) and *P*(1) of CN yield (2.47 ± 0.22) °K at 2.6 mm wavelength. The absence of lines at points marked *a* provides the upper limits on I_λ in Fig. 1.

been done by CLAUSER and THADDEUS [61] who now have data for 11 stars (Table I, and Fig. 2*b*) *c*)). The agreement between the various rotational temperatures is strong evidence for a universal excitation mechanism and so for a radiation field that fills at least the Galaxy with fairly constant intensity. The agreement of the rotational temperature with the microwave temperature then supports the black-body hypothesis as may be seen from the point marked CN at 2.6 mm in Fig. 1.

The other point marked CN in Fig. 1 (at 1.3 mm) is an upper limit which follows from Bortolot, Clauser and Thaddeus's [62] failure to observe absorption lines arising from electronic transitions starting from CN in the second rotational state. Similarly they derived the upper limits marked CH (0.56 mm) and CH^+ (0.36 mm) from their failure to observe absorption lines related to these molecules in the first rotational state.

It is this latter evidence which conflicts with the far infra-red rocket measurement of SHIVANANDAN, HOUCK and HARWIT which we have already dis-

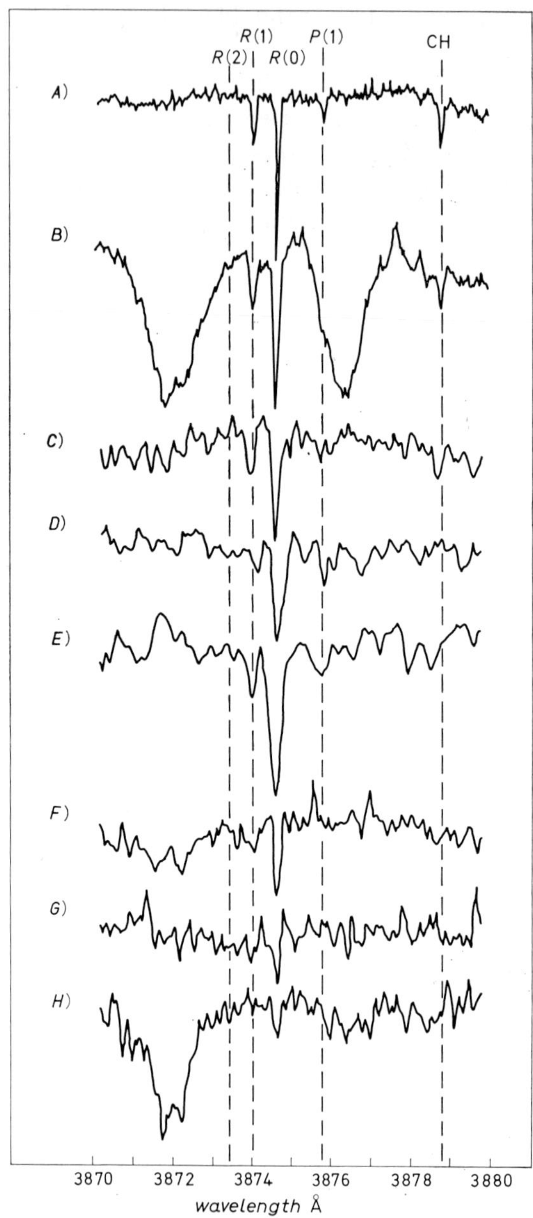

Fig. 2c). – Confirmation of the universal nature of the CN excitation in seven stars in addition to ζ Ophiuchi. The absorption line at $R(1)$ is always found, yielding a rotational temperature of about 3 °K. *A*) ζ Oph $\times$ 10, 24 spectra; *B*) ζ Per $\times$ 10, 7 spectra; *C*) X Per $\times$ 5, 2 spectra; *D*) BD $+66°$ 1674 $\times$ 2, 2 spectra; *E*) BD $+66°$ 1675 $\times$ 2, 2 spectra; *F*) 20 Aql $\times$ 2, 1 spectrum; *G*) AE Aur $\times$ 5, 3 spectra; *H*) 55 Cyg $\times$ 5, 3 spectra.

TABLE I. – *The rotational temperature* T_{10} *of the* CN (0, 0) *violet band against eleven stars, corrected for the finite optical depth of the lines.* Uncertainties are estimated from the plate grain. V is the visual magnitude and Sp the spectral class of the stars. The heliocentric radial velocities of CN and HI 21 cm are included in those instances where 21 cm peaks are found to be well defined.

Star	V	Sp	T_{10}(CN) (°K)	v(CN) (km/s)	V(HI) (km/s)
ζ Oph	2.56	O9.5V	2.74 ± 0.22	−14.8	−12.7
ζ Per	2.83	B1Ib	2.82 ± 0.30	12.6	13.4
55 Cyg	4.83	B3Ia	< 5.5		
AE Aur	5.3	O9.5V	3.5 ± 2.3		
20 Aql	5.37	B3IV	2.5 ± 1.8	−12.2	− 2.0, −11.5
HD 12953 ([c])	5.68	A1Ia	3.7 ± 0.7		
13 Ceph ([d])	5.79	B8Ib	2.8 ± 0.4		
HD 26571 ([a])	6.10	B8II-III	~ 3		
X Per	6.08	O	2.8 ± 0.8	13.5, 23.2	13.3, 22.8
BD+66° 1675 ([b])	9.05	O7	2.39 ± 0.4	−17.2	−14.2, −19.5
BD+66° 1674 ([b])	9.5	O	2.45 ± 0.6	−17.4	−14.2, −19.5

(*a*) G. HERBIG: Lick 120 inch Coude, private communication.
(*b*) From spectra taken by G. MÜNCH with the 200 inch telescope.
(*c*) M. PEIMBERT: 120 inch Coude, private communication.
(*d*) V. BORTOLOT and P. THADDEUS: 120 inch Coude.

cussed (p. 193). The conflict is illustrated in Fig. 3. One could resolve it by assuming that the rocket measurement actually refers to one or more lines which happen to lie outside the CH wavelength. In that case the rocket measurement need not be incompatible with the hot-big-bang picture and would refer to some additional source of radiation (*).

Another point that should be made is that one cannot be quite certain

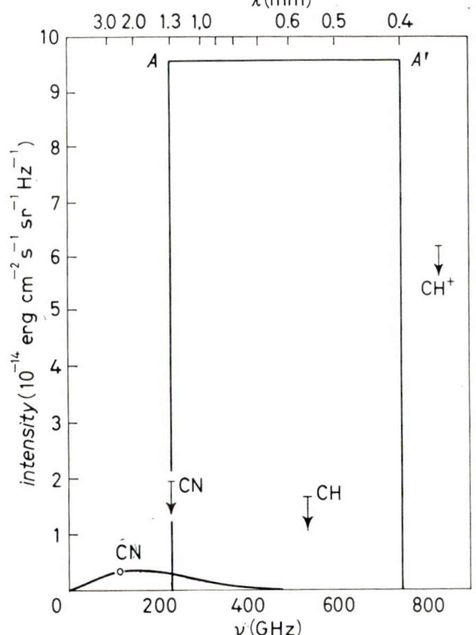

Fig. 3. – Comparison of the results of a rocket flight with evidence from molecular lines in ζ Ophiuchi (Fig. 2 *b*)): ——— 2.83 °K black-body.

(*) *Note added in proofs.* – See the footnote on p. 193.

that the failure to observe absorption lines leads to a rigorous upper limit to the radiation temperature at the appropriate wavelength. It has recently been discovered from the microwave lines of interstellar formaldehyde that this molecule is actually *colder* than 2.7 °K at the wavelength of the lines (PALMER, ZUCKERMAN, BUHL and SNYDER [63]). Presumably we are dealing here with a population inversion (TOWNES and CHEUNG [64]). The question then arises: could the observed temperature of CN or CH at the appropriate wavelengths actually be less than that of the incident radiation at these wavelengths because of a similar population inversion? The level structure of these molecules makes it unlikely (SOLOMON [65]) but this point deserves further investigation.

Our conclusion then is that the molecular points marked on Fig. 1 are probably correct, and serve to strengthen the hypothesis that we are dealing with radiation that has a black-body spectrum with a temperature ~ 2.7 °K.

8. – The thermal history of the universe.

In this Section we shall accept the hot-big-bang theory of the universe and study its thermal history, assuming for the most part that the universe has the isotropy and homogeneity of the Robertson-Walker models. The observational evidence in favour of these symmetry assumptions will be discussed later (p. 200).

The thermal history is much simplified by the easily proved fact that the time scale for free-free scattering to thermalize the radiation field is far less than the expansion time scale in the early stages of the universe. There is no doubt, therefore, that we are dealing with black-body radiation in these early stages. Now as discussed by ELLIS the energy density ϱ_{rad} of this radiation depends on the scale factor $R(t)$ in the following way:

$$\varrho_{\text{rad}} \propto R^{-4}\,.$$

By contrast the energy-density of ordinary matter in the absence of radiation behaves like

$$\varrho_{\text{mat}} \propto R^{-3}\,.$$

It follows that, unless there is no radiation whatsoever, at early enough times there was far more energy density in radiation than in matter. This conclusion is not altered if we allow for the interaction between matter and radiation. However, if we go back to sufficiently early times that the temperature exceeds the threshold for pair creation (leptonic or baryonic), we would then be dealing with relativistic gases (except very close to the threshold temperature (*)) whose behaviour differs little from that of radiation. We shall therefore refer to the early stages as being radiation-dominated. During these early stages we have the relations

$$R \propto t^{\frac{1}{2}}, \qquad T = \frac{10^{10}}{t_{\text{sec}}^{\frac{1}{2}}}\,{}^\circ\text{K}.$$

It is not always convenient to use the ratio $\varrho_{\text{rad}}/\varrho_{\text{mat}}$ to characterize the relative amounts of radiation and matter in the universe, as this quantity is time-dependent. It is often better to work in terms of the entropy density S of the radiation, which satisfies

$$S \propto T^3 .$$

Since

$$T^3 \propto \frac{1}{R^3} \propto n_{\text{mat}} ,$$

we have

$$\frac{S}{n_{\text{mat}}} = \text{const} .$$

We may express this by saying that the entropy per baryon is independent of time, and so is a universal constant. In an Einstein-de Sitter universe this constant would have the value $\sim 10^8 k$, where k is Boltzmann's constant. Since S/k is approximately the number density of photons we can also write

$$\frac{n_{\text{ph}}}{n_{\text{baryon}}} \sim 10^8 .$$

It is here that we meet a rather unsatisfactory feature of the hot big-bang cosmology. The number of photons per baryon must be regarded as an initial condition specified at $t=0$, and the question of accounting for the particular value it has simply does not arise. It would be more satisfactory if we could account for its observed value in terms of more primitive assumptions. One possibility of doing this is provided by what I may call the Misner programme (MISNER [67, 67*a*]). In this programme one starts out with a

(*) A curious thermodynamic point arises here. A partially relativistic gas cannot remain in thermal equilibrium in an expanding system, and has a nonvanishing bulk viscosity (ANDERSON and STEWART [66]). This point is explained in Ehlers' lectures. Its practical consequences are not important for us here.

generic universe containing no symmetries and indulging in random motions of a white noise type. Dissipation processes are then invoked to suppress disturbances on length scales not now observed. In these processes heat is generated which becomes rapidly thermalized. MISNER believes that if one starts from a singularity an infinite number of possible starting points may lead to a finite number of possible end results, even if the system is governed by differential equations which are regular everywhere except at the singular origin. In this way he hopes to understand the high degree of symmetry the universe is observed to possess (p. 200). It may also be possible to understand the photon-baryon ratio along similar lines, or at least to reduce the complete arbitrariness this quantity possesses in the Gamow theory.

We must now consider the thermal history of the gas in more detail. We have

$$T_{\text{rad}} \propto \frac{1}{R}. \tag{2}$$

If we assume that the matter behaves like a perfect gas with the ratio of its specific heats equal to $\frac{5}{3}$, then *if uncoupled to the radiation*, its temperature T_{mat} would behave like

$$T_{\text{mat}} \propto \frac{1}{R^2}. \tag{3}$$

Thus noninteracting matter cools down more rapidly than does radiation, but its energy-density decreases more slowly than that of radiation. Of course, for black-body radiation T and ϱ determine each other, while for nonrelativistic matter T and ϱ are completely independent.

In the early stages of the universe the matter was ionized and closely coupled to radiation by Thomson scattering. At that time the heat capacity of the radiation far exceeded that of matter (so long as the matter was nonrelativistic) and so it would be a good approximation to write

$$T_{\text{rad}} = T_{\text{mat}} \propto \frac{1}{R},$$

until the temperature drops to the point (~ 3000 °K) where the matter recombines (*) and becomes only weakly coupled to the radiation. From that point on we will have (2) and (3) holding. We would then expect that the temperature of intergalactic matter at a redshift ~ 2 should be far less than a few degrees absolute. This is in flagrant contradiction with our conclusion

(*) The recombination radiation may be a significant constituent of the present background (PEEBLES [68], ZEL'DOVICH and SUNYAEV [69]).

(p. 190) that such matter must be highly ionized. We must therefore introduce the *re-heating* processes discussed on p 190.

It is interesting that we can place an upper limit on the redshift at which this re-heating occurred. There are two reasons for this. First the resulting free electrons scatter the black-body radiation and so distort its spectrum. This distortion arises because although photon number is conserved by the scattering, the photons on average gain energy from the hot electrons, so that the relative number of high-frequency photons is increased. To keep the distortion of the spectrum down to an acceptable level the concentration of free electrons must be suitably low, and so the intergalactic gas can only become re-ionized at a relatively late stage in the expansion, when its density has

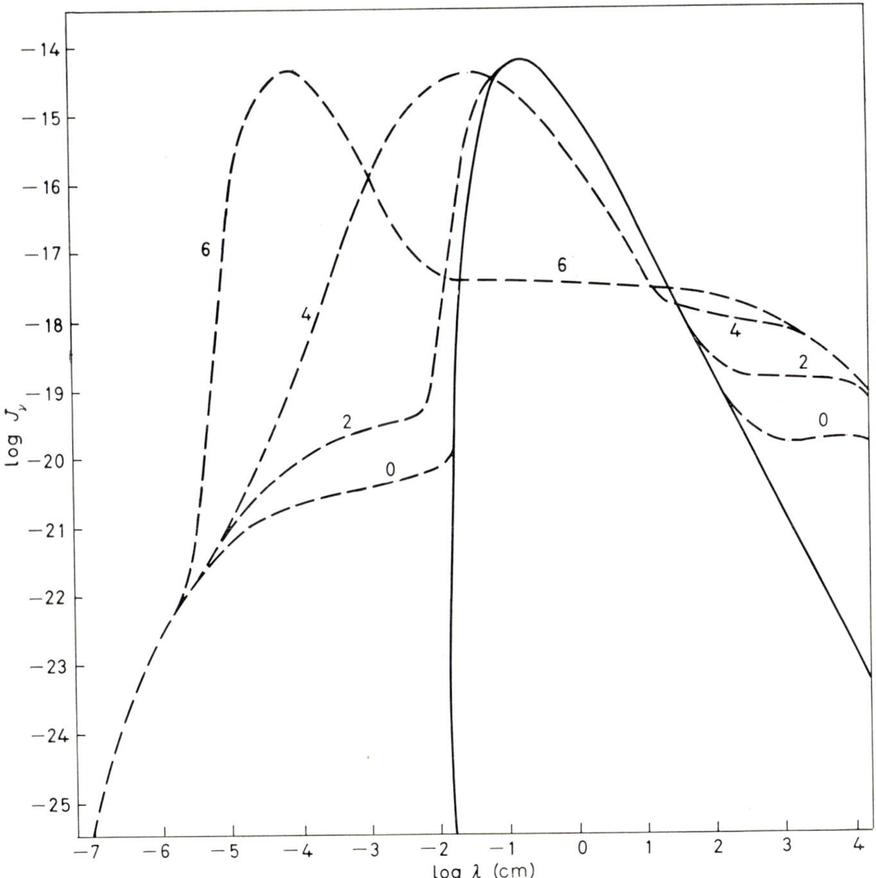

Fig. 4. – Distortion of the Planck function produced by an isothermally expanding gas with $\log T = 6.5$. Continuous curve: Planck function of 3.5°; broken curves: distortion produced by the heating; numbers attached to these curves: log of the density at which heating commenced.

dropped sufficiently. The second reason is that the radiation emitted by the re-ionized gas must not exceed the observed background radiation at any wavelength. The emissivity of the gas is proportional to the square of the electron density, and so again this density must not be too great. Both these considerations have been worked out quantitatively by WEYMANN [70] and by SUNYAEV and ZEL'DOVICH [71]. WEYMANN finds that the « dangerous » wavelengths where the background may be exceeded are in the radio and X-ray regions. For a universe with the magic density, the upper limit on the redshift at which significant re-ionization can occur is about 200. The results of Weymann's calculations are shown in Fig. 4.

The effects of scattering are more complicated if the expansion of the universe is not quite isotropic (REES [72]). In fact one can set more stringent limits on any possible anisotropy from measurements with 25 per cent accuracy of the spectrum of the microwave background near its peak than from the direct measurements of the isotropy of the background, which have already come close to 0.1 per cent accuracy. Scattering in an anisotropic universe would also lead to a linear polarization ε of the radiation. REES finds in typical cases that

$$\varepsilon \sim \left(\frac{1}{10} \text{ to } 3\right) \times \text{anisotropy} .$$

The best observed upper limit on the linear polarization comes from recent work of BARON and WILKINSON [73] at 3 cm. They find an upper limit of 1 per cent.

9. – Isotropy of the background.

The original measurements of PENZIAS and WILSON already showed that the microwave background was isotropic to a precision of a few per cent, indicating a probable extragalactic origin for the radiation. The precision achievable is much greater than this since one is here concerned with comparing the radiation from two different parts of the sky, rather than making an absolute measurement of the radiation from one direction. Consequently we now have available far more accurate results, which have the profoundest significance for cosmology.

The results are of two types, corresponding to angular scales measured in minutes of arc (CONKLIN and BRACEWELL [74, 74*a*], EPSTEIN [75], PENZIAS, SCHRAML and WILSON [55]) and in degrees or tens of degrees (PARTRIDGE and WILKINSON [76], WILKINSON and PARTRIDGE [77], PARTRIDGE [45], CONKLIN [78]). No significant intensity variations with direction have been claimed, with the exception of the recent work of CONKLIN [78] which is discussed later. (WILKINSON and PARTRIDGE [77]) found provisional evidence for a peak when the Milky Way was in the beam, and a trough in another direction. Their further observations (PARTRIDGE [45]) showed that these variations are not significant). As an example of the precision of the small-scale measurements we quote the result of CONKLIN and BRACEWELL [74] that $\Delta T/T$ is less than $2\cdot10^{-3}$ on an angular scale of 1° over the 4 hours of right ascension which they studied. We have already mentioned (p. 194) how results of this sort severely limit possible discrete source models for the origin of the microwave radiation.

The latest measurements by the Princeton group on the larger scale fluctuations along the celestial equator are shown in Fig. 5 (taken from PARTRIDGE [45]).

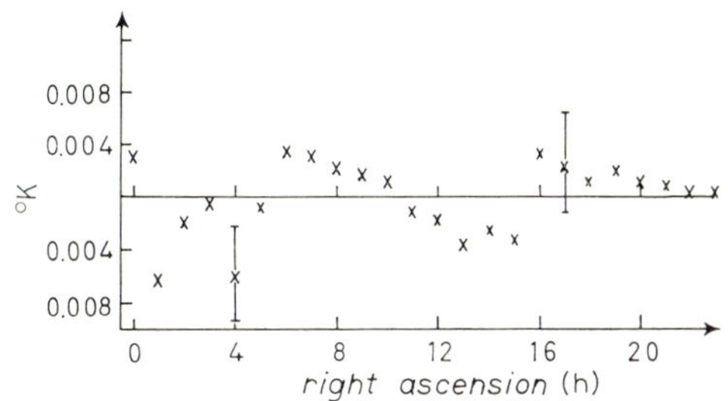

Fig. 5. – The temperature distribution along the celestial equator is shown. More than two years of data, obtained at Princeton and Yuma, have been averaged. Since the number of raw data points in each bin varies, the indicated error bars (standard deviations of the means) are only approximate. $N \approx 450$, $\sigma \approx 0.0037$ °K.

From the theoretical point of view we are interested in possible peaks or troughs with an angular scale of several degrees (p. 204), and also with 12 hour and 24 hour variations. Fourier analysis of the data in Fig. 5 shows that

$$\frac{\Delta T}{T} = (0.55 \pm 0.59)\cdot 10^{-3} \qquad \text{12 h period,}$$

$$\frac{\Delta T}{T} = (0.40 \pm 0.56)\cdot 10^{-3} \qquad \text{24 h period.}$$

Thus no variations are observed for either period to a precision of about 0.1%. As has now often been said these are by far the most accurate measurements ever made in cosmology.

We must now consider the theoretical implications of these measurements. For this purpose it is helpful to think in terms of the *immediate* sources of the black-body radiation. One meets a similar situation in connection with the visible radiation from the sun. The ultimate source of this radiation is the central region of the sun where the energy-producing thermonuclear reactions are occurring. However, the γ-rays released by these reactions are scattered many times on their way to the sun's surface, and are degraded in energy into optical photons. The multiple scattering serves to destroy any detailed information the γ-rays may contain about conditions near the centre of the sun. We therefore usually think in terms of the sun's radiation as coming from the layer in the sun after which there is little further scattering. This surface of last scattering, which is at about unit optical depth inside the sun, is, of course, the sun's photosphere, and it is the temperature of the photosphere which determines the intensity and the quality of the radiation we receive from the sun.

In a similar way it is helpful to think of the cosmic black-body radiation as coming, not from its ultimate source in or near the hot big-bang, but from the surface of last scattering at about unit optical depth away from us. The redshift of this surface is determined by eq. (1) and is 7 for a universe which contains the magic density of ionized gas. For a low-density universe we may have to look right back to before the decoupling era ($T\sim 3000$ °K) in order to reach unit optical depth. In that case the redshift of the last scattering surface would be about 1000 (since $T\propto 1+z$). We may thus think of ourselves as immersed in a heat bath, namely, at the centre of a sphere whose surface has a redshift z_0 ($7<z_0<1000$) and a temperature $T_0\sim 3(1+z_0)$ °K. Of course this picture has its limitations, as is shown by Rees's discussion of multiple scattering in an anisotropic universe. However, for a first orientation it is very useful.

We can now return to the problem of analysing the significance of the isotropy measurements. To avoid tiresome repetition we will now suppress the phrase « with a precision of 0.1 per cent », and shall also assume that this precision holds in all directions, not just in those so far observed. In terms of the last-scattering surface we see that the isotropy implies that

i) z_0 is the same in all directions, that is, the universe is expanding at the same rate in all directions, at least out to z_0. The isotropy of the Hubble constant could also be studied directly from a detailed analysis of the Hubble diagram, but the precision attainable by this method is hardly better than $\sim$20 per cent, and applies only to relatively small red shifts. The absence of a 12 hour anisotropy shows in particular that the universe out to z_0 cannot be one of a large class of homogeneous anisotropic models (THORNE [79]).

ii) T_0 is the same everywhere on the last-scattering surface (we exclude here a close cancellation in possible variations of T_0 and z_0).

We might try and explain i), without simply appealing to initial conditions at $t=0$, by invoking dissipative processes to smooth out initial anisotropies in the expansion (MISNER [80, 67], DOROSHKEVICH, ZEL'DOVICH and NOVIKOV [81], STEWART [82], HAWKING [83], MATZNER [84]). The most important such process appears to be neutrino viscosity, which arises from the interaction between electrons and neutrinos thermally excited at temperatures exceeding 10^{10} °K (see p. 214). The extent to which an arbitrarily large initial anisotropy can be smoothed away is still under dispute, and more calculations are needed.

As regards ii) the situation is also not straightforward. There is a limit to the extent to which transport processes can reduce initial temperature gradients, because of the existence of particle horizons. For instance if $z_0 \sim 7$, two regions on the last-scattering surface which are more than 30° apart have not been in causal contact since $t = 0$ (if the Robertson-Walker models are good approximations right back to $t = 0$). Yet their temperatures differ by less than 0.1 per cent. MISNER [67*a*] has proposed that the early stages of the universe differed so much from the Robertson-Walker models that particle horizons did not exist, and that smoothing processes could have occurred on any required scale. This is a very attractive suggestion which should undoubtedly stimulate further work on the properties of generic solutions of Einstein's field equations near $t = 0$ (cf. MACCALLUM [85]). A closely related problem is that of the singular behaviour of solutions at $t = 0$, which is discussed by ELLIS.

iii) A third consequence of the isotropy is that we can set limits on the properties of large-scale irregularities in the distribution of matter in the universe. This was first pointed out by SACHS and WOLFE [86], and further studies of this problem have been made by REES and SCIAMA [86*a*], WOLFE [87], DAUTCOURT [88], LONGAIR and SUNYAEV [89], and CHIBISOV and OZERNOY [90]. The main point is that if a region on the last-scattering surface is observed through an irregularity, the redshift of that region would be changed from z_0 to, say, $z_0 + \delta z$. Now as explained in Ellis's lectures we would expect the simple relation

$$T' = \frac{T_0}{1 + z_0 + \delta z} \,. \tag{4}$$

However, we must be a little careful because, owing to the change in propagation-time for a ray traversing the irregularity, we would be observing the last-scattering surface at a slightly earlier epoch, when its temperature was slightly greater. Thus T_0 in (4) must contain a correction factor for this effect.

SACHS and WOLFE give a rigorous (linearized) calculation for a Fourier component of an arbitrary density perturbation, while REES and SCIAMA, and WOLFE discussed the effects of localized irregularities such as would be produced by clusters of quasars, if they exist. The effects of incipient galaxies on the last-scattering surface have also been considered by some of the abovementioned authors. Smaller-scale irregularities such as galaxy clusters as they now exist would produce effects too small to observe. Since no fluctuations in the background have yet been established, we will not go into further detail here, except to illustrate what type of large-scale irregularity can now be ruled out. As an example we quote the case of a single spherical irregularity with proper diameter 750 Mpc, contrast density ~ 3 at a redshift of 1.5 in an Einstein-de Sitter universe. Such an irregularity would lead to a decrease of ~ 0.3 per cent in the radiation temperature in the direction of its centre, and ~ 0.9 per cent somewhat away from the centre. The angular extent of the profile would be ~ 20 degrees. A similar irregularity in a low-density universe with $q_0 \sim \frac{1}{10}$ would give rise to a decrease in temperature of ~ 0.2 per cent near the edge, but a comparable increase at the centre.

It is clear that back to the last-scattering surface the universe is highly homogeneous on a large scale. In addition it is unreasonable to suppose that the high isotropy we observe is a special property of our own location in space-time. It is therefore plausible that any observer at the present epoch would see high isotropy around him. It would then follow that the universe as a whole is highly homogeneous at the present epoch. Thus the symmetry assumptions underlying the Robertson-Walker models probably hold true to greater accuracy than even the boldest cosmologist (except perhaps MILNE) would have dared to hope. Whether the same holds true beyond a redshift of z_0 is an open question; as we have seen, the breaking of the symmetry beyond z_0 may be necessary to explain the existence of the symmetry within z_0.

iv) *The peculiar velocity of the earth.* A fourth consequence of the high isotropy of the microwave background is the limit it places on the motion of the Earth through this background. The reason is that an isotropic radiation field defines a *rest frame* at each point, namely that frame in which the radiation *is* isotropic. An observer who moves relative to the rest frame would, by virtue of the Doppler effect, see an increased intensity in front of him and a decreased intensity behind him. In any one direction the radiation would continue to have a black-body spectrum, but with a temperature corresponding to the appropriate redshift factor (cf. our previous remarks on the relation between temperature and redshift and Ellis's lectures). For a peculiar velocity $\ll c$ we can use the classical Doppler formula to obtain for the variation of temperature with direction in any plane

$$T(\theta) = T(0)\left(1 + \frac{v}{c}\cos\theta\right),$$

where v is the component of peculiar velocity in the plane and θ is the angle between this component and the direction of observation. Thus if observations of T are made at a constant declination we would expect to observe a 24 hour (dipole) variation in temperature with right ascension.

The most recent Princeton measurements (PARTRIDGE [45]) gave an upper limit for the component of the peculiar velocity of the Earth along zero declination of $\sim 300\ \text{km s}^{-1}$. Subsequently CONKLIN [78] has claimed to find a positive result, but before examining his work it will be helpful to consider the predictions that have been made of the Earth's peculiar velocity. Since the last-scattering surface has a very large redshift, the rest frame defined by the microwave background is the rest frame defined by the universe as a whole. This remark would remain true whatever the origin of the background, so long as its sources were distributed over a large region of the universe (as is the case for all the theories so far proposed). We are therefore concerned with the motion of the Earth relative to the universe as a whole. This situation immediately reminds us of Mach's principle. To simplify the discussion we shall defer our comments on the relevance of this principle until later (p. 215). For the moment we shall simply assume without question that a local inertial frame has zero angular velocity with respect to the sources of the microwave background.

The best-known peculiar velocity of the Earth arises, of course, from its motion around the Sun. However, this velocity is only about 30 km s^{-1}, and is dwarfed by the velocity of the Sun in its motion around the centre of the Galaxy. This motion is also driven by gravitation, and according to our assumption, we expect it to be a motion relative to the distant parts of the universe. The circular velocity at the sun's location is not known very accurately, but is believed to be in the vicinity of 250 km s^{-1} (SCHMIDT [91]). The uncertainty in this value is about 15 per cent. In addition the sun has a peculiar motion of about 20 km s^{-1} relative to the local standard of rest (as defined by a suitable average of the motions of nearby stars).

We must now consider the motion of our Galaxy as a whole. We are usually regarded as belonging to a Local Group of twenty or so galaxies, of which the Andromeda galaxy *M*31 is the brightest and most massive. It is still an unsettled question whether the Local Group is gravitationally bound, but in any case solutions have been made for the peculiar motion of our Galaxy relative to the Group as a whole. The method is the same as that used for finding the peculiar motion of the sun relative to the nearby stars, except that in the galactic case one has only radial motions available (no proper motions). The method essentially consists of determining a common reflex motion in the galaxies of the Local Group as observed from our own Galaxy. It assumes that the objects concerned have a negligible common (streaming) motion of their own, which may not be correct. It is, perhaps, comforting that the solutions so far obtained (MAYALL [92], HUMASON and WAHLQUIST [93], BYRNES [94], DE VAUCOULEURS and PETERS [95]) give a result not very different from that of the solar motion around our own Galaxy (which, must, of course, contribute to the observed effect). The resulting velocity of our Galaxy as a whole through the Local Group is about (80 ± 20) km s^{-1}, and the total peculiar motion of the Sun relative to the Local Group is about (315 ± 15) km s^{-1} towards $l_{II} \sim 95°$, $b_{II} \sim -8°$.

We now come to the most uncertain part of our problem: what is the peculiar velocity of the Local Group as a whole? We cannot ignore this question because the Local Group may well be under the gravitational influence of some large-scale irregularity in the universe. Indeed in our earlier discussion of the effect of such an irregularity on the black-body background we should have included its gravitational effect on the Galaxy, which would give rise to a further peculiar velocity. As stressed by SACHS and WOLFE [86], the division of the total change in temperature into a part due to the gravitational action of the irregularity on the radiation passing through it and a part due to our peculiar velocity in the gravitational field of the irregularity (and a still further part due to peculiar motions on the last-scattering surface) does not have an invariant significance. The division depends on the co-ordinate system used in the calculation. In physical terms we can say that in an irregular universe there is no unique standard of rest defined at a point. Such a standard of rest could be defined by some averaging procedure (*e.g.* by minimizing the anisotropy in the background) but many equally natural averaging procedures could be constructed with differing results. By contrast, in a homogeneous and isotropic universe the symmetry itself defines a natural rest frame at each point (in relativistic language one would use the timelike geodesics orthogonal to the hypersurfaces of homogeneity).

Despite these relativistic difficulties it is possible to discuss in a straightforward way the question whether a nearby irregularity is influencing the Local Group if the gravitational potential of the irregularity (in units of c^2) is much less than the velocity conferred on the Local Group (in units of c). The corresponding criterion is well satisfied within our own Galaxy and in the Local Group itself, so that our discussion of the peculiar motions on those scales does not run into any relativistic difficulties. The nearest major irregularity to the Local Group is the Virgo cluster of galaxies, about 10 Mpc away. Now some astronomers (RUBIN [96], DE VAUCOULEURS [97], OGORODNIKOFF [98]) believe that the Local Group belongs to a flattened rotating supercluster dominated by the Virgo group of galaxies. The gravitational potential of this supercluster is about $10^{-5}\,c^2$, whereas, as we shall see, the velocity induced in the Local Group has been estimated as about $10^{-3}c$. We may thus study this part of the Earth's peculiar motion without having to pay attention to relativistic niceties.

It is still an open question whether the Local Group does belong to a Virgo supercluster. However, in view of the imminent possibility that the net peculiar velocity of the Earth would be measured by the radio astronomers, it became necessary to re-assess the question and to redetermine our motion, if any, in this system (SCIAMA [99], STEWART and SCIAMA [100], DE VAUCOULEURS and PETERS [95]). The primary evidence is the distribution of bright galaxies in the sky (Fig. 6) which shows a flattened distribution,

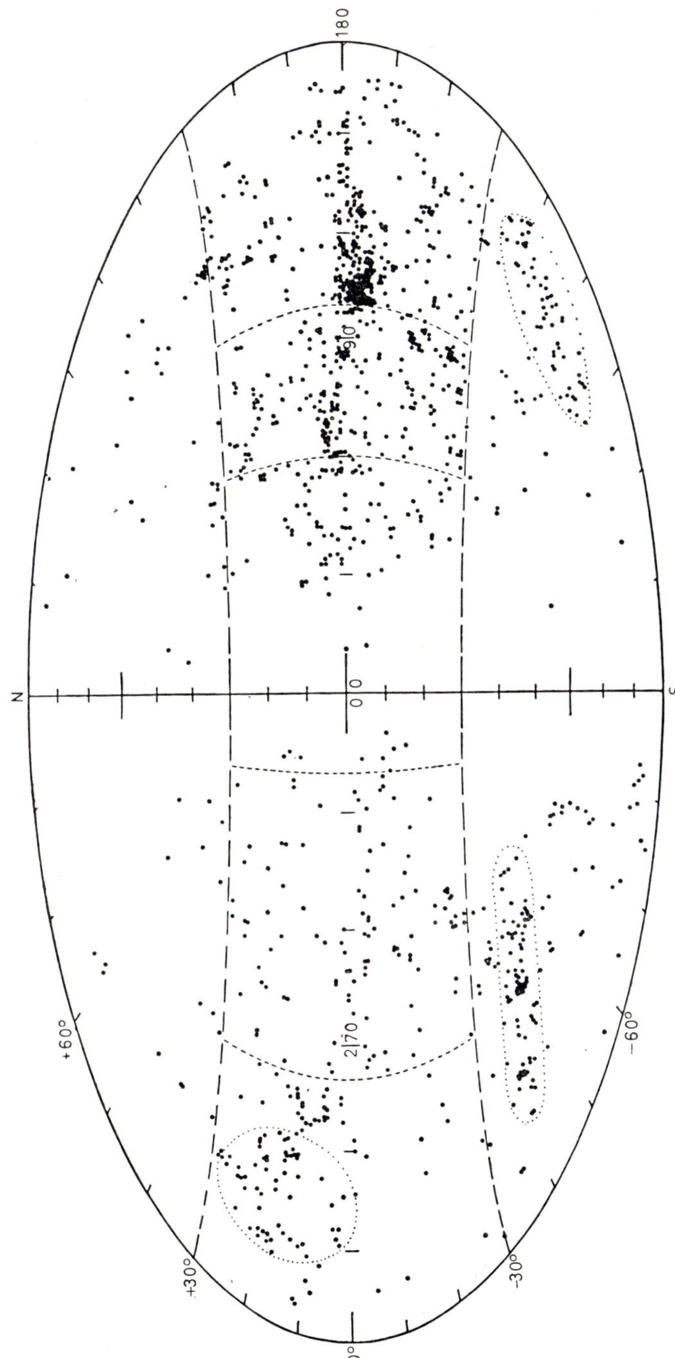

Fig. 6. – Distribution of galaxies of 8-th to 13-th magnitude in supergalactic co-ordinates.

centred on the Virgo cluster, that cannot arise from selection effects. The flattening is suggestive of rotation, although difficult dynamical problems are involved (for instance, the inferred rotation period is about 10^{11} years, so that only $\frac{1}{10}$ of a rotation could have occurred in a Hubble time. However, the flattening may have been produced in primordial turbulence that could have preceded the formation of the supercluster.) One can test for rotation, for differential expansion, and for motion perpendicular to the plane of the system, by studying the distribution with direction of the redshifts of the nearby galaxies in the supercluster. (Similar analyses of stellar motions reveal the rotation of our own Galaxy.) The most detailed discussion is that of DE VAUCOULEURS and PETERS [95]. They find for the motion of the Local Group in the supercluster:

a) a rotational velocity of (400 ± 100) km s^{-1};

b) a velocity towards the Virgo cluster (after subtraction of the linear Hubble expansion, with a Hubble constant based on galaxies more distant than the Virgo group) of (300 ± 100) km s^{-1};

c) a perpendicular component of velocity of (300 ± 100) km s^{-1}. The vector sum of these velocities is (590 ± 170) km s^{-1} towards $l_{\text{II}}\sim 283°$, $b_{\text{II}}\sim 16°$. When we combine this with our peculiar motion relative to the Local Group, we obtain a net velocity $\sim$300 km s^{-1} towards $l_{\text{II}}\sim 290°$, $b_{\text{II}}\sim 24°$.

The next question concerns the peculiar velocity of the supercluster as a whole. If it were under the influence of a still larger irregularity we would be in danger of running into relativistic ambiguities. In fact we can say very little about this question. Apart from the lack of any major anisotropy in the Hubble diagram for galaxies much more distant than the scale of the supercluster, all we can appeal to is the isotropy of the black-body measurements themselves. The upper limit of 300 km s^{-1} in the equatorial direction given by PARTRIDGE [45] does at least suggest that the supercluster (if it exists) does not have a peculiar velocity of an order greater than the velocities we have been discussing so far. This result also enables us to place limits on possible large-scale irregularities in the universe.

We must now consider the recent claim by CONKLIN [78] to have detected the peculiar motion of the Earth through the microwave background. Working at a wavelength of 3.7 cm, and observing at a fixed declination of $+32°$, CONKLIN found the variation of background temperature with right ascension depicted in Fig. 7. The dotted line indicates the range of right ascension for which no useful measurements were possible because of solar radiation in the sidelobes of the detector. It would appear from these results that in this experiment the Milky Way has been successfully detected at 3.7 cm; the peak A and the trough B correspond to the times when the Milky Way was in the beam (there were in fact two beams about 5 hours apart whose responses were differenced. That is why at B the Milky Way produced a trough.) The galactic radiation must clearly be corrected for before a 24 hour Fourier component representing our peculiar motion can be extracted from the data. CONKLIN estimated this correction factor by taking a galactic map at 75 cm (PAULINY-TOTH and SHAKESHAFT [101]) and transferring it to 3.7 cm with a spectral index of -0.8. The corrected data are shown in Fig. 7 by the ongdashed line.

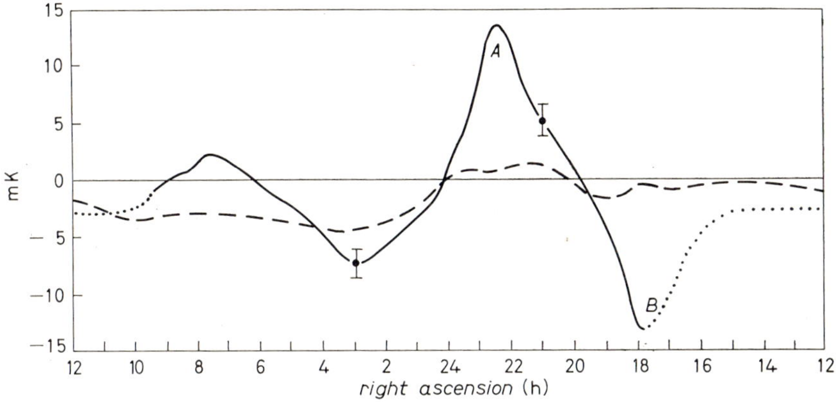

Fig. 7. – Mean of twenty-three observations; the data have been averaged over 1 h of right ascension. The dotted line indicates estimated values: the broken line shows the observations after subtraction of nonthermal contributions. Zero level is arbitrary; the total length of error bars is two standard deviations.

Fourier analysis then gave a 24 hour component of amplitude (1.6 ± 0.75) millidegrees with a maximum at 13 hours right ascension. This result is shown as the vector OB in Fig. 8. For comparison we note the result in the absence of any correction for galactic radiation: 2.05 millidegrees at 19 hours. Had the correction been made with a spectral index of -0.9 the point B in Fig. 8 would have moved to B'.

CONKLIN interprets his result in the following way. The vector OA in Fig. 8 represents the expected result if our peculiar motion were due entirely to the motion of the Earth with respect to the Local Group, as estimated by DE VAUCOULEURS and PETERS [95]. The vector AC represents the additional effect expected from the motion of the Local Group in the Virgo supercluster, also as estimated by DE VAUCOULEURS and PETERS. The final point C lies close to the points B and B'. Accordingly CONKLIN concludes that our motion in the supercluster is real and that the supercluster as a whole has a low-velocity component in the plane of observation.

This result is of such importance that it is necessary to scrutinize Conklin's procedure with great care. Two objections can be made to this procedure. The first, a relatively minor one, is that CONKLIN resolved all the relevant velocities along a plane with declination 32°. However his observations were not made in this plane, but at all right ascensions with this declination, that is, his directions of observation lay on a cone whose axis is the Earth's rotation axis. A comparison of the intensities received in this cone of directions

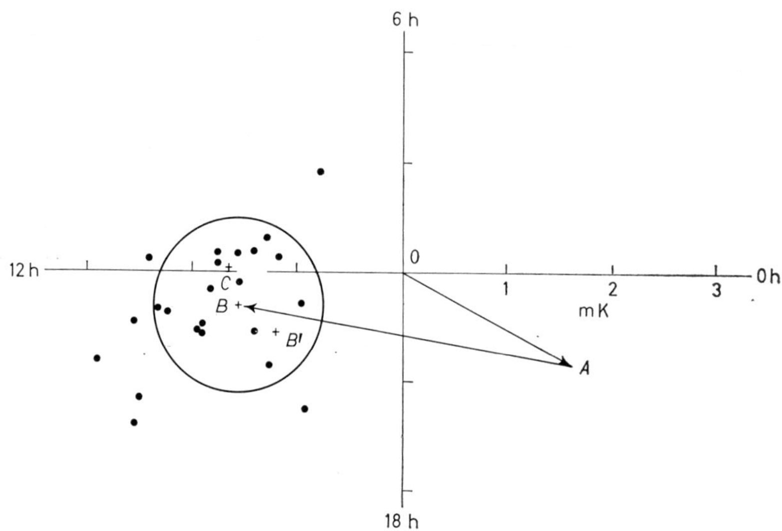

Fig. 8. – Polar diagram of the 24 h component of the background radiation. Corrected values for individual days are shown; the large circle gives the probable error of the mean.

gives the velocity component in the plane perpendicular to the Earth's axis (PARTRIDGE [45]). Thus Conklin's results should be divided by cos 32°, yielding (1.9±0.9) millidegrees at 11 hours, or in terms of velocity (210±100) km s^{-1}. For the same reason, the predicted peculiar velocities should also be projected on to the equatorial plane. The points *A* and *C* in Fig. 8 then become 1.75° mK at 21 h 30 and 2.35° mK at 12 h. The agreement is still impressive, perhaps fortuitously so in view of the uncertainties in the analysis of the supercluster motions.

The second objection is more serious. The correction due to radiation from the Milky Way is clearly very important, but it is quite uncertain whether Conklin's extrapolation from the 75 cm data assuming a nonthermal spectral index is correct. One would expect the HII regions of the Galaxy to make a relatively important thermal contribution at a short wavelength like 3.7 cm; indeed they may make the dominant contribution. Before a reliable result can be obtained for our peculiar motion through the background it is absolutely necessary to have a reliable estimate of the galactic contributions to this background. This will require an accurate map of the Galaxy made at a series of wavelengths as close to 3 cm as possible. Until this has been done we must keep an open mind about the motion of the Local Group in the Virgo supercluster, and indeed about the existence of the supercluster itself.

10. – Relation to Mach's principle.

We now discuss in more detail our basic assumption that a local inertial frame is one relative to which the universe as a whole is nonrotating. The most stringent form of this assumption is Mach's Principle (SCIAMA [102, 103]), one statement of which asserts that the local inertial frame is actually *determined* by the universe as a whole. The nature of the causal relationship involved will be discussed in detail later, but it is clear that not all cosmological solutions of Einstein's field equations satisfy this principle. There are, for instance, expanding models of the Gödel type, in which the universe is homogeneous but rotating. In such models an inertial observer would see the universe as a whole rotating around him. One might expect on dimensional grounds that the angular velocity of this rotation would not greatly exceed the inverse time-scale $10^{-10}\,y^{-1}$ of the expansion, although this restriction is not a logical necessity. It would correspond to a limiting angular velocity of about 10^{-3} s of arc per century. According to Mach's Principle this angular velocity should be zero (apart from a possible small contribution from nearby irregularities, see later).

It is now of interest to ask the purely empirical question: with what precision is a local inertial frame nonrotating with respect to distant matter? This question has been discussed by SCHIFF [104] who concludes that studies of planetary motion are not as accurate for this purpose as consideration of the rotation of the Galaxy. We know that this rotation is relative to a local inertial frame because the Galaxy possesses the flattening associated with the action of centrifugal forces. We also know that the rotation is relative to the outermost stars in the Galaxy which have a much longer period than the bulk of the Galaxy and so are essentially « fixed » (eventually external galaxies and quasars will be used for the comparison). Thus we know that to a precision somewhat greater than the angular velocity of the sun in the Galaxy, the two definitions of nonrotation are in agreement. This precision is somewhat better than one second of arc per century. We certainly expect therefore that the rotational velocity of the sun will also be a velocity through the microwave background.

We could improve on the precision of the comparison if future measurements confirm Conklin's claim that our rotational velocity in the supercluster is also a velocity through the background. For the supercluster (if it exists) also appears to be flattened, so that our rotational velocity is relative to an inertial frame centred on the Virgo cluster. The precision we could then claim would be somewhat greater than our angular velocity in the supercluster, which according to the analyses we have described is about 10^{-3} second of arc per century. This is comparable with the upper limit expected on dimensional considerations.

We can also discuss the problem in terms of arguments that do not rely on our first establishing a local inertial frame. For example, if the universe as a whole were homogeneous but were rotating relative to an inertial frame we would expect to see transverse Doppler shifts in the spectra of distant galaxies, except along the axis of rotation. Since the transverse shift is of second order in v/c this method is not very sensitive, and we cannot do more than rule out a transverse velocity of order c for a source a Hubble radius away. This argument has been used by KRISTIAN and SACHS [105] to set an upper limit to the angular velocity, or vorticity, of the universe of $\sim 1.5 \cdot 10^{-3}$ seconds of arc per century. This limit is comparable with the one we would obtain if the supercluster argument were successful.

A much more powerful limit can be derived along these lines from the upper limit of 300 km s^{-1} for our motion through the microwave background (HAWKING [83]). For in a homogeneous rotating universe we can consider ourselves as rotating about an axis a Hubble radius away. Our rotational velocity must then not exceed 300 km s^{-1} rather than the $3 \cdot 10^5$ km s^{-1} of the previous argument. We thus obtain an upper limit on the vorticity of the universe of about 10^{-6} seconds of arc per century. Even this limit can be sharpened, as HAWKING pointed out. The reason is that the sources on the last-scattering surface would also have a peculiar velocity because of their rotation, and this velocity would exceed our own by the inverse ratio of the cosmic scale factors involved. The result would be a Doppler shift of the radiation temperature which would depend on direction in a manner characteristic of the cosmological model adopted. In a high-density closed model this dependence would be similar to that in flat space and would lead to the limit

$$\omega < \frac{10^{-6}}{z_0} \text{ s of arc per century}.$$

In a low density open model the dependence of Doppler shift on direction would be more complicated, but one can derive the upper limit

$$\omega < \frac{3 \cdot 10^{-39}}{\varrho_0} \text{ s of arc per century},$$

where ϱ_0 is the present density of matter in gm cm^{-3}.

In this last type of model HAWKING obtains a very interesting dependence of radiation temperature on direction. The vorticity in the model leads to an angular dependence $\operatorname{ctg}(\theta/2)\cos\varphi$ and the shear to a dependence $\operatorname{ctg}^2(\theta/2)\cdot\cos 2\varphi$. These relations only hold for $\theta > 5$ minutes of arc, where θ is the angle between the direction of observation and a certain preferred direction defined by the symmetry of the model. The angle φ is the azimuthal angle relative to this preferred direction. It is clear that one would get a hot spot near $\theta = 0$ (cf. NOVIKOV [106], SAUNDERS [107]), and if such a hot spot were found it would be evidence in favour of a low-density open model. The hot spot could presumably be distinguished from one due to a localized irregularity by the φ-dependence.

Finally, HAWKING notes that the upper limit on the vorticity of the universe is much smaller than might be expected on dimensional grounds, and investigates whether this could be due to dissipative process of the type we have already discussed in connexion with the low shear of the universe. The result is negative: viscosity is not effective in damping out rotational velocities very much less than c. Rather reluctantly he concludes that the explanation for the low vorticity may after all be Machian initial conditions on the universe.

Hawking's reluctance is related to the fact that hitherto no clear detailed and rigorous statement has been made of the way distant matter would determine the local inertial frame in accordance with Mach's Principle. How is the contribution of each piece of matter to be weighted, especially in a nonlinear theory like general relativity or any likely reasonable modification of it? A possible answer to this question has recently been found by SCIAMA, WAYLEN and GILMAN [108] following the pioneering work of ALTSHULER [109] and LYNDEN-BELL [110]. The idea is to express the metric at a point of space-time in terms of a Kirchoff-type volume integral over the sources inside a four-volume surrounding the point and an integral of boundary data over the surface. The surface integral replaces the contribution of the sources outside the volume, and also of any source-free contribution which of course should vanish in a Machian solution. This integral representation is possible despite the nonlinearity of Einstein's field equations because the Green's function that acts as the kernel of the representation is actually a *functional* of the metric itself. Each element of source thus contributes linearly to the gravitational potential in the background space determined by all the other sources.

To obtain a unique (retarded) Green's function we must add the requirement that when a first-order change is made in the sources, the new potential may be derived to first-order by using the Green's function of the unperturbed space-time. This is the nearest we can get to regarding Einstein's equations as linear. The linear differential operator defining the Green's function is

$$\Box - 2R^{(\varrho\sigma)}_{(\mu\nu)}, \tag{5}$$

where $\Box$ is the covariant d'Alembertian operator, $R^{\varrho}{}_{\mu}{}^{\sigma}{}_{\nu}$ is the Riemann-Christoffel curvature tensor, and round brackets denote symmetrization. The Kirchoff-type representation then has the form

$$g^{\alpha'\beta'}(x') = 2K\int_{\Omega} G^{-\alpha'\beta'\nu}{}_{\mu}(T^{\mu}{}_{\nu} - \tfrac{1}{2}T^{\lambda}{}_{\lambda}\delta^{\mu}{}_{\nu})[-g(x)]^{\frac{1}{2}}\,\mathrm{d}^4x + \int_{\partial\Omega} G^{-\alpha'\beta'\nu}{}_{\nu}{}^{;\sigma}[-g(x)]^{\frac{1}{2}}\,\mathrm{d}S_{\sigma}\,,$$

where Ω is a four-volume containing the point x', $\partial\Omega$ is its boundary and $G^{-\alpha'\beta'\nu}{}_{\mu}$ is the retarded Green's function associated with the operator (5). The requirement that the surface integral should have a suitable limiting behaviour as Ω approaches the whole of space-time then gives us a well-defined statement of Mach's principle (GILMAN [111]). It is intuitively plausible that in this form Mach's principle would require that the vorticity of a homogeneous universe should vanish, but this remains to be verified. This formulation has the advantage that it can be applied in an irregular universe, so that the influence of nearby sources on the choice of inertial frames can be completely specified.

11. – The helium problem.

The present state of the helium problem is highly confused, and so we shall give only a brief account of it in these lectures (cf. TAYLER [112]). The problem itself consists of two parts: *a*) to determine the abundance of helium in different parts of the universe, *b*) to discover what proportion of this helium has a cosmological origin, that is, was formed by thermonuclear processes in the hot big-bang. The reason why attention is focussed on helium rather than all the elements heavier than hydrogen is two-fold: i) the helium abundance in some regions of the Galaxy is difficult to account for by formation processes occurring after the birth of the Galaxy. The difficulty is not so great for heavier elements. ii) A straightforward calculation of helium formation in the hot big-bang leads to an abundance of the same order as referred to in i). By contrast, a negligible amount of heavier elements (with one or two possible exceptions) is formed at this stage.

The confusion arises when one looks in more detail both at abundance measurements and at the cosmological formation process. Confining oneself to our Galaxy alone there is now substantial evidence that the helium abundance may vary appreciably with position in the Galaxy, in some places to values far below the amounts expected from the straightforward cosmological calculation. This situation is described by G. R. BURBIDGE. On the other hand one can think of many complicating factors which would change substantially the amount of helium formed cosmologically. It would even be possible to have fluctuations in the amount of helium produced in different parts of the early universe, fluctuations which might conceivably survive to some extent until after the formation of our Galaxy. To indicate the possibilities here it is sufficient to mention that in the Robertson-Walker models (which are used for the straightforward calculation of the helium formation) the size of a causally connected region at the time of helium formation is so small that the amount of material it contains is of the order of one solar mass. Thus our present Galaxy would contain material coming from about 10^{11} unrelated regions. It would not be surprising then to find substantial small-scale helium variations in the Galaxy even if most of it were of cosmological origin. We could avoid this conclusion only if mixing were a highly efficient process at some stage after the helium was formed, or if the early universe deviated from a Robertson-Walker model so much that it contained essentially no particle horizons, and transport processes smoothed out any inhomogeneities (as MISNER [67*a*] has advocated).

With these warnings we shall now describe briefly the straightforward calculation, and then indicate how the results would be changed by introducing some of the complicating factors. We first recall our approximate relation between temperature and time which holds in the early radiation-dominated stages of a Robertson-Walker model:

$$T = \frac{10^{10}\ {}^\circ\mathrm{K}}{t^{\frac{1}{2}}\ (\mathrm{s})}.$$

Thus for times less than 1 s the temperature would exceed 10^{10} °K. This is beyond the threshold for the creation of electron-positron pairs. Moreover, as HAYASHI [113] pointed out, the weak interactions would then have been strong enough to maintain thermal equilibrium between neutrons, protons, electron pairs and electron-type neutrino pairs, via the reactions

$$\mathrm{p} + \bar{\nu} \leftrightarrow \mathcal{N} + \mathrm{e}^+ ,$$

$$\mathrm{p} + \mathrm{e}^- \leftrightarrow \mathcal{N} + \nu .$$

Below 10^{10} °K the weak interactions can no longer maintain the neutrons in statistical balance with the protons because the concentration of electron pairs is beginning to drop abruptly. The $\mathcal{N}/\mathrm{p}$ ratio is then frozen in, until a few hundred seconds have passed and neutron decay begins to be appreciable. This frozen-in ratio, corresponding to thermal equilibrium at a temperature somewhat below 10^{10} °K, is about 15 per cent (as follows from the Boltzmann relation and the neutron-proton mass-difference).

The first nuclear reaction of importance occurs about 100 s later when the temperature has dropped to 10^9 °K. This reaction is

$$\mathcal{N} + P \rightarrow D + \gamma . \tag{6}$$

It proceeds at earlier times too, but then the radiation temperature is so high that the deuterons are rapidly photodisintegrated. The reaction (6) is followed by a series of rapid reactions, of which the end-product is mainly helium Typical examples of such reactions are

$$\begin{aligned} D + D &\rightarrow {}^3\text{He} + \mathcal{N} , \\ &\searrow {}^3\text{H} + p , \\ {}^3\text{He} + \mathcal{N} &\rightarrow {}^3\text{H} + p , \\ {}^3\text{H} + D &\rightarrow {}^4\text{He} + \mathcal{N} . \end{aligned}$$

All these reactions are much faster than (6), so the amount of helium formed depends essentially on how many deuterons are formed by (6). If the material density is very low at $t \sim 100$ s this reaction never has a chance to get going, and so little helium is formed. For higher densities at $t \sim 100$ s the reaction rate of (6) increases and so more helium is formed. The amount of helium formed soon levels off, however, because once the reaction rate of (6) exceeds the expansion rate of the universe at that time essentially all the frozen-in neutrons are incorporated into helium nuclei. If we plot the helium abundance as a function of s, the entropy per baryon introduced on p. 198, we will obtain a curve that first rises as s decreases and then has a plateau. This plateau is only departed from when s is so low that the universe is still matter-dominated at 10^{10} °K. In this case most of the matter would be turned into helium.

The dependence of helium abundance on s is shown in Fig. 9. Detailed calculations show that the plateau abundance continues for a range of about six powers of 10 in s, a range which includes the values of s expected for the actual universe (with a present black-body temperature of 2.7 °K and $2 \cdot 10^{-29} > \varrho > 2 \cdot 10^{-31}$ g cm^{-3}). Even without detailed calculations we can estimate the height of the plateau from our estimated 15 per cent for the frozen-in ratio of neutrons. Since there are as many protons as neutrons in a helium nucleus, and since on the plateau all neutrons are incorporated into helium nuclei, the final helium abundance on the plateau should be 30 per cent by mass. This is comparable to the observed helium abundance in many parts of the Galaxy.

A determination of the exact value of the plateau abundance and the (small) dependence on s in this range requires accurate knowledge of the weak

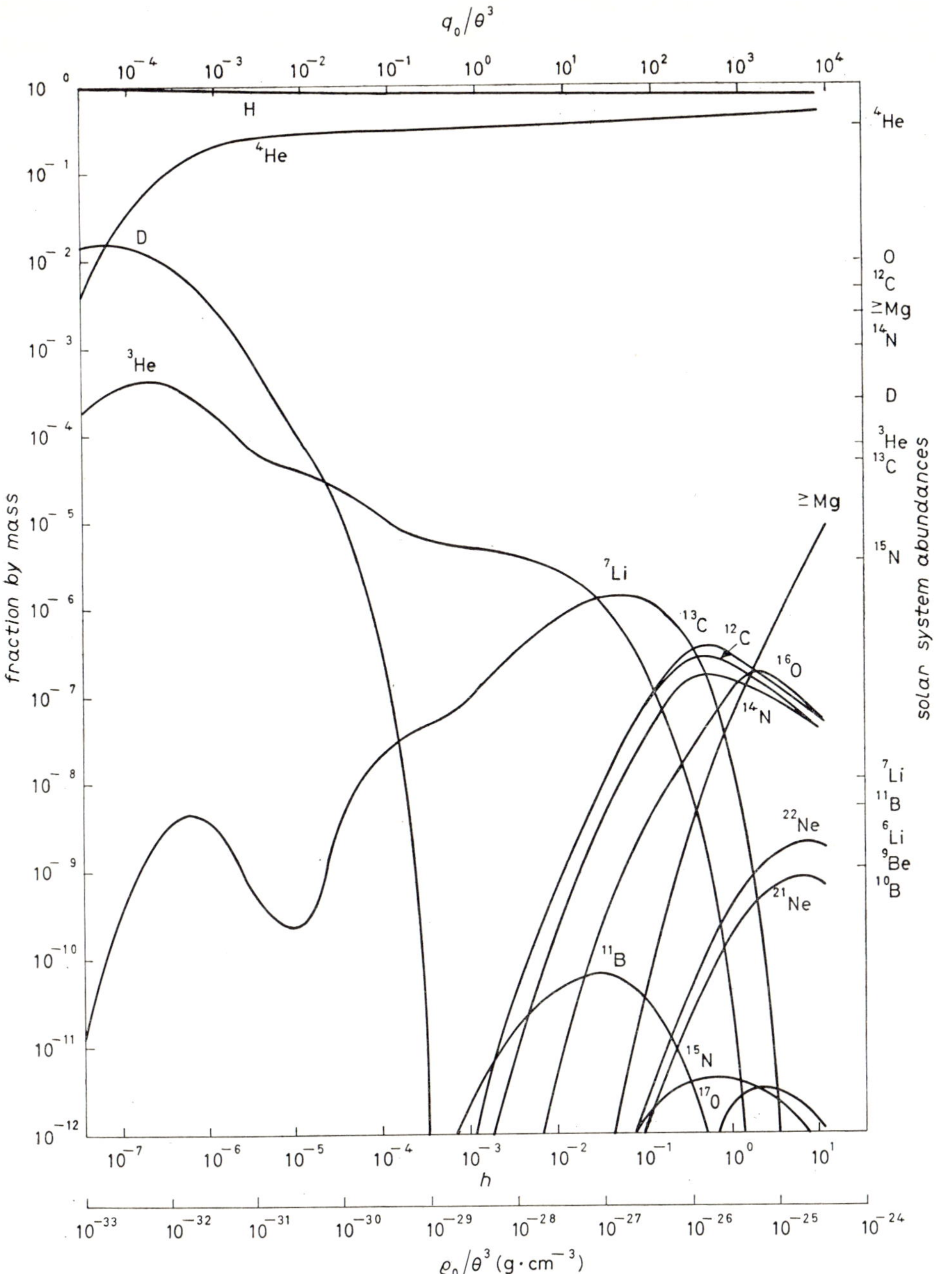

Fig. 9. – Element production in the hot big-bang. The parameter h is inversely proportional to the entropy per baryon s.

interactions and a numerical calculation of all the reactions between the light particles, only a few of which were listed earlier. Hayashi's [113] original value for the plateau abundance was 40 per cent, which is rather higher than the observations permit. A later series of calculations by ALPHER, FOLLIN and HERMAN [114], HOYLE and TAYLER [115], and SMIRNOV [116] gave about 35 per cent, which is still rather high. If we descend below the plateau to avoid this difficulty, we run into another one, namely an unacceptably high value for the present day black-body temperature.

This was the situation until PEEBLES [117] published another calculation in which he improved the accuracy of the numerical integrations. He then obtained 27 per cent for the plateau abundance, in good agreement with many of the measurements in the Galaxy. His results were confirmed by WAGONER, FOWLER and HOYLE [118] who made a much more elaborate calculation. This calculation took into account all the reactions that can occur between the light elements and used the most up-to-date values available for their cross-sections. The build-up of heavier elements was also considered (144 different reactions being included) with results that will be described later. The final helium abundance as a function of s is plotted in Fig. 9. The region above the plateau, where the universe is matter-dominated during the nuclear reactions, is not included because the computer programme would have required considerable modification to handle this regime.

The plateau begins at $s \sim 10^4 k$ with a helium abundance of 40 per cent and ends at $s \sim 2.5 \cdot 10^{10} k$ at a level of 24 per cent. The material density at a given temperature can thus change by a factor $2.5 \cdot 10^6$ for a change in the helium abundance of less than a factor 2. The observed helium abundance (even if it were constant throughout the Galaxy) would thus not be a sensitive indicator of s.

WAGONER, FOWLER and HOYLE considered that their calculation of the helium abundance was accurate to 1 per cent. However, recent experimental evidence (CHRISTENSEN, NIELSEN, BAHNSEN, BROWN and RUSTAD [119]) suggests that the half-life of the neutron is about 10 per cent less than was previously believed. The resulting increase in the weak interaction coupling constant would imply that neutrons remain in thermal equilibrium for a longer time, so that the frozen-in abundance of neutrons would be reduced. The final helium abundance would also be reduced—by about 10 per cent according to TAYLER [112].

We must also mention briefly the results of WAGONER, FOWLER and HOYLE for the build-up of other elements. Their complete results are shown in Fig. 9. It will be seen that for $s \sim 10^{10} k$ the D and ^{3}He abundances agree well with observation and that even ^{7}Li could be brought into line if one admitted localized irregularities in the distribution of matter in which s were reduced by a factor ~ 100. This consideration would support a low-density universe.

However, as WAGONER, FOWLER and HOYLE point out, there are other, non-cosmological mechanisms for making (and destroying) these light elements, and the present situation is quite unclear.

Figure 9 also gives the results for the primeval build-up of the elements heavier than lithium. The main difference from earlier calculations is that previously the absence of stable nuclei of atomic weights 5 and 8 completely prevented the build-up of heavier elements. Reactions are now known that bridge these gaps, but even so far too few heavy elements are formed. We must therefore still appeal to noncosmological processes for the formation of these elements, processes such as nuclear reactions in hot stars or massive objects. This is not really surprising because there is now considerable evidence that the concentration of heavy elements in the Galaxy has changed with time. Old stars, for instance, tend to have a lower metal content than young stars. The correlation of element abundance, stellar type, and motion and position in the Galaxy has become an important branch of astrophysics which should lead to a better understanding of the formation of the heavier elements. Since this is probably not a cosmological problem we shall have to pass it by.

So far we have been discussing what I have called the straightforward calculation of helium formation. We must now consider some complications which might arise. The first is the possibility that the universe is filled with electron-type neutrinos or anti-neutrinos with about 10^{10} neutrinos per baryon. Such a neutrino flux could not be detected to-day, so that there is no empirical objection to introducing this hypothesis. At first sight it might seem that there should be no theoretical objection either, since we have already accepted the idea that there might be about 10^{9} photons per baryon in the universe. There is one important difference, however, in the two situations. Irreversible processes can occur that create entropy and this entropy can be in the form of photons; photon number is not conserved. By contrast, there is believed to be a conservation law for leptons (neutrinos, electrons, muons) which in the present context would mean that no process occurring in the universe can change the net number of neutrinos per baryon. This number is fixed once and for all. Admittedly we do not understand what determines this number but most physicists, I believe, would be reluctant to assume that it is of the order of 10^{10} rather than of the order of unity (or zero), unless it were absolutely necessary.

We have still to explain why a preponderance of neutrinos would affect the helium production. The reason is that reactions of the type

$$\nu + \mathcal{N} \rightarrow p + e^-$$

would go very fast, converting most of the neutrons into protons. There would then be very few neutrons left to undergo the key reaction

$$\mathcal{N} + p \rightarrow D + \gamma ,$$

and very little helium would be formed.

Similarly if there were a preponderance of antineutrinos, the protons would be rapidly removed by the reaction

$$\bar{\nu} + p \rightarrow \mathcal{N} + e^{+} ,$$

and there would be few protons left to undergo the key reaction. Moreover in this case the neutrons could not decay at first because the antineutrinos that would be emitted in this decay are subject to the exclusion principle and would not find an empty level to occupy (if the initial preponderance of antineutrinos were sufficiently high). Since the anti-neutrinos would be diluted by the expansion of the universe the time would come when the neutrons could decay, but by then the temperature would be too low for the key reaction to occur. Again the result would be very little helium.

Another method of suppressing helium is to introduce anisotropy into the universe at the formation stage. The main effect of the anisotropy would be to change the *time-scale* of the expansion. There are two separate problems here. The first is the essentially nuclear problem of calculating the dependence of the helium abundance on the time-scale. This problem has been solved by PEEBLES [117] with the result shown in Fig. 10. The second problem is

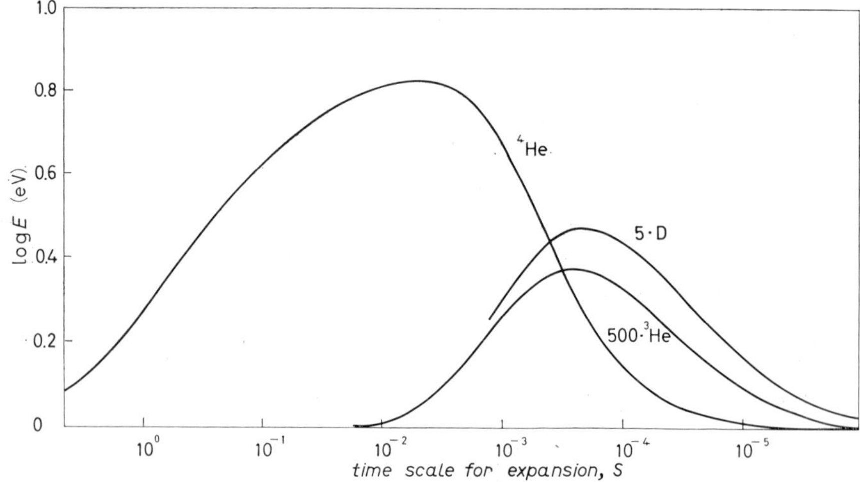

Fig. 10. – Element abundances by mass, produced in the early Universe, assuming present mean density $1.8 \cdot 10^{-29}$ gm/cm³ and fireball radiation temperature 3 °K.

the cosmological one of relating assumptions about the anisotropy to the resulting change of time-scale (HAWKING and TAYLER [120], THORNE [79], CARSWELL [121], GREENSTEIN [122]). In this connexion it will also be necessary to understand thoroughly the dissipation of anisotropy by viscous processes, a problem we have already touched on in discussing the isotropy of the black-body background. A further question is whether one can have substantially different expansion time-scales in regions outside each others' particle horizons (if particle horizons do in fact exist in the actual universe), and if so how well the resulting gradient in the helium abundance would survive later mixing processes. It would be particularly valuable to know the helium abundance in the intergalactic medium, a question that may be resolved when systematic ultra-violet observations become possible from satellites orbiting the Earth above the atmosphere (SCHEUER [123]).

Finally we should mention that some cosmologists have studied the effect on helium formation of modifications of the general theory of relativity made originally for other purpose (DICKE [124], GREENSTEIN [125]). In view of all the other uncertainties in the helium problem we shall not go into such questions here.

12. – High-energy interactions with the black-body radiation.

From a laboratory point of view 3 °K is a very low temperature. Indeed to measure it the microwave observers had to use a reference terminal immersed in liquid helium. Nevertheless from an astrophysical point of view 3 °K is a very high temperature. A universal black-body radiation field at this temperature would contribute an energy density everywhere $\sim 1\ \text{eV}\,\text{cm}^{-3}$. This is comparable with the energy density in our Galaxy of the various modes of interstellar excitation—starlight, cosmic rays, magnetic fields and turbulent gas clouds. So even in our Galaxy the cosmological background radiation would be for many purposes as important as the well-known energy modes of local origin. In intergalactic space, however, these localized energy densities probably drop off by a factor between 100 and 1000, whereas the black-body component would maintain its energy density at $1\ \text{eV}\,\text{cm}^{-3}$. It would thus become the dominant form of energy density in intergalactic space, with the possible exception of the rest-energy density of matter itself. Even then, in a low density universe with $n \sim 10^{-7}\ \text{cm}^{-3}$ the present energy density in the radiation field would be as much as 1 per cent of the rest-energy density in matter.

Under these circumstances we might expect the black-body radiation field to exert a noticeable influence on astrophysical processes, especially on those involving high-energy particles or radiation that can interact with the black-body photons. This is the question we shall study in this Section. While no known effect can definitely be attributed to this interaction, it is possible to make well-defined predictions that are amenable to future test. At the same time the absence of any clear effect to date puts an upper limit of about 10 °K on the temperature of a possible universal black-body radiation field.

It will be useful for what follows to state the following properties (in round numbers) of a 3 °K radiation field. The number density of photons is 10^3 cm^{-3}, and since the total energy density is 1 eV cm^{-3}, the mean energy per photon is 10^{-3} eV. These estimates are useful for making quick calculations of many of the effects of the radiation field, without having to consider in detail the full energy range of the photons in the black-body spectrum.

The phenomena with which we shall be concerned involve interactions with cosmic-ray protons (and heavier nuclei), γ-rays and electrons. We shall consider these in turn.

12·1. *Cosmic-ray protons.* – The energy spectrum of cosmic rays is shown in Fig. 11. It is of power-law type with a roughly constant spectral index for energies in the range between 10^9 and 10^{20} eV. It is not yet quite certain that the highest energy cosmic rays, which are of concern here, are mainly protons, but for convenience of exposition we shall assume it. A proton of 10^{20} eV has a Lorentz factor γ $(=(1-v^2/c^2)^{-\frac{1}{2}})$ of 10^{11}. In the rest frame of such a proton one of our black-body photons of energy 10^{-3} eV becomes a photon of energy 10^8 eV. Such an energetic photon striking a stationary proton would be close to the threshold for producing a pion (rest mass $\sim$137 MeV). This means that from the terrestrial point of view a cosmic-ray proton of 10^{20} eV could collide with a black-body photon, produce a pion, and so be degraded in energy. Now cosmic rays with energies in the range $(10^{18}\div10^{20})$ eV are usually thought to have an extragalactic origin because

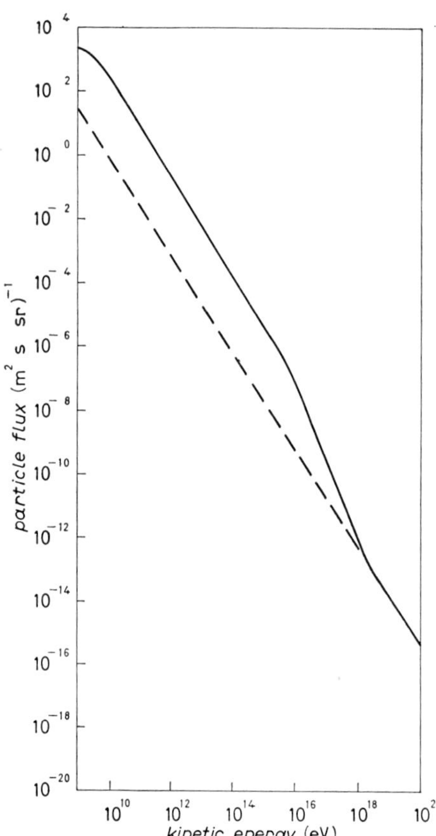

Fig. 11. – The energy spectrum of the total cosmic radiation. The dashed line represents a possible extragalactic component.

they cannot be confined by the magnetic field of the Galaxy, and so their flux near the Earth cannot be built up for the storage time of a few million years that is characteristic of lower-energy cosmic rays. Thus if all the cosmic-rays originated in the Galaxy we might expect a sharp drop in the flux at about 10^{18} eV. This sharp drop is not found (Fig. 11) and indeed the spectrum appears to flatten slightly in this region. This suggests, although it does not prove, that there exists an intergalactic flux of cosmic rays with the flatter spectrum, that dominates the galactic flux locally at energies above 10^{18} eV.

Now the observed spectrum shows no signs of a cut-off right out to cosmic-ray energies of 10^{20} eV, which is the maximum energy that existing techniques can detect. This limit is tantalizingly close to the threshold for attenuation by the black-body background. Once it sets in this attenuation is very severe (GREISEN [126]). Recent calculations by STECKER [127] give the following lifetimes t for protons of various energies (STECKER [128] has also made calculations for the disruption of heavier nuclei):

$$E < 6\cdot 10^{19}\ \text{eV}\,, \qquad t > 10^{10}\ \text{y}\,,$$

$$E \sim 10^{20}\ \text{eV}\,, \qquad t \sim 10^{9}\ \text{y}\,,$$

$$E > 10^{20}\ \text{eV}\,, \qquad t > 5\cdot 10^{7}\ \text{y}\,.$$

Experiments are now being conducted that should enable the spectrum to be extended to the range $(10^{21} \div 10^{22})$ eV. If there are still no signs of attenuation found, and if the black-body radiation really does fill intergalactic space, it would mean that these ultra-high-energy cosmic rays have a lifetime less than 50 million years, and so could have come from no further than 50 million light-years (for comparison, the Virgo cluster of galaxies is 30 million light-years away). On the other hand if severe attenuation is found it would strongly suggest that these cosmic rays come from still greater distances.

12·2. *Cosmic γ-rays.* – We shall now see that high-energy cosmic γ-rays would also be degraded by interaction with black-body photons. We cannot now look at the collision from the rest frame of the high-energy object, but we can use a frame in which the γ-ray and the black-body photon have the same energy. If the energy of the γ-ray is E relative to the Earth, and we transform to a frame moving away from the γ-ray with velocity v, then the energy of the γ-ray becomes E/γ $(\gamma = (1 - v^2/c^2)^{-\frac{1}{2}})$ and the energy of a black-body photon moving towards the γ-ray becomes $10^{-3}\gamma$ eV. Setting these equal we have

$$\frac{E}{\gamma} = 10^{-3}\,\gamma\,.$$

Let us now choose E so that the common energy of each photon is just the rest energy of an electron. We would then be at the threshold for electron-pair creation. This requires that

$$\frac{E}{\gamma} = 10^{-3}\gamma = m_e c^2 .$$

Eliminating γ we have

$$10^{-3}E = (m_0 c^2)^2 ,$$

or

$$E = 2.5 \cdot 10^{14}\ \text{eV} .$$

At this energy and above, cosmic γ-rays would be degraded by the black-body photons through pair creation.

The effect is a large one (JELLEY [129], GOULD and SCHREDER [130]). The cross-section σ for the process near threshold is roughly the square of the classical electron radius $\sim 10^{-25}$ cm², and so the mean free path, which would be about $1/n\sigma$ where n is the density of the black-body photons, becomes $\sim 10^{22}$ cm. This is smaller than the size of the Galaxy. This cut-off should continue right out to 10^{21} eV as Fig. 12 shows. Similarly, discrete γ-ray sources would be subject to this strong absorption.

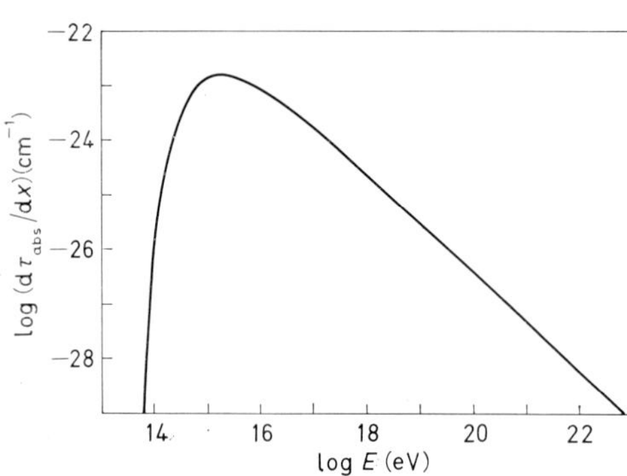

Fig. 12. – Absorption probability per unit path-length as a function of energy by means of the process $\gamma + \gamma' \to e^+ + e^-$ for high-energy photons traversing a black-body photon gas at 3.5 °K.

Unfortunately no γ-rays above an energy of 100 MeV have yet been reliably detected from cosmical objects or in the form of a diffuse background. The energy range above 10^{14} eV is under experimental study at the moment but it is not clear whether the unattenuated flux would be large enough to be detectable.

12'3. *Relativistic electrons.* – This is the most interesting of the interaction processes because many of the relativistic electrons concerned emit a substantial flux of radio noise through the synchrotron process (magnetic bremsstrahlung), and so can be studied by the methods of radio astronomy.

The dominant interaction process is inverse Compton scattering (inverse because in the terrestrial frame of reference the photon gains energy and the electron loses energy). By working, as before, in the rest frame of the cosmic-ray particle, we see that on average an incident photon of energy E in the terrestrial frame gains an energy E' given by

$$E' \sim \gamma^2 E ,$$

where γ is the Lorentz factor of the electron.

Consider in the first place the electrons that are responsible for the diffuse radio emission of our Galaxy. A typical energy for such an electron would be 1 GeV. Its γ would then be 2000, and with $E \sim 10^{-3}$ eV we see that the scattered photon would be raised in energy to about 4 kilovolt. This takes us right into the X-ray region at a wavelength ~ 5 Å. The Galaxy would thus be an X-ray source (HOYLE [131], GOULD [132]). We must ensure, of course, that its strength is compatible with the observed X-ray background. Now it turns out that the inverse Compton effect and the synchrotron mechanism are essentially the same process of a relativistic electron emitting radiation while under the influence of an electromagnetic field. (GINZBURG and SYROVATSKY [133], FELTEN and MORRISON [134]). Consequently they lead to a similar rate of energy transfer from the electrons if the energy density in the photon field and the static magnetic field are comparable. This is just the case for our Galaxy, as we have seen. Thus the known energy flux in the galactic radio background should be of the same order as the energy flux in the emitted X-rays. This flux would be about 1 per cent of the observed isotropic X-ray background (Fig. 13).

The proposed black-body radiation field thus passes this test. The margin by which it does so is not as large as it might seem, however, because the rate of energy transfer to the X-rays is proportional to the Compton collision rate, and so to the photon density in the black-body radiation field, that is, to the third power of its temperature. Moreover, if the radiation field had a higher temperature, the mean energy of its photons would be greater. Accordingly an electron of lower energy would suffice to produce a given X-ray energy. We must therefore allow for the fact that in our Galaxy there are more electrons at the lower energies. The net result of all this is that if the black-body background had a temperature of only 10 °K, the X-ray flux from the galactic disc would exceed the observed X-ray flux. This would be inconsistent with the fact that this X-ray flux is isotropic to a precision of about 10 per cent (SEWARD, CHODIL, MARK, SWIFT and TOOR [135], BOLDT, DESAI, HOLT and SERLEMITSOS [136]). However, recently (COOKE, GRIFFITHS and POUNDS [137]) a 100 per cent increase in the X-ray background has been detected along the galactic equator. The origin of this increase may have

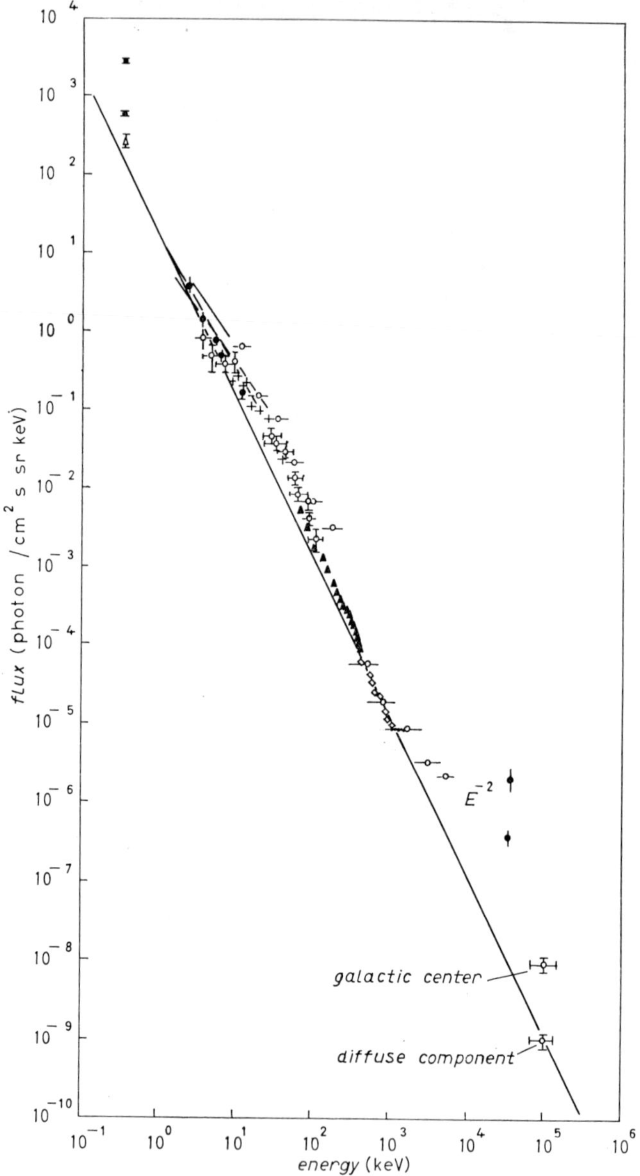

Fig. 13. – The diffuse X-ray background.

nothing to do with our present discussion, but the data make it clear that a black-body radiation field with $T > 10$ °K in the Galaxy would be hard to accept. The question whether the excess 8 °K far infra-red radiation discovered by SHIVANANDAN, HOUCK and HARWIT [46] and by HOUCK and HARWIT [47], and discussed on p. 193, could fill the Galaxy is unclear (O'CONNELL and VARMA [138]); much depends on the angular scale of the excess X-ray flux.

There is another important aspect to this problem and that is the resulting energy drain on the electrons. For $T \sim 3$ °K the drain due to the production of X-rays is of the same order as that due to the radio emission (or to Compton collisions with starlight photons which also have an energy density ~ 1 eV cm^{-3}). If we tried to increase T however, this drain would increase very rapidly, roughly like T^4. It would soon become so high that it would be very difficult to understand where the electrons got their energy from in the first place. For $T \sim 10$ °K, for instance, a 1 GeV electron would lose half its energy in 10 million years, instead of the billion years characteristic of the synchrotron process alone. A further point is that the observed energy spectrum of the electrons shows no sign of significant energy losses out to at least 300 GeV, at which energy the lifetime of an electron in the 3 °K radiation field would be only about $3 \cdot 10^4$ y. This tells us that these electrons must leak out of the local trapping region in a shorter time than this. In a 10 °K radiation field the leakage time would have to be less than 300^4 y. This would have very severe implications also for the propagation of cosmic-ray protons and heavier nuclei.

These considerations take on even greater importance when we apply them to a radio source with a substantial redshift z. The temperature of the black-body radiation at such a source would be increased by the factor $1+z$ over its local value, and the all-important energy density by $(1+z)^4$. The balance between the synchrotron and Compton effects that holds in our Galaxy would then be disturbed, unless the magnetic fields in radio galaxies are much greater than in our Galaxy. In any case the Compton lifetime of a relativistic electron of given energy would be 81 times shorter in a radio source with a redshift of 2 than in our own Galaxy. This effect may contribute significantly to the evolution of radio sources which is so important for analysing the source counts. The situation would be even more extreme for a radio source with a redshift of, say, 4. At such a redshift the Compton lifetime of an electron would be 625 times less than that of a contemporary electron. These interactions would also give rise to X-rays, which might make the dominant contribution to the observed X-ray background, especially if one allows in addition for the evolutionary increase in the power or number of radio sources with large red-shifts (BERGAMINI, LONDRILLO and SETTI [139], BRECHER and MORRISON [140], REES and SETTI [141], LONGAIR [142], LONGAIR and SUNYAEV [143]).

It is also interesting to consider the Compton collisions that might be occurring in intergalactic space (FELTEN and MORRISON [134]). To estimate the resulting X-ray flux we observe that the path-length of a line of sight through the universe is about 10^5 times that through the Galaxy. Since the galactic X-ray flux due to the Compton process is expected to be about one per cent of the observed background, an intergalactic flux of relativistic electrons which is 10^{-3} of the galactic flux would suffice to account for the observed X-ray background. This looks an attractive explanation at first sight. However, we must not overlook the losses arising from the inverse Compton process itself. These losses lead to a lifetime for the electrons of only about a billion years, one tenth of the expansion time-scale. It would therefore be necessary to maintain the intergalactic electron flux at 10^{-3} of the galactic flux against this heavy loss-rate. The energetic problem is very severe, and no convincing resolution of it has yet been forthcoming.

At the time of writing up these lectures (September 1969) the most likely explanation of the X-ray background, at any rate for energies above 1 keV, is inverse Compton collisions between relativistic electrons in radio galaxies and the black-body radiation field, the majority of the X-rays coming from redshifts in the range from 3 to 5. However, several other explanations have been proposed, and a full survey of the problem has been given by SETTI and REES [144] in their report to the IAU Symposium No. 37 on nonsolar gamma- and X-ray astronomy, which was held in Rome in May 1969. The Symposium proceedings contain many papers about the X-ray background which are worth studying. In particular the question is discussed whether the soft X-ray background at $\frac{1}{4}$ keV is an extrapolation of the higher-energy background or represents a new component that could be bremsstrahlung from a hot intergalactic gas with the magic density (cf. pp. 190-191, and BERGERON [33], BUNNER, COLEMAN, KRAUSHAAR, MCCAMMON, PALMIERI, SHILEPSKY and ULMER [145]). Intergalactic absorption processes may also be important in this connection. It seems likely that in the future X-ray astronomy, radio astronomy and the black-body background will become inextricably woven with astrophysical cosmology, to the ultimate benefit of all four subjects.

REFERENCES

[1] A. SANDAGE: *Astrophys. Journ.*, **152**, 149 (1968); *Observatory*, **88**, 91 (1968).
[2] E. P. HUBBLE: *The Realm of the Nebulae* (Yale, 1936).
[3] A. EINSTEIN and W. DE SITTER: *Proc. Nat. Acad. Sci.*, **18**, 213 (1932).
[4] A. EINSTEIN: *Sitz. Ber. Preuss. Akad. Wiss.*, 142 (1917).
[5] W. DE SITTER: *Mont. Not. Roy. Astr. Soc.*, **78**, 3 (1917).
[6] H. BONDI and T. GOLD: *Mont. Not. Roy. Astr. Soc.*, **108**, 252 (1948).
[7] F. HOYLE: *Mont. Not. Roy. Astr. Soc.*, **108**, 372 (1948).
[8] W. RINDLER: *Mont. Not. Roy. Astr. Soc.*, **116**, 662 (1956).
[9] M. RYLE and P. A. G. SCHEUER: *Proc. Roy. Soc.*, A **230**, 448 (1955).

[10] P. A. G. SCHEUER: *Stars and Stellar Systems.* - Vol. **9**: *Galaxies and the Universe,* to be published.
[11] M. RYLE: *Ann. Rev. Astr. and Astr.*, **6**, 249 (1968).
[12] G. G. POOLEY and M. RYLE: *Mont. Not. Roy. Astr. Soc.*, **139**, 515 (1968).
[13] W. DAVIDSON: *Mont. Not. Roy. Astr. Soc.*, **32**, 389 (1966).
[14] M. S. LONGAIR: *Mont. Not. Roy. Astr. Soc.*, **133**, 421 (1966).
[15] M. ROWAN-ROBINSON: *Mont. Not. Roy. Astr. Soc.*, **138**, 445 (1968).
[16] D. W. SCIAMA and M. J. REES: *Nature*, **211**, 1283 (1966).
[17] M. SCHMIDT: *Astrophys. Journ.*, **151**, 393 (1968).
[18] M. SCHMIDT: *Ann. Rev. Astr. and Astr.*, **7**, 527 (1969).
[19] R. J. GOULD: *Ann. Rev. Astr. and Astr.*, **6**, 195 (1968).
[20] P. A. G. SCHEUER: *Nature*, **207**, 963 (1965).
[21] J. E. GUNN and B. A. PETERSON: *Astrophys. Journ.*, **142**, 1633 (1965).
[22] G. R. BURBIDGE and E. M. BURBIDGE: *Quasi-Stellar Objects*, Chap. 12 (San Francisco and London, 1967).
[23] A. A. PENZIAS and R. W. WILSON: *Astrophys. Journ.*, **156**, 799 (1969).
[24] R. J. GOULD and D. W. SCIAMA: *Astrophys. Journ.*, **140**, 1634 (1964).
[25] M. J. REES, D. W. SCIAMA and G. SETTI: *Nature*, **217**, 326 (1968).
[26] G. B. FIELD, P. SOLOMON and E. J. WAMPLER: *Astrophys. Journ.*, **145**, 351 (1966).
[27] J. N. BAHCALL and E. E. SALPETER: *Astrophys. Journ.*, **142**, 1677 (1965); **144**, 847 (1966).
[28] M. J. REES and D. W. SCIAMA: *Astrophys. Journ.*, **145**, 6 (1966); *Nature*, **212**, 1001 (1966); *Astrophys. Journ.*, **147**, 353 (1967).
[29] D. W. SCIAMA: *Quart. Journ. Roy. Astr. Soc.*, **5**, 196 (1964).
[30] R. J. GOULD and G. R. BURBIDGE: *Astrophys. Journ.*, **138**, 696 (1963).
[31] G. B. FIELD and R. C. HENRY: *Astrophys. Journ.*, **140**, 1002 (1964).
[32] R. J. WEYMANN: *Astrophys. Journ.*, **147**, 887 (1967).
[33] J. BERGERON: *Astr. and Astr.*, **3**, 42 (1969).
[34] R. A. SUNYAEV: *Astrophys. Lett.*, **3**, 33 (1969).
[35] D. W. SCIAMA: *Rendiconti S.I.F.*, Course XXXV (New York, 1966). p. 428.
[36] S. W. HAWKING and G. F. R. ELLIS: *Astrophys. Journ.*, **152**, 25 (1968).
[37] J. N. BAHCALL and R. M. MAY: *Astrophys. Journ.*, **152**, 37 (1968).
[38] F. T. HADDOCK and D. W. SCIAMA: *Phys. Rev. Lett.*, **14**, 1007 (1965).
[39] P. MORRISON: *Astrophys. Journ.*, **157**, L73 (1969).
[40] M. J. REES: *Astrophys. Lett.*, **4**, 61 (1969).
[41] A. A. PENZIAS and R. W. WILSON: *Astrophys. Journ.*, **142**, 419 (1965).
[42] R. H. DICKE, P. J. PEEBLES, P. G. ROLL and D. T. WILKINSON: *Astrophys. Journ.*, **142**, 414 (1965).
[43] G. GAMOW: *Rev. Mod. Phys.*, **21**, 367 (1949).
[44] G. B. FIELD: *Riv. Nuovo Cimento* (Report of Inaugural Meeting of the European Physical Society, Florence, 1969).
[45] R. B. PARTRIDGE: *Amer. Sci.*, **57**, 37 (1969); private communication (1969).
[46] K. SHIVANANDAN, J. R. HOUCK and M. HARWIT: *Phys. Rev. Lett.*, **21**, 1460 (1968).
[47] J. R. HOUCK and M. HARWIT: *Astrophys. Journ.*, **157**, L45 (1969).
[48] D. W. SCIAMA: *Nature*, **211**, 277 (1966).
[49] F. HOYLE and N. C. WICKRAMASINGHE: *Nature*, **218**, 1126 (1968).
[50] J. V. NARLIKAR and N. C. WICKRAMASINGHE: *Nature*, **217**, 1236 (1969).
[51] J. R. SHAKESHAFT and A. S. WEBSTER: *Nature*, **217**, 339 (1968).
[52] T. GOLD and F. PACINI: *Astrophys. Journ.*, **152**, L 115 (1968).

[53] Y. N. Pariiski: *Sov. Astr.*, **12**, 219 (1968).
[54] A. M. Wolfe and G. R. Burbidge: *Astrophys. Journ.*, **156**, 345 (1969).
[55] A. A. Penzias, J. Schraml and R. W. Wilson: *Astrophys. Journ.*, **157**, L 49 (1969).
[56] M. G. Smith and R. B. Partridge: *Astrophys. Journ.*, **159**, 737 (1970).
[57] C. Hazard and E. E. Salpeter: *Astrophys. Journ.*, **157**, L87 (1969).
[58] G. B. Field and J. L. Hitchcock: *Astrophys. Journ.*, **146**, 1 (1966); *Phys. Rev. Lett.*, **16**, 817 (1966).
[59] P. Thaddeus and J. F. Clauser: *Phys. Rev. Lett.*, **16**, 819 (1966).
[60] A. McKellar: *Publ. Dom. Ap. Obs.*, **7**, 251 (1941).
[61] J. F. Clauser and P. Thaddeus: *Topics in Relativistic Astrophysics*, edited by S. P. Maran and A. G. W. Cameron (New York, 1969).
[62] V. J. Bortolot, J. F. Clauser and P. Thaddeus: *Phys. Rev. Lett.*, **22**, 307, 806 (E) (1969).
[63] P. Palmer, B. Zuckerman, D. Buhl and L. E. Snyder: *Astrophys. Journ.*, **156**, L147 (1969).
[64] C. H. Townes and A. C. Cheung: *Astrophys. Journ.*, **157**, L103 (1969).
[65] P. Solomon: private communication (1969).
[66] J. L. Anderson and J. M. Stewart: to be published.
[67] C. W. Misner: *Astrophys. Journ.*, **151**, 431 (1968).
[67*a*] C. W. Misner: *Phys. Rev.*, **186**, 1328 (1969); *Phys. Rev. Lett.*, **22**, 1071 (1969).
[68] P. J. E. Peebles: *Astrophys. Journ.*, **153**, 1 (1968).
[69] Y. B. Zel'dovich and R. A. Sunyaev: *Astrophys. and Space Sci.*, **4**, 301 (1969).
[70] R. J. Weynmann: *Astrophys. Journ.*, **145**, 560 (1966).
[71] R. A. Sunyaev and Y. B. Zel'dovich: *Nature*, **223**, 721 (1969).
[72] M. J. Rees: *Astrophys. Journ.*, **153**, L1 (1968).
[73] W. Baron and D. T. Wilkinson: to be published.
[74] E. K. Conklin and R. N. Bracewell: *Phys. Rev. Lett.*, **18**, 614 (1967).
[74*a*] E. K. Conklin and R. N. Bracewell: *Nature*, **216**, 777 (1968).
[75] E. E. Epstein: *Astrophys. Journ.*, **148**, L157 (1967).
[76] R. B. Partridge and D. T. Wilkinson: *Phys. Rev. Lett.*, **18**, 557 (1967).
[77] D. T. Wilkinson and R. B. Partridge: *Nature*, **215**, 719 (1967).
[78] E. K. Conklin: *Nature*, **222**, 971 (1969).
[79] K. S. Thorne: *Astrophys. Journ.*, **148**, 51 (1967).
[80] C. W. Misner: *Nature*, **214**, 40 (1967); *Phys. Rev. Lett.*, **19**, 533 (1967).
[81] A. G. Doroshkevich, Y. B. Zel'dovich and I. D. Novikov: *Žurn. Ėksp. Teor. Fiz.*, **53**, 644 (1967).
[82] J. M. Stewart: *Astrophys. Lett.*, **2**, 133 (1968); *Mont. Not. Roy. Astr. Soc.*, **145**, 347 (1969).
[83] S. W. Hawking: *Mont. Not. Roy. Astr. Soc.*, **142**, 129 (1968).
[84] R. A. Matzner: *Astrophys. and Space Sci.*, **4**, 459 (1969); *Astrophys. Journ.*, **157**, 1085 (1969).
[85] M. A. H. MacCalum: to be published, *Comm. Math. Phys.* (1970).
[86] R. K. Sachs and A. M. Wolfe: *Astrophys. Journ.*, **147**, 73 (1967).
[86*a*] M. J. Rees and D. W. Sciama: *Nature*, **217**, 1511 (1968).
[87] A. M. Wolfe: *Astrophys. Journ.*, **156**, 803 (1969).
[88] G. Dautcourt: *Mont. Not. Roy. Astr. Soc.*, **144**, 255 (1969).
[89] M. S. Longair and R. A. Sunyaev: *Nature*, **223**, 719 (1969).
[90] G. V. Chibisov and L. M. Ozernoy: *Astrophys. Lett.*, **3**, 189 (1969).
[91] M. Schmidt: *Stars and Stellar Systems V Galactic Structure* (Chicago, 1965).

[92] N. U. Mayall: *Astrophys. Journ.*, **104**, 319 (1946).
[93] M. L. Humason and H. D. Wahlquist: *Astron. Journ.*, **60**, 254 (1955).
[94] D. V. Byrnes: *Publ. Astr. Soc. Pacific*, **78**, 46 (1966).
[95] G. de Vaucouleurs and W. L. Peters: *Nature*, **220**, 868 (1968).
[96] V. C. Rubin: *Astron. Journ.*, **56**, 47 (1951).
[97] G. de Vaucouleurs: *Astron. Journ.*, **63**, 253 (1958).
[98] K. F. Ogorodnikoff: *Dynamics of Stellar Systems* (New York, 1965), p. 95.
[99] D. W. Sciama: *Phys. Rev. Lett.*, **18**, 1065 (1967).
[100] J. M. Stewart and D. W. Sciama: *Nature*, **216**, 748 (1967).
[101] I. I. K. Pauliny-Toth and J. R. Shakeshaft: *Mont. Not. Roy. Astr. Soc.*, **124**, 61 (1962).
[102] D. W. Sciama: *The Unity of the Universe* (London, 1959).
[103] D. W. Sciama: *The Physical Foundations of General Relativity* (New York, 1969).
[104] L. I. Schiff: *Rev. Mod. Phys.*, **36**, 510 (1964).
[105] J. Kristian and R. K. Sachs: *Astrophys. Journ.*, **143**, 379 (1966).
[106] I. D. Novikov: *Sov. Astr.*, **12**, 427 (1969).
[107] P. T. Saunders: *Mont. Not. Roy. Astr. Soc.*, **142**, 213 (1969).
[108] D. W. Sciama, P. C. Waylen and R. C. Gilman: *Phys. Rev.*, **187**, 1762 (1969).
[109] B. L. Altshuler: *Akad. Nauk. USSR*, **55**, 1311 (1966); *Žurn. Èksp. Teor. Fiz.*, **51**, 1143 (1966).
[110] D. Lynden-Bell: *Mont. Not. Roy. Astr. Soc.*, **135**, 413 (1967).
[111] R. C. Gilman: Ph. D. thesis (Princeton University).
[112] R. J. Tayler: *Quart. Journ. Roy. Astr. Soc.*, **8**, 313 (1967); *Nature*, **217**, 433 (1968).
[113] C. Hayashi: *Progr. Theor. Phys.*, **5**, 224 (1950).
[114] R. A. Alpher, J. W. Follin and R. C. Herman: *Phys. Rev.*, **92**, 1347 (1953).
[115] F. Hoyle and R. J. Tayler: *Nature*, **209**, 1108 (1964).
[116] Y. N. Smirnov: *Sov. Astr.*, **8**, 864 (1965).
[117] P. J. E. Peebles: *Astrophys. Journ.*, **146**, 542 (1966).
[118] R. V. Wagoner, W. A. Fowler and F. Hoyle: *Astrophys. Journ.*, **148**, 3 (1967).
[119] C. J. Christensen, A. Nielsen, A. Bahnsen, W. K. Brown and B. N. Rustad: *Phys. Lett.*, **26** B, 11 (1967).
[120] S. W. Hawking and R. J. Tayler: *Nature*, **209**, 1278 (1966).
[121] R. F. Carswell: *Mont. Not. Roy. Astr. Soc.*, **144**, 279 (1969).
[122] G. S. Greenstein: *Nature*, **223**, 938 (1969).
[123] P. A. G. Scheuer: *Observatory*, **87**, 204 (1967).
[124] R. H. Dicke: *Astrophys. Journ.*, **152**, 1 (1968).
[125] G. S. Greenstein: *Astrophys. and Space Sci.*, **2**, 155 (1968).
[126] K. Greisen: *Phys. Rev. Lett.*, **16**, 748 (1966).
[127] F. W. Stecker: *Phys. Rev., Lett.*, **21**, 1016 (1968).
[128] F. W. Stecker: *Phys. Rev.*, **180**, 1264 (1969).
[129] J. V. Jelley: *Phys. Rev. Lett.*, **16**, 479 (1966).
[130] R. J. Gould and G. Schreder: *Phys. Rev. Lett.*, **16**, 252 (1966); *Phys. Rev.*, **155**, 1404, 1408 (1967).
[131] F. Hoyle: *Phys. Rev. Lett.*, **15**, 131 (1965).
[132] R. J. Gould: *Phys. Rev. Lett.*, **15**, 511 (1965).
[133] V. L. Ginzburg and S. I. Syrovatsky: *The Origin of Cosmic Rays* (London, 1964).
[134] J. E. Felten and P. Morrison: *Astrophys. Journ.*, **146**, 686 (1966).
[135] F. Seward, G. Chodil, H. Mark, C. Swift and A. Toor: *Astrophys. Journ.*, **150**, 845 (1967).

[136] E. A. Boldt, U. D. Desai, S. S. Holt and P. J. Serlemitsos: *I.A.U. Symposium*, No. 37, *Nonsolar gamma- and* X-*ray astronomy*, to be published.
[137] B. A. Cooke, B. E. Griffiths and K. A. Pounds: *Nature*, **224**, 134 (1969).
[138] R. F. O'Connell and S. D. Varma: *Phys. Rev. Lett.*, **22**, 1443 (1969).
[139] R. Bergamini, P. Londrillo and G. Setti: *Nuovo Cimento*, **52** B, 495 (1967).
[140] K. Brecher and P. Morrison: *Astrophys. Journ.*, **150**, L61 (1967).
[141] M. J. Rees and G. Setti: *Nature*, **219**, 127 (1968).
[142] M. S. Longair: *Observatory*, **88**, 199 (1968).
[143] M. S. Longair and R. A. Sunyaev: *Astrophys. Lett.*, **4**, 65 (1969).
[144] G. Setti and M. J. Rees: *I. A. U. Symposium*, No. 37, *Nonsolar gamma and* X-*ray astronomy*, to be published.
[145] A. N. Bunner, P. C. Coleman, W. L. Kraushaar, D. McCammon, T. M. Palmieri, A. Shilepsky and M. Ulmer: *Nature*, **223**, 1222 (1969).

Relativistic Stars, Black Holes and Gravitational Waves (Including an in-Depth Review of the Theory of Rotating, Relativistic Stars) (*).

K. S. THORNE

California Institute of Technology - Pasadena, Cal.

1. – Overview.

1·1. *The relevance of general relativity to Astronomy.* – In 1965, at Course XXXV of this summer school, I presented a series of lectures on roughly the same topic as this series: « The General Relativistic Theory of Stellar Structure and Dynamics. » At that time astronomers were beginning to suspect that general-relativistic effects *might* be important for astronomy, not solely on the cosmological scale, but also on the scale of isolated astrophysical bodies. Quasars had just recently been discovered and were posing puzzles, which suggested that they might be associated with (relativistic) gravitational collapse, or with (relativistic) supermassive stars. Also, detailed models of supernovae were suggesting more strongly than ever that (relativistic) neutron stars should be formed in supernova outbursts.

Since 1965 two additional astronomical discoveries have made the case for the astrophysical relevance of general relativity almost irrefutable: the discovery of pulsars by HEWISH and his group at Cambridge University in 1967, and the possible discovery of bursts of cosmic gravitational waves by WEBER at the University of Maryland in 1969.

Studies of pulsars since 1967 have yielded strong evidence that the clock mechanism which regulates the pulses is the rotation of a neutron star, and that some of the neutron stars associated with pulsars were, indeed, formed in supernova outbursts.

(*) Supported in part by the National Science Foundation [GP-9433, GP-9114] and the Office of Naval Research [Nonr-220(47)].

The characteristics of Weber's gravitational-wave events agree with the characteristics that one would expect for waves from the collapse of a very massive star ($\gtrsim 50 M_\odot$) to near its gravitational radius—with one exception: WEBER sees too many events, by several orders of magnitude, to agree with current ideas about stellar evolution. This situation is simultaneously tantalizing and frustrating. However, we can be sure of at least one thing: No object in which Newtonian gravity is even roughly valid can be a source of the types of gravitational-wave bursts that WEBER believes he is seeing!

1'2. *The nature of these lectures.* – Pulsars and Weber's observations are of only motivational relevance to these lectures. The lectures will *not* discuss them in detail. The reasons are: i) there is not enough space [poor excuse!]; ii) by the time the lectures are published, our whole viewpoint on pulsars and on Weber's experiment may have changed [better excuse!]; iii) I am not really competent to discuss the fine points of pulsar theory or of Weber's experiment [best excuse!]. Fortunately, in other chapters of this volume BERTOTTI gives an excellent review of Weber's experiment and related matters, and PACINI gives a fine review of pulsar observations and theory.

What, then, is the nature of these lectures? They are intended to be a review of recent (post-1965) developments in the theory of relativistic stars and gravitational waves. In a sense, they are companion lectures to the ones which I gave here in Varenna four years ago [1], and which I shall refer to henceforth as « RSSD-65 » (*i.e.* « Relativistic Stellar Structure and Dynamics 1965 »). They are also companions to my 1966 Les Houches lectures on the same subject ([2]; referred to henceforth as « RSSD-66 »).

RSSD-65 was written in such a style that it could be understood by an astrophysicist with no prior knowledge of general relativity theory. In RSSD-65 I remarked: « Although this "poor man's" approach to general relativity is extremely useful and powerful in the present context, it is so *only* because we restrict our attention to situations with spherical symmetry and thereby strongly limit the types of dynamical effects which are considered ». In the present lectures we shall concentrate heavily on nonspherical, rotating stars, so we will be forced to abandon the « poor man's » approach, and we will have to assume of the reader an understanding of the fundamentals of general relativity theory.

On all of the topics discussed in the oral version of my lectures, with one exception—rotating stars, I have already published fairly detailed reviews elsewhere. Rather than republish those reviews here, I have chosen to refer the reader to them in their original settings, as well as to other recent review articles. As a consequence, this written version of my lectures is little more than a list of references on a variety of topics, plus a detailed review of the theory of rotating, relativistic stellar models.

These lectures are divided into five Sections. Section **1** is this Introduction. Section **2**, on *nonrotating, relativistic stars*, consists primarily of a list of references to places in the literature where this subject has already been reviewed in depth. Section **3** is a comprehensive review of the theory of *rotating, relativistic stars*, most of which has been worked out since 1965. Section **4**, on *gravitational collapse and black holes*, is largely a list of references to the literature, as is Sect. **5**, on recent developments in the theory of *gravitational waves.*

The notational conventions used here are the same as those of RSSD-65 and RSSD-66, except that I drop the tiresome practice of putting a «*» on quantities measured in geometrized units ($c = G = 1$). A few words of warning are in order about the differences between my conventions and those used by EHLERS and ELLIS in this book: i) I set $c = G = 1$; they set $c = 8\pi G = 1$, where c is the speed of light and G is Newton's gravitational constant. ii) I use «timelike conventions» (the Minkowskii metric is $\eta_{00} = +1, \eta_{11} = \eta_{22} = \eta_{33} = -1$); they use «spacelike conventions» ($\eta_{11} = \eta_{22} = \eta_{33} = +1, \eta_{44} = -1$). iii) I let Greek indices run from 0 to 3, and Latin from 1 to 3; they let Greek run from 1 to 3 and Latin from 1 to 4. iv) I use the symbol ϱ to represent total density of mass-energy; they use ϱ for rest-mass density, and μ for total density of mass-energy.

In these lectures I shall often write down the basis vectors associated with a co-ordinate system, using the notation of differential geometers: $\partial/\partial x^\alpha = \boldsymbol{e}_\alpha$ will denote a contravariant basis vector, and $\mathbf{d}x^\alpha = \boldsymbol{e}^\alpha$ will denote a covariant basis vector. Hence, a vector which relativists usually denote by its contravariant components A^α or its covariant components A_α will often be written

$$\boldsymbol{A} = A^\alpha \partial/\partial x^\alpha \qquad \text{or} \qquad \boldsymbol{A} = A_\alpha \mathbf{d}x^\alpha .$$

I am grateful to J. DEMARET, J. IPSER, and J. MELVIN for assistance in the preparation of these lecture notes.

2. – Nonrotating, relativistic stellar models.

OPPENHEIMER and VOLKOFF [3] laid the fundations for the modern theory of nonrotating, relativistic stars; but the theory did not develop into its present, detailed form until the period 1958-1965. Since 1965 there have been few developments in the fundamental aspects of the theory, though there has been fairly vigorous activity on applications of it to neutron stars and supermassive stars.

The oral version of these lectures concentrated on the fundamentals of the theory, following quite faithfully chapters **1** and **2** of RSSD-65, and chapters **3** and **4** of RSSD-66.

The reader interested in applications to astrophysical problems may find the following reviews helpful: Chapter **6** of RSSD-66 for supermassive stars; chapter **7** of RSSD-66 and chapter **6** of RSSD-65 for neutron stars; HARRISON, THORNE, WAKANO, and WHEELER [4] for dense white dwarfs and neutron stars; WHEELER [5] for neutron stars, with an emphasis on observational aspects; and ZEL'DOVICH and NOVIKOV [6] for all types of applications, from a strongly astrophysical, not-too-mathematical viewpoint.

Readers will find the next Section much easier to follow if they have already read chapters **1** and **2** of RSSD-65 or chapters **3** and **4** of RSSD-66.

3. – Rotation of relativistic stellar models.

3'1. *General properties of rotating, relativistic stellar models.*

3'1.1. Co-ordinate systems. In these lectures we shall restrict ourselves to *axially symmetric* rotating stars, *i.e.* we shall ignore Jacobi-ellipsoid-like configurations; and we shall assume that our stars are all made of perfect fluid, *i.e.* we shall ignore stresses due to magnetic fields and crystallization of matter.

In analysing an axially symmetric, rotating star it is helpful to introduce a co-ordinate system which, so far as possible, embodies the symmetries of the star's spacetime geometry. Such a co-ordinate system has the following pro-properties (see Appendix to this Section):

i) It possesses very particular time and angular co-ordinates, $x^0 \equiv t$ and $x^3 \equiv \varphi$, of which the metric coefficients and all matter variables are independent:

$$g_{\alpha\beta} = g_{\alpha\beta}(x^1, x^2)\,, \qquad \varrho = \varrho(x^1, x^2)\,, \qquad p = p(x^1, x^2)\,, \text{ etc.} \tag{3.1a}$$

(ϱ is density of total mass-energy; p is pressure). Independence of t corresponds to the fact that the star is stationary in time; independence of φ corresponds to the fact that it is axially symmetric.

ii) Since φ is an angular co-ordinate about the rotation axis, the co-ordinate values (t, x^1, x^2, φ) and $(t, x^1, x^2, \varphi + 2\pi)$ correspond to the same point in spacetime

$$(t, x^1, x^2, \varphi) \equiv (t, x^1, x^2, \varphi + 2\pi)\,. \tag{3.1b}$$

iii) The star's matter rotates in the φ direction, so its 4-velocity has components

$$u^t = u^t(x^1, x^2) \neq 0\,, \qquad u^\varphi = \Omega(x^1, x^2)\, u^t(x^1, x^2) \neq 0\,, \qquad u^1 = u^2 = 0\,, \tag{3.1c}$$

where $\Omega(x^1, x^2)$ is the angular velocity measured in units of co-ordinate time

$$\Omega = u^\varphi/u^t = \mathrm{d}\varphi/\mathrm{d}t \,. \tag{3.1d}$$

iv) Because the star rotates in the φ direction, it and its space-time geometry are *not* invariant under time reversal $(t \to -t)$; such a reversal would change the direction of rotation. Nor are they invariant under inversion of the φ direction $(\varphi \to -\varphi)$; that would also change the direction of rotation. However, they *are* invariant under a simultaneous inversion of t and φ $(t \to -t; \varphi \to -\varphi)$. As a result, the metric coefficients $g_{t1}, g_{t2}, g_{\varphi 1}$, and $g_{\varphi 2}$, which would have to change sign under the simultaneous inversion of φ and t, must vanish

$$g_{t1} = g_{t2} = g_{\varphi 1} = g_{\varphi 2} = 0 \,, \tag{3.1e}$$

leaving the line element in the form

$$\mathrm{d}s^2 = g_{tt}\,\mathrm{d}t^2 + 2g_{t\varphi}\,\mathrm{d}t\,\mathrm{d}\varphi + g_{\varphi\varphi}\,\mathrm{d}\varphi^2 + g_{AB}\,\mathrm{d}x^A\,\mathrm{d}x^B \,. \tag{3.1f}$$

Here capital Latin indices run over 1 and 2. The presence of an off-diagonal $g_{t\varphi}$ term in the metric leads, we shall see, to a « dragging of inertial frames ».

v) Far from the star the co-ordinates become inertial so that, in particular,

$$g_{tt} \to 1 \,, \qquad g_{t\varphi} \sim 1/r \,, \qquad g_{\varphi\varphi} \to r^2 \sin^2\theta \qquad \text{as } r \to \infty \,. \tag{3.1g}$$

Here r and θ (not necessarily the same as x^1 and x^2) are the usual spherical co-ordinates of the nearly-Euclidean space far from the star.

Co-ordinates systems of the above type are uniquely determined up to transformations of the two auxiliary coordinates, x_1 and x^2:

$$x^1_{\text{new}} = f(x^1_{\text{old}}, x^2_{\text{old}}) \,, \qquad x^2_{\text{new}} = h(x^1_{\text{old}}, x^2_{\text{old}}) \,. \tag{3.2}$$

Such transformations can be used to impose additional constraints on the metric coefficients ... *e.g.*,

$$g_{12} = 0 \,, \qquad g_{33} = g_{22}(\sin x^2)^2 \ldots \,, \tag{3.3}$$

or to impose a constraint such as $g_{12} = 0$ and to guarantee that the surfaces of constant pressure, p, are surfaces of constant x^1. However, such specializations of the (x^1, x^2) co-ordinates are of a very different type from the specializations of t and φ which lead to conditions (3.1): the specializations of t and φ are dictated by the geometric symmetries of space-time and embody *all* such symmetries. Specializations of x^1 and x^2 are made only so as to simplify the mathematics when solving Einstein's field equations. For this reason, we shall avoid specializing x^1 and x^2 until we discuss the actual procedures for constructing rotating stellar models.

The metric coefficients g_{tt}, $g_{t\varphi}$, and $g_{\varphi\varphi}$ are invariant under the transformations (3.2) of x^1 and x^2. This suggests that they must be geometric invariants associated with the symmetries of space-time. Their invariant significance is most elegantly expressed in terms of the Killing-vector fields, $\partial/\partial t$ and $\partial/\partial\varphi$, which generate the invariant time translations and rotations of space-time: $\partial/\partial t$ is defined as the unique Killing-vector field which is timelike at $r=\infty$ and which is normalized so that

$$(\partial/\partial t)\cdot(\partial/\partial t)=1 \quad \text{at} \quad r=\infty\,. \tag{3.4a}$$

Similarly, $\partial/\partial\varphi$ is defined as the unique Killing-vector field which is orthogonal to $\partial/\partial t$ at $r=\infty$,

$$(\partial/\partial\varphi)\cdot(\partial/\partial t)=0 \quad \text{at} \quad r=\infty \tag{3.4b}$$

and which is normalized so that a motion through $\Delta\varphi=2\pi$ brings the space back to where it started. The metric coefficients g_{tt}, $g_{t\varphi}$, and $g_{\varphi\varphi}$ are just the invariant scalar products of these Killing fields:

$$g_{tt}=(\partial/\partial t)\cdot(\partial/\partial t)\,,\qquad g_{t\varphi}=(\partial/\partial t)\cdot(\partial/\partial\varphi)\,,\qquad g_{\varphi\varphi}=(\partial/\partial\varphi)\cdot(\partial/\partial\varphi)\,. \tag{3.4c}$$

Appendix to Subsection 3·1.1 (*).

Proof that for the space-time manifold of an axially symmetric, rotating star there exist co-ordinates with the properties i) to v).

The assumptions which go into the proof are *a*) that the space-time manifold of the star is asymptotically flat at radial infinity; *b*) that the star and its manifold are stationary, so that there exists a Killing-vector field $\boldsymbol{\xi}_{(t)}$ which,

(*) I thank B. Carter for a very helpful discussion of the material in this Appendix, particularly for pointing out to me the applicability of his theorem on orthogonal transitivity, and for informing me of his recent results on commutivity when there are more than two Killing vectors present.

at radial infinity, is timelike and has unit length:

$$\xi_{(t)} \cdot \xi_{(t)} \to +1 \text{ at radial infinity}; \tag{3.5a}$$

c) that the star and its manifold are axially symmetric; *i.e.* that there exists a Killing-vector field, $\xi_{(\varphi)}$, with closed orbits, which, at radial infinity, is spacelike and is orthogonal to $\xi_{(t)}$ and has length $r \sin\theta$:

$$\xi_{(\varphi)} \cdot \xi_{(\varphi)} \to -r^2 \sin^2\theta\,, \qquad \frac{\xi_{(\varphi)} \cdot \xi_{(t)}}{|\xi_{(\varphi)}|\,|\xi_{(t)}|} \to 0 \text{ at radial infinity}; \tag{3.5b}$$

d) that the star is made of a perfect fluid which rotates in the $\xi_{(\varphi)}$ direction—*i.e.* which has a 4-velocity of the form

$$\boldsymbol{u} = u^t \xi_{(t)} + u^\varphi \xi_{(\varphi)}\,. \tag{3.5c}$$

The first step in the proof is to show that the Killing vectors $\xi_{(t)}$ and $\xi_{(\varphi)}$ commute. It is well known that the commutator of any two Killing vectors is itself a Killing vector. Suppose, as case i), that the only Killing vectors present are $\xi_{(t)}$ and $\xi_{(\varphi)}$. Then

$$[\xi_{(\varphi)}, \xi_{(t)}] \equiv a\xi_{(\varphi)} + b\xi_{(t)}\,,$$

where a and b are constants. At radial infinity the commutator must vanish because of the asymptotically-flat nature of the geometry; this is possible if and only if $a = b = 0$ ..., *i.e.* if and only if $\xi_{(\varphi)}$ and $\xi_{(t)}$ commute. In the less general case ii), where there are other Killing vectors present, a theorem due to CARTER [7] guarantees that $\xi_{(t)}$ can be chosen in such a manner that it commutes with $\xi_{(\varphi)}$. ([7] also gives very general conditions for commutivity which do not rely on asymptotic flatness.)

The next step in the proof is to show that the group of motions generated by $\xi_{(\varphi)}$ and $\xi_{(t)}$ is *orthogonally transitive*—*i.e.* that there exists a family of 2-dimensional surfaces which are orthogonal to the surfaces of transitivity of the group. To prove orthogonal transitivity, we invoke theorem 2 of CARTER [8]. Our star and its manifold satisfy the hypotheses of that theorem by virtue of the following: *a*) The nonempty subset $\mathscr{F}$ is the axis of symmetry, $\theta = 0$, near radial infinity (Carter's condition II). *b*) The star's stress-energy tensor, $T = (\varrho + p)\,\boldsymbol{u} \otimes \boldsymbol{u} - pg$, with the 4-velocity (3.5*c*), is invertible in the group, as is the metric tensor g; consequently, by virtue of Einstein's field equations, the Ricci tensor is also invertible in the group (Carter's condition I). Thus, Carter's theorem is applicable, and orthogonal transitivity is proved (*).

(*) Prior to Carter's work, so far as I know, the most general proof of orthogonal transitivity for axially symmetric systems—and hence the most general proof that one can introduce co-ordinate systems of the form used in the text—was Papapetrou's [9] proof for the *vacuum case.*

With this knowledge of the properties of the group of motions, we are ready to construct the co-ordinate system. Because the Killing vectors $\boldsymbol{\xi}_{(\varphi)}$ and $\boldsymbol{\xi}_{(t)}$ commute, we can choose our co-ordinates ($x^0 = t, x^1, x^2, x^3 = \varphi$) in such a manner that

$$\boldsymbol{\xi}_{(\varphi)} = \partial/\partial\varphi\,, \qquad \boldsymbol{\xi}_{(t)} = \partial/\partial t\,. \tag{3.6a}$$

And by virtue of the orthogonal transitivity, we can simultaneously adjust our co-ordinates so that

$$\begin{cases} g_{\varphi A} = (\partial/\partial\varphi)\cdot(\partial/\partial x^A) = 0\,, \\ g_{tA} = (\partial/\partial t)\cdot(\partial/\partial x^A) = 0\,, \end{cases} \qquad \text{for } A = 1, 2\,. \tag{3.6b}$$

This co-ordinate system has the properties claimed in the text: Property i) follows from the fact that $\partial/\partial\varphi$ and $\partial/\partial t$ are Killing vectors; property ii) follows from the hypothesis that $\partial/\partial\varphi$ has closed orbits and from its normalization near radial infinity where spacetime is asymptotically flat; property iii) follows from hypothesis (3.5*c*); property iv) follows from condition (3.6*b*); and property v) follows from the hypothesis of asymptotic flatness, together with the field equations (to be discussed later), which guarantee the stated radial dependence for $g_{t\varphi}$.

3˙1.2. Differential rotation and rigid rotation. *If the co-ordinate angular velocity $\Omega = \mathrm{d}\varphi/\mathrm{d}t$ is a constant throughout the configuration, then the configuration is in rigid rotation* in the following sense: the distance on a hypersurface $t =$ const between any two baryons which are close together is independent of time, t. Equivalently stated, to a distant observer with X-ray vision, the configuration appears to rotate rigidly; moreover, as measured by the clock of the distant observer, for which t is proper time, the angular velocity of the matter is Ω. Of course, if Ω is not constant throughout the star, then its matter is in differential rotation. Although our identification of $\Omega =$ constant with rigid rotation should be obvious, it is interesting to verify it formally by calculating the expansion, Θ, the shear, $\sigma_{\alpha\beta}$, and the rotation, $\omega_{\alpha\beta}$, associated with the fluid motion (3.1*c*) in the geometry (3.1*f*). (For a discussion of the concepts of expansion, shear and rotation, see the lectures by Ellis in this volume.) Straightforward computation reveals that:

i) The expansion is always zero

$$\Theta = 0\,, \tag{3.7a}$$

independently of whether Ω is constant.

ii) The shear vanishes at a given point if and only if Ω is constant ($\partial\Omega/\partial x_A = 0$) there—*i.e.* if and only if the rotation is rigid (cf. BOYER [10])

$$\text{(3.7}b\text{)}\qquad \begin{cases} \sigma_{AB} = \sigma_{tt} = \sigma_{\varphi\varphi} = \sigma_{t\varphi} = 0\,, \qquad \sigma_{tA} = -\,\Omega\sigma_{\varphi A}\,, \\ \sigma_{\varphi A} = \dfrac{1}{2}\left(\dfrac{g_{tt}g_{\varphi\varphi} - g_{t\varphi^2}}{g_{tt} + 2g_{t\varphi}\Omega + g_{\varphi\mu}\Omega^2}\right)\dfrac{\partial\Omega}{\partial x^A} \end{cases}$$

(note: $g_{tt}g_{\varphi\varphi} - g_{t\varphi^2}$ cannot be zero anywhere in the star since, if it were, the fluid there would be moving along a spacelike or null world line).

iii) The rotation, which is given by

$$\text{(3.7}c\text{)}\qquad \begin{cases} \omega_{AB} = \omega_{tt} = \omega_{\varphi\varphi} = \omega_{t\varphi} = 0\,, \qquad \omega_{tA} = -\,\Omega\omega_{\varphi A}\,, \\ \omega_{\varphi A} = \frac{1}{2}\,[u^t g_{\varphi\varphi}(\Omega - \omega)]_{,A} - \frac{1}{4}\,(u^t)^3 g_{\varphi\varphi}(\Omega - \omega)(g_{tt,A} + 2g_{t\varphi,A}\Omega + g_{\varphi\varphi,A}\Omega^2)\,, \\ \omega \;\;\; = -\,g_{\varphi t}/g_{\varphi\varphi}\,, \end{cases}$$

vanishes everywhere if and only if $\Omega = \omega \equiv -\,g_{\varphi t}/g_{\varphi\varphi}$ everywhere. We shall later (Subsect. **3**˙1.7) call ω the « angular velocity of cumulative dragging ». For bounded stars in asymptotically flat spacetime the mean value of Ω is generally greater than the mean value of ω.

3˙1.3. Injection energy (*). We now turn our attention to the first of several calculations and theorems related to the properties of rotating relativistic stellar models. A more complete version of these calculations and theorems is given in Subsect. **10**˙7 of [6]. Our first calculation is an extension to rotating stars of the concept of *injection energy*, which was introduced for nonrotating stars in Subsect. **3**˙5.7 of RSSD-65.

Let an astrophysicist far from a rotating equilibrium configuration create δA baryons and drop them, along with an additional mass-energy W_0, down an idealized pipe which is inserted into the star, to a colleague situated at a point $\boldsymbol{x}$ in the star. Let the nuclear abundances of the δA baryons be the same as the abundances in the star at $\boldsymbol{x}$, so the total mass-energy dropped is

$$W_{\text{initial}} = \mu_B(\boldsymbol{x})\,\delta A + W_0\,,$$

where $\mu_B(\boldsymbol{x})$ is the rest mass per baryon at $\boldsymbol{x}$. The distant astrophysicist is at rest with respect to the star's gravitational field, and he drops the baryons radially so that they have initial 4-momenta

$$p_{t_{\text{initial}}} = \mu_B(\boldsymbol{x})\,\delta A + W_0\,, \qquad p_{j_{\text{initial}}} = 0 \qquad \text{for } j = 1, 2, 3\,.$$

(*) Much of Subsect. **3**˙1.3 and **3**˙1.4 is based on RSSD-66.

As the baryons fall, the gravitational field accelerates them, thereby making p_1 and p_2 nonzero; but since the metric is independent of t and φ, geodesic motion guarantees that p_φ will remain zero and p_t will remain constant (*). (The pipe down which the baryons fall is carefully shaped so that the baryons never touch its walls as they fall.)

The astrophysicist's colleague inside the star is rotating with the fluid in his neighborhood and therefore has a 4-velocity $u^\alpha(\boldsymbol{x})$ with the form (3.1*c*). As the rotation carries him under the mouth of the pipe, the colleague catches the baryons and excess mass-energy, converting their kinetic energy of fall into mass. The total mass-energy which he then has is

$$W_{\text{initial at }\boldsymbol{x}} = p_\alpha u^\alpha = p_{t\,\text{initial}}\, u^t(x) = [\mu_B(x)\,\delta A + W_0]\, u^t(x)\,. \tag{3.8}$$

The colleague then uses some of this mass-energy to heat the δA baryons to the temperature, T, of the surrounding fluid, to compress them to the pressure, p, of the surrounding fluid, and to make a space for them in the surrounding fluid and inject them into it. The total mass-energy required for this process—including the rest mass of the baryons injected—is (cf. Subsect. **3**.5.7 of RSSD-65)

$$W_{\text{injected}} = (\varrho + p)\, n^{-1}\,\delta A$$

where n is the number density of baryons in the surrounding fluid. The remaining mass-energy,

$$W_{\text{excess at }\boldsymbol{x}} = (\mu_B\,\delta A + W_0)\, u^t - (\varrho + p)\, n^{-1}\,\delta A\,,$$

(*) This is a consequence of the following useful theorem: *If the metric tensor, $g_{\mu\nu}$, relative to a particular co-ordinate system is independent of one of the co-ordinates, x^β,*

$$g_{\mu\nu} \neq \text{function of } x^\beta \text{ for all } \mu, \nu \text{ and for a particular } \beta\,,$$

then the numerical value of the β-th covariant component of the four-momentum of any particle or photon is conserved along its geodesic path,

$$p_\beta = \text{constant}\,.$$

Proof of theorem: the geodesic equation reads

$$\frac{\mathrm{D}u_\beta}{\mathrm{d}s} = 0 = \frac{\mathrm{d}u_\beta}{\mathrm{d}s} - \Gamma^\nu_{\beta\sigma} u^\sigma u_\nu =$$

$$= \frac{\mathrm{d}u_\beta}{\mathrm{d}s} - \frac{1}{2}\left(\frac{\partial g_{\beta\lambda}}{\partial x^\sigma} + \frac{\partial g_{\sigma\lambda}}{\partial x^\beta} - \frac{\partial g_{\beta\sigma}}{\partial x^\lambda}\right) u^\sigma u^\lambda = \frac{\mathrm{d}u_\beta}{\mathrm{d}s} - \frac{1}{2}\left(\frac{\partial g_{\beta\lambda}}{\partial x^\sigma} - \frac{\partial g_{\beta\sigma}}{\partial x^\lambda}\right) u^\sigma u^\lambda = \frac{\mathrm{d}u_\beta}{\mathrm{d}s}\,.$$

Consequently, μ_β is conserved. But for a particle of nonzero rest mass, μ, we have $p_\beta = \mu u_\beta$; and for a photon—with appropriate choice of the affine parameter—we have $p_\beta = u_\beta$. Hence, p_β is also conserved. *q.e.d.*.

the colleague throws back up the pipe to the astrophysicist in just the reverse motion with which it fell down. Some of the excess mass-energy goes into the kinetic energy of the throw and is lost in climbing out of the gravitational field. Consequently, the excess mass-energy received by the distant astrophysicist,

$$W_{\text{excess at }\infty} = W_{\text{excess at }x}/u^t = (\mu_B\,\delta A + W_i) - [(\varrho + p)/(nu^t)]\,\delta A\;,$$

is less than the excess mass-energy at x. The mass-energy is red-shifted by $1/u^t$, the inverse of the factor by which it was blue-shifted in falling down the pipe (cf. eq. (3.8)). (Recall that u^t is the contravariant time component of the fluid's 4-velocity at the injection point.)

The distant astrophysicist calculates the injection energy at x—*i.e.* (by conservation of mass-energy), the increase in the star's total mass M which results from injection—by subtracting $W_{\text{excess at }\infty}$ from W_{initial}:

$$\delta M = [(\varrho + p)/nu^t]_{\text{at }x}\,\delta A = [(\partial\varrho/\partial n)_s(1/u^t)]_{\text{at }x}\,\delta A\;. \tag{3.9}$$

(Recall that M is the mass which governs the Keplerian motions of distant Newtonian planets.)

3˙1.4. Convection in a rigidly rotating star. We now turn our attention from the computation of injection energy to some applications of it. In discussing applications we confine our attention to equilibrium configurations in *rigid rotation* because a key equation in the analysis, eq. (3.10), is only valid in the rigid case. (For a generalization to the differentially rotating case see [10]; also Subsect. **10˙7** of [6].)

As our first application of injection energy, we discuss convection in rigidly rotating configurations with homogeneous chemical compositions, from the same point of view as was used in Subsect. 3˙5.7 of RSSD-65 for nonrotating configurations. It is evident from physical considerations that *such a rigidly rotating star is in marginal convective equilibrium if and only if small fuid elements can be moved about arbitrarily, but slowly, inside it without any work being done or energy being released.* Equivalently stated: *A rigidly rotating star is marginally convective if and only if its injection energy is independent of position.*

As for nonrotating stars, so also here, constant injection energy is equivalent to isentropy (*); and, consequently: *A rigidly rotating star is marginally convective if and only if it is isentropic.* To prove the equivalence of insentropy and constant injection energy for rigidly rotating configurations, proceed as follows:

(*) The equivalence of isentropy and constant injection energies for rigidly rotating configurations was first deduced by Boyer [11] and has also been discussed by Boyer and Lindquist [12] and by Hartle and Sharp [13].

First show that the equations of stress balance, $T_{\mu;\nu}^{\ \nu} = 0$, for a rigidly rotating configuration can be put into the form

$$(\varrho + p)^{-1}(\partial p/\partial x^\mu) = \partial(\ln u^t)/\partial x^\mu \,, \tag{3.10}$$

when a co-ordinate system of the form described at the beginning of Subsect. **3**˙1.1 is used. (Note: this form of the equation of stress balance is not valid in a differentially rotating star.) Then show that eq. (3.10) and the constancy of the injection energy,

$$(\varrho + p)/nu^t = \text{const} \,,$$

are compatible if and only if the configuration is isentropic—*i.e.*, if and only if

$$(\varrho + p)^{-1}(\partial \varrho/\partial x^\mu) = n^{-1}(\partial n/\partial x^\mu) \,. \tag{3.11}$$

The above isentropy criterion for the marginal convective equilibrium of a rigidly rotating configuration has the same form in relativity theory as in Newtonian theory. As was noted in Subsect. **3**˙2.8 of RSSD-65 this should not be surprising. Convective instability is a purely local phenomenon. Consequently, it cannot be affected by nonlinearities in the gravitational field, which act only over finite distances.

3˙1.5. Von Zeipel's theorem in general relativity. Another useful theorem about rigidly rotating stars is this: *In a rigidly rotating star the following sets of surfaces coincide—the surfaces of constant density, ϱ, the surfaces of constant pressure, p, the surfaces of constant redshift factor, $1/u^t$, and* (if n is a unique function of p and ϱ) *the surfaces of constant baryon number density, n, and of constant injection energy, $(\varrho + p)/nu^t$.* *Proof*: from the equation of stress balance (3.10) for a rigidly rotating star we see that the gradients of p and $1/u^t$ are parallel, so their level surfaces must coincide. Also from eq. (3.10) we learn that

$$0 = \varepsilon^{\alpha\beta\gamma\delta}[(\varrho + p)^{-1}p_{,\delta} - (\ln u^t)_{,\delta}]_{;\gamma} = (\varrho + p)^{-2}\varepsilon^{\alpha\beta\gamma\delta}\,\varrho_{,\gamma}\,p_{,\delta} \,,$$

where a comma denotes a partial derivative, a semi-colon denotes a covariant derivative, and $\varepsilon^{\alpha\beta\gamma\delta}$ is the completely antisymmetric tensor. This equation has the geometric interpretation that the gradients of ϱ and p are parallel so that their level surfaces coincide. That the level surfaces of n and $(\varrho + p)/nu^t$ also coincide with those of ϱ, p, and $1/u^t$ then follows trivially from the fact that n and $(\varrho + p)/nu^t$ are unique functions of ϱ, p, and $1/u^t$. *q.e.d.*

The above theorem is the general-relativistic version of Von Zeipel's theorem, which has played a crucial role in Robert Dicke's interpretation of his solar-oblateness observations. It is instructive to take the Newtonian limit of this theorem and get the familiar equation for the level surfaces. In general relativity the level surfaces are surfaces of constant redshift factor, $1/u^t$. From the normalization condition $u^\alpha u_\alpha = 1$ for the 4-velocity, one discovers that the square of this constant redshift factor is

$$(1/u^t)^2 = g_{tt} + 2g_{t\varphi}\Omega + g_{\varphi\varphi}\Omega^2 = \text{constant on level surfaces.} \tag{3.12}$$

In the Newtonian limit the metric coefficients become

$$g_{tt} = 1 - 2U\,, \qquad g_{t\varphi} = 0\,, \qquad g_{\varphi\varphi} = -r^2\sin^2\theta\,,$$

where U is the Newtonian potential (with the sign convention $U > 0$). Hence, the Newtonian level surfaces are given by

$$(1/u^t)^2 = \text{const} = 1 - 2U - (r^2\sin^2\theta\Omega^2)\,,$$

or, equivalently,

$$U + \tfrac{1}{2}r^2\sin^2\theta\,\Omega^2 = \text{const}\,. \tag{3.12 N}$$

This is the familiar Newtonian equation, which describes a bulging of the level surfaces at the equator (rotational flattening).

A special case of a level surface is the outside boundary of a rigidly rotating star. That this boundary must be a surface of constant redshift factor $1/u^t$ one can see not only from the above analysis, but also from this simple fact: if the boundary did not have a constant redshift factor, then the star could liberate energy by moving some of its surface particles from regions of small redshift (small u^t) to regions of large redshift (large u^t).

The constancy of $1/u^t$ over the boundary of a rigidly rotating star is a useful calculational tool, as BOYER [10, 11] first pointed out, Suppose that one knows the external gravitational field for a special rigidly rotating configuration—*e.g.*, the Kerr field (see Subsect. 3·1.8)—, but that one knows little about the configurations which could generate that gravitational field. Then one can determine as follows the locations, in the known external field, of the boundaries of all rigidly rotating configurations that could generate it: 1) specify those parameters which fix the external field uniquely (*e.g.*, for the Kerr field, specify the total mass-energy, M, and the total angular momentum, $J = aM$). 2) Specify the constant angular velocity, $\Omega = u^\varphi/u^t$, as measured by a distant observer. 3) Specify the constant value, $1/u^t$, of the redshift factor at the boundary of the equilibrium configuration. 4) Then those points in the external gravitational field at which eq. (3.12) is satisfied form the boundary of the configuration.

3'1.6. Redshift of light from the surface of a rotating star. One should not make the mistake of thinking that the «redshift factor» $1/u^t$ at a star's boundary measures the redshift of all photons emitted from the boundary. On the contrary, $u^t = 1 + z$ is the redshift only for photons which have zero angular momentum relative to the star's rotation axis. (We derived it originally as the factor by which the energy of a particle of finite rest mass is shifted if it is dropped *from rest*, *i.e.* with zero angular momentum at infinity, and is caught by an observer on the star who is moving with the fluid.)

For a photon with finite angular momentum the redshift can be derived as follows: the emitting atom sees the photon to have an energy which is the projection of the photon's 4-momentum on the atom's 4-velocity

$$E_{\text{emitted}} = p_t u^t + p_\varphi u^\varphi = p_t u^t [1 + \Omega p_\varphi / p_t]. \tag{3.13a}$$

The components p_t and p_φ of the photon's 4-momentum are conserved along its trajectory; and far from the star, where the co-ordinates are nearly inertial, they are the photon's energy and the projection of its angular momentum on the star's rotation axis:

$$E_{\text{obs}} = p_t \,, \qquad j_{\text{obs}} = p_\varphi \,. \tag{3.13b}$$

The ratio $p_\varphi / p_t = j_{\text{obs}} / E_{\text{obs}}$ is the impact parameter of the photon relative to the star's rotation axis:

$$\frac{p_\varphi}{p_t} \equiv l = \begin{pmatrix} \text{impact parameter of photon} \\ \text{relative to rotation axis} \end{pmatrix} = \alpha r \,. \tag{3.14}$$

Here r is the separation between the star and a distant, stationary observer who observes the photon, and α is the angle which the observer measures between the photon's trajectory and the radial plane that passes through the rotation axis. (The radial plane is determined, far from the star, by the axially symmetric, Newtonian gravitational field of forces, ∇U, *not* by the optical appearance of the star's disk. The dragging of inertial frames—see next Section—deforms the trajectories of photons, thereby displacing the center of the star's apparent disk away from the gravitationally determined radial direction.)

Combining eqs. (3.13) and (3.14), we obtain for the photon's redshift

$$\frac{E_{\text{emitted}}}{E_{\text{observed}}} = 1 + z = u^t [1 + \Omega l] = \frac{1 + \Omega l}{[g_{tt} + 2 g_{t\varphi} \Omega + g_{\varphi\varphi} \Omega^2]^{\frac{1}{2}}_{\text{boundary}}} \,. \tag{3.15}$$

Note that the Doppler broadening of a spectral line is by the factor

$$z_{max} - z_{min} = u^t \Omega(l_{max} - l_{min}) = (d\varphi/d\tau)_{boundary}(l_{max} - l_{min}) , \tag{3.16}$$

where $(d\varphi/d\tau)_{boundary}$ is the angular velocity which an observer on the star's boundary measures himself to have relative to the distant stars; $l_{max} > 0$ is the maximum impact parameter (corresponding to a photon emitted from that equatorial limb of the star's disk which is moving toward the observer); and $l_{min} < 0$ is the minimum impact parameter (corresponding to a photon from that limb which moves away from the observer). Although $l_{min} = - l_{max}$ in Newtonian theory, this is not true in general relativity. The dragging of inertial frames, which deforms the photon trajectories, destroys this relation, and also causes an asymmetry in the Doppler-broadened spectral line. The dragging also produces a slightly different redshift in a right-hand circularly polarized photon than in a left-hand photon, thus causing a very small splitting of spectral lines (ZEL'DOVICH [14]).

3˙1.7. Dragging of inertial frames. One of the most intriguing of the relativistic effects produced by the rotation of a star is the dragging of inertial frames (also called the « Lense-Thirring effect »; cf. THIRRING [15]). There are two ways to describe the dragging of inertial frames—by means of its *local* effects on gyroscopes and the motion of particles, and by means of its *cumulative* effects on the motions of particles. The simpler of these is the cumulative description, so we shall present it first.

Let an observer far from a rotating star throw a particle of finite or zero rest mass directly toward the star's rotation axis, so that it has zero angular momentum relative to the axis, $j = p_\varphi = 0$, and zero impact parameter, $l = p_\varphi/p_t = 0$. As the particle falls freely inward, its angular momentum p_φ remains zero; but, due to the dragging of inertial frames, its angular velocity does *not* remain zero. The angular velocity at any point along the trajectory is

$$\omega = d\varphi/dt = p^\varphi/p^t = (g^{\varphi t} p_t)/(g^{tt} p_t) = g^{\varphi t}/g^{tt} = - g_{\varphi t}/g_{\varphi\varphi} \,. \tag{3.17a}$$

Notice that this angular velocity depends only on the local geometric invariants $g_{\varphi t}$ and $g_{\varphi\varphi}$; it does not depend on the particle's rest mass, or on the initial inward velocity with which the particle is thrown (so long as it is thrown toward the rotation axis). Throughout the gravitational field of a slowly rotating star (Subsect. 3˙3), and in the nearly Newtonian region far from a rapidly rotating star, this *angular velocity of « cumulative dragging »* is given by

$$\omega = 2J/r^3 , \tag{3.17b}$$

where J is the star's total angular momentum, Thus, the dragging is in the same direction as the star's rotation. If J is positive (right-handed rotation of the star), ω is positive (right-handed rotation of the particle).

As a result of this cumulative dragging, the falling particle—which is assumed to not collide with the star's matter if it passes through the star—will be deflected away from the rotation axis. Although the distant observer who throws it in, and a distant observer who sees it come out, will measure it to have zero impact parameter as it passes them, they will find that it was deflected by the star's rotation through an angle

$$\alpha = \pi - \int\limits_{\text{along trajectory}} [-g_{\varphi t}/g_{\varphi\varphi}]\,\mathrm{d}t\,. \tag{3.18a}$$

(See Fig. 1.) In the more realistic case of a photon, which flies past a rotating star with its point of nearest approach at distance $s \gg M$ from the star's center, the cumulative dragging produces a small change in the usual light-deflection formula. The angle of deflection is change from $\alpha = 4M/s$ to

$$\alpha = (4M/s)[1-(\boldsymbol{J}\cdot\boldsymbol{n}/Ms)] = (4M/s)[1-J/Ml] \tag{3.18b}$$

(see COHEN and BRILL [16]). Here $\boldsymbol{J}$ is the star's angular momentum, M is its mass, $\boldsymbol{n}$ is the direction of the photon's angular momentum, and l (as above) is the photon's impact parameter relative to the star's rotation axis.

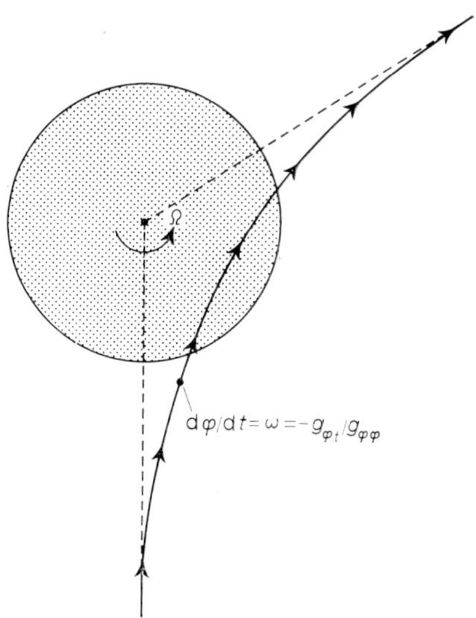

Fig. 1. – Trajectory of a particle with zero impact parameter relative to the rotation axis, which is deflected by the dragging of inertial frames. The particle's angular velocity—the « angular velocity of cumulative dragging »—is given, at each point on its orbit, by equations (3.17a), (3.17b). The total deflection is through an angle given by equation (3.18a).

A second way to describe the dragging of initial frames involves the difference in travel times for light beams that circle a star in different directions, bouncing off a series of mirrors as they go. This way was devised by BARDEEN [63] after these lectures were written, so it is not described here.

The third way to discuss the dragging of inertial frames is in terms of its local effects on gyroscopes and particles. Consider an event q in or near the star. Focus attention on two local frames (*i.e.*

two tetrads of basis vectors being carried by observers) near that event. The first frame is that of a static observer (*i.e.* an observer who is at rest relative to the « distant stars »). Since the co-ordinates (t, x^1, x^2, φ) become inertial at infinity, the space co-ordinate grid is itself at rest relative to the « distant stars, » so the static observer must be at rest relative to it. In other words, he moves along a world line with $(x^1, x^2, \varphi) = \text{const}$; his 4-velocity is

$$\boldsymbol{u} = (g_{tt})^{-\frac{1}{2}} (\partial/\partial t) . \tag{3.19}$$

We require that this static observer keep his frame—*i.e.* the little tetrad of basis vectors that he carries with himself—fixed relative to the « distant stars, » or equivalently, fixed relative to the space co-ordinate grid. The static observer performs local physical experiments in the frame that he carries with him (though he must always keep in mind that his frame is being accelerated, since he is not falling freely). In performing experiments he, like any ordinary physicist or engineer, distinguishes between the (locally) flat, 3-dimensional Euclidean space of his frame, which is measured by means of rods, and the time direction, which is measured by means of clocks. The basis vector associated with his time direction is just his 4-velocity, $\boldsymbol{e}_0 = \boldsymbol{u}$. His little Euclidean 3-space is orthogonal to $\boldsymbol{u}$, so as basis vectors in it he can take any set of three vectors that are orthogonal to $\boldsymbol{u}$ and fixed relative to the « distant stars » —*i.e.* relative to the (x^1, x^2, φ) co-ordinate grid. If he were an experimentalist, he would choose his space vectors to be orthogonal to each other and have unit length; but if he were a theoretician, he might prefer to choose them in the following nonorthonormal manner, which ties in nicely with the (x^1, x^2, φ) grid:

$$\boldsymbol{e}_1 = \partial/\partial x_1 , \qquad \boldsymbol{e}_2 = \partial/\partial x^2 , \qquad \boldsymbol{e}_3 = \partial/\partial\varphi - (g_{\varphi t}/g_{tt})(\partial/\partial t) . \tag{3.20a}$$

With this choice of basis vectors, the metric of his little, locally flat Euclidean 3-space would have components h_{jk}, which are related to the components g_{jk} of the metric of spacetime and u_j of this 4-velocity by

$$h_{jk} = \boldsymbol{e}_j \cdot \boldsymbol{e}_k = g_{jk} - u_j u_k = g_{jk} - g_{tj} g_{tk}/g_{tt} , \qquad h^{jk} = g^{jk} . \tag{3.20b}$$

The second frame we consider is a local inertial frame which is momentarily at rest relative to the static observer at the event q. The dragging of inertial frames by the rotating star causes the space axes of this local inertial frame to be rotating relative to the space basis vectors $\boldsymbol{e}_1, \boldsymbol{e}_2, \boldsymbol{e}_3$ of the static observer. The angular velocity with which the static observer sees the inertial axes to rotate is

$$\boldsymbol{\omega}_{IA} = \tfrac{1}{2} (g_{tt})^{\frac{1}{2}} (\nabla \times \boldsymbol{\gamma}) . \tag{3.21a}$$

This equation is written in the elementary vector notation of the static observer's little, 3-dimensional, local Euclidean space; in that space g_{tt} is a scalar, and the vector $\boldsymbol{\gamma}$ is

$$\boldsymbol{\gamma} = g^{tj}\boldsymbol{e}_j = g^{t\varphi}\boldsymbol{e}_3\,. \tag{3.21b}$$

This formula for the angular velocity of the inertial axes can be inferred from problem 1 at the end of Sect. **89** of LANDAU and LIFSHITZ [17]. (Note that $\boldsymbol{\omega}_{IA} = -\boldsymbol{\Omega}$, where $\boldsymbol{\Omega}$ is the quantity calculated there; also note that because of a difference in signature conventions, our formula for $\boldsymbol{\omega}_{IA}$ is the negative of the formula which would be inferred from there.) An alternative method of calculating $\boldsymbol{\omega}_{IA}$ will be outlined in the Appendix to this Section. A third method can be based on eq. (3.7c).

In the particular case of our rotating star, the angular velocity of the inertial axes is orthogonal to the φ-direction

$$\boldsymbol{\omega}_{IA} = -\frac{g_{tt}}{2\sqrt{(-g)}}\left\{\frac{\partial(g_{t\varphi}/g_{tt})}{\partial x^1}\frac{\partial}{\partial x^2} - \frac{\partial(g_{t\varphi}/g_{tt})}{\partial x^2}\frac{\partial}{\partial x^1}\right\}. \tag{3.22}$$

(To derive this equation from eqs. (3.21a), (3.21b) one must keep in mind that the basis vectors $\boldsymbol{e}_1, \boldsymbol{e}_2, \boldsymbol{e}_3$ as defined in (3.20a) are not orthonormal.) Far from a rotating relativistic star, in a nearly spherical coordinate system, $g_{t\varphi}$ becomes

$$g_{t\varphi} \approx g_{t\varphi}/g_{tt} \approx 2(J/r)\sin^2\theta \tag{3.23}$$

(cf. eqs. (3.17) and (3.18); also Sect. **3**.3.4 below). Consequently, the angular velocity of the inertial axes relative to a static observer is

$$\boldsymbol{\omega}_{IA} \simeq \frac{J}{r^3}\left\{\sin\theta\frac{\partial}{r\partial\theta} + 2\cos\theta\frac{\partial}{\partial r}\right\} = (1/r^3)\{-\boldsymbol{J} + 3(\boldsymbol{J}\cdot\partial/\partial r)(\partial/\partial r)\}\,. \tag{3.24}$$

Notice that near the poles of the star ($\theta \approx 0$; $\boldsymbol{J} \approx J(\partial/\partial r)$), the inertial axes rotate in the same direction as the fluid of the star rotates ($\boldsymbol{\omega}_{IA}$ is parallel to $\boldsymbol{J}$); but near the equator ($\theta \approx \pi/2$; $\boldsymbol{J}\cdot\partial/\partial r \approx 0$), the inertial axes rotate in the opposite direction ($\boldsymbol{\omega}_{IA}$ is antiparallel to $\boldsymbol{J}$).

While this might seem paradoxical at first, an analogy which is attributed to Schiff makes it seem more reasonable: Consider a rotating, solid sphere immersed in a viscous fluid. As it rotates, the sphere will drag the fluid along with it in qualitatively the same way as a rotating star drags freely falling particles. At various points in the fluid, set down little rods, and watch how the fluid rotates them as it flows past them. Near the poles the fluid clearly will rotate the rods in the same direction as the star rotates. But near the equator, because the fluid is dragged more rapidly at small radii than at large, the end of a rod closest to the sphere is dragged by the fluid more rapidly than the far end of the rod. Consequently, the rod rotates in the opposite direction to the rotation of the sphere (*)!

Let us return to the gravitational problem. A static observer discovers that the inertial axes near him are rotating relative to his axes—and, hence, relative to the distant stars—by examining the behavior of gyroscopes and freely falling particles in his frame. A gyroscope which he carries with him, applying the necessary forces to its center of mass (no torques!), will keep its axis at rest relative to the momentarily comoving local inertial frame. Consequently, the static observer will see the gyroscope's axis precess with the angular velocity of the local inertial frame; he will see a rate of change of angular momentum $\boldsymbol{S}$ for the gyroscope, which is given by

$$\mathrm{d}\boldsymbol{S}/\mathrm{d}\tau = \boldsymbol{\omega}_{IA}\times\boldsymbol{S}\,. \tag{3.25}$$

Similarly, the static observer will notice that a Coriolis force acts on any freely falling particle that moves through his frame; more particularly, he will measure, for a particle with small ordinary velocity $\boldsymbol{v}$ ($|\boldsymbol{v}| \equiv v \ll 1$) relative to him, an acceleration given by

$$\boldsymbol{a} \equiv \mathrm{d}\boldsymbol{v}/\mathrm{d}\tau = -\nabla[\ln\sqrt{(g_{tt})}] + 2\boldsymbol{\omega}_{IA}\times\boldsymbol{v}\,. \tag{3.26}$$

The first term is just the acceleration which results from the observer's refusal to fall freely; the second is the Coriolis acceleration which results from his refusal to attach his frame to a set of gyroscopes—*i.e.* to an inertial frame. For a derivation of eq. (3.26) see problem 1 at the end of Sect. **89** of LANDAU and LIFSHITZ [17]. (See also the Appendix to this Subsection).

A group of physicists at Stanford University, under the direction of FAIRBANK, is preparing an experiment to measure the dragging of inertial frames produced by the rotation of the Earth. They will put a super-conducting gyroscope in orbit around the Earth and measure its tendency to precess relative to the distant stars. In addition to the precession produced by the dragging of inertial frames, they should see a « curvature precession » due to the

---

(*) This analogy can be made fully rigorous. Recall (eq. (3.7*c*)) that a family of world lines with 4-velocity

$$\boldsymbol{u} = u^t[(\partial/\partial t) + \omega(\partial/\partial\varphi)]\,,$$

$\omega =$ angular velocity of cumulative dragging $= -g_{\varphi t}/g_{\varphi\varphi} = 2J/r^3$ at large r is irrotational in the sense that the antisymmetric part, $\omega_{\alpha\beta}$, of $u_{\alpha;\beta}$ vanishes. This means physically that a gyroscope which is carried along one of these world lines does *not* precess relative to the neighboring world lines. Consequently, an imaginary fluid with the above 4-velocity plays the same role for the gravitational problem as does the real fluid for Schiff's analogy.

fact that the gyroscope is moving along an orbit for which

$$\text{d(Circumference)}/\text{d(Proper radial distance)} = 2\pi(1 - M/r) < 2\pi\,. \tag{3.27}$$

The curvature precession is one and a half orders of magnitude larger than the frame-dragging precession, but both should be measurable. Notice that for this experiment the angular velocity of the inertial axes is

$$\omega_{IA} \sim J/R^3 \sim 0.5MR^2\Omega/R^3 \sim 0.5(M/R)\,\Omega \sim \sim 0.5\cdot 10^{-9}\,\Omega \sim 3\cdot 10^{-9}\,\text{rad/day} \sim 0.2\text{ s of arc per year}\,. \tag{3.28}$$

Appendix to Subsection 3'1.7.

In the text we discussed the angular velocity of the inertial axes, $\boldsymbol{\omega}_{IA}$, from the standpoint of a $3+1$ space-time split of the space-time geometry. An alternative calculation of $\boldsymbol{\omega}_{IA}$ makes use of the full 4-dimensional notation (*). Let a stationary observer near a rotating star carry a gyroscope at its center-of-mass. The spin, $\boldsymbol{S}$, of the gyroscope then rotates with the inertial axes. Mathematically speaking, the spin is Fermi-transported along the observer's world line. If we take the Killing-vector field $\boldsymbol{\xi} \equiv \partial/\partial t$ as a tangent (but not a unit tangent!) to the world line, then the fact that the gyroscope is at rest in the stationary observer's frame means that

$$\boldsymbol{\xi}\cdot\boldsymbol{S} = 0\,,$$

and the Fermi-transport law reads

$$\nabla_{\boldsymbol{\xi}}\boldsymbol{S} = -\boldsymbol{\xi}(\boldsymbol{S}\cdot\nabla_{\boldsymbol{\xi}}\boldsymbol{\xi})/(\boldsymbol{\xi}\cdot\boldsymbol{\xi})\,.$$

Here $\nabla_{\boldsymbol{\xi}}$ denotes the invariant derivative along $\boldsymbol{\xi}$. The stationary observer does not Fermi-transport his space axes; rather, he keeps them fixed with respect to the stationary grid of our co-ordinate system. Mathematically speaking, this means that he Lie-drags them with the Killing-vector field $\boldsymbol{\xi} = \partial/\partial t$. As a consequence, the rate of change of $\boldsymbol{S}$ relative to his space grid, as measured in units of the group-theoretically defined time co-ordinate t, is the Lie-derivative of $\boldsymbol{S}$ with respect to $\boldsymbol{\xi}$:

$$(\mathrm{d}\boldsymbol{S}/\mathrm{d}t)_{\text{relative to stationary observer's grid}} = \mathscr{L}_{\boldsymbol{\xi}}\boldsymbol{S} \equiv \nabla_{\boldsymbol{\xi}}\boldsymbol{S} - \nabla_{\boldsymbol{S}}\boldsymbol{\xi}\,. \tag{3.29a}$$

(*) This approach was pointed out to me by R. Geroch.

It is a straightforward computation, which makes use of Killing's equation $\xi_{\alpha;\beta}+\xi_{\beta;\alpha}=0$ and of the orthogonality relation $\boldsymbol{\xi}\cdot\boldsymbol{S}=0$, to put the Lie-derivative into the form

$$(\mathscr{L}_{\xi}\boldsymbol{S})^{\alpha}=\xi_{\nu}\varepsilon^{\nu\alpha\sigma\lambda}\omega_{IA\sigma}S_{\lambda}\,, \tag{3.29b}$$

where

$$\omega^{\alpha}_{IA}=-\frac{\xi_{\nu}\varepsilon^{\nu\alpha\sigma\lambda}\nabla_{\sigma}\xi_{\lambda}}{2(\boldsymbol{\xi}\cdot\boldsymbol{\xi})}\,. \tag{3.29c}$$

Equations (3.29*a*), (3.29*b*) are the 4-dimensional version of the 3-dimensional equation (3.25) for the transport of the spin. From this it is clear that the 4-vector $\boldsymbol{\omega}_{IA}$ is orthogonal to the stationary observer's world line, and that its nonvanishing space-part represents the angular velocity of the spin $\boldsymbol{S}$ relative to the stationary observer's frame, as measured in units of the observer's *proper* time. (The difference between proper time and co-ordinate time cancels out on the two sides of eq. (3.29*b*) because each side is proportional to $\boldsymbol{\xi}$.) Thus, the $\boldsymbol{\omega}_{IA}$ of equation (3.29*c*), when regarded as a 3-vector in the space orthogonal to $\boldsymbol{\xi}$, must be identical to the $\boldsymbol{\omega}_{IA}$ of equation (3.21*a*). One can verify, by straightforward manipulations in our special co-ordinate system, that this is so.

3'1.8. Methods of constructing a relativistic, rotating stellar model. When actually constructing stellar models, one should specialize the co-ordinates x^1 and x^2 in such a way as to simplify the task (cf. Subsect. **3'**1.1). A particular choice of x^1 and x^2 which has been used widely in the past—but not necessarily the most useful choice—is the one which makes $g_{12}=0$ and $g_{22}=g_{\varphi\varphi}/\sin^2 x^2$. In the resulting co-ordinate system, with the notation $x^1\equiv r$ and $x^2\equiv\theta$, and with special symbols introduced for the metric coefficients, the line element reads

$$\mathrm{d}s^2=N^2\,\mathrm{d}t^2-e^{2\gamma}\,\mathrm{d}r^2-r^2K^2[\mathrm{d}\theta^2+\sin^2\theta\,(\mathrm{d}\varphi-\omega\,\mathrm{d}t)^2]\,. \tag{3.30}$$

Notice that the line element depends on four « gravitational potentials »

$$N=N(r,\theta)\,,\qquad \gamma=\gamma(r,\theta)\,,\qquad K=K(r,\theta)\,,\qquad \omega=\omega(r,\theta)\,. \tag{3.31}$$

For the special case of zero rotation these potentials become

$$N\to\exp\,[2\Phi(r)]\,,\qquad \gamma\to-\tfrac{1}{2}\ln\,[1-2m(r)/r]\,,\qquad K\to 1\,,\qquad \omega\to 0\,, \tag{3.32}$$

leaving the line element in the standard form

$$\mathrm{d}s^2=\exp\,[2\Phi]\,\mathrm{d}t^2-(1-2m/r)^{-1}\,\mathrm{d}r^2-r^2(\mathrm{d}\theta^2+\sin^2\theta\,\mathrm{d}\varphi^2)$$

for a spherically symmetric, nonrotating star. Notice also that the function $\omega(r, \theta)$ in the line element (3.30) is precisely equal to the angular velocity of cumulative dragging (eq. (3.17)) discussed in the last Section.

The Einstein field equations which govern the structure of a rotating star —*i.e.* which link the metric functions $N(r, \theta)$, $\gamma(r, \theta)$, $K(r, \theta)$, $\omega(r, \theta)$ to the thermodynamic variables $\varrho(r, \theta)$, $p(r, \theta)$, and to the angular velocity $\Omega(r, \theta)$— have been worked out by COHEN and BRILL [16] and by others. There is known only one exact analytic solution to these equations for the interior of a nonsingular, bounded rotating body (WAHLQUIST [18]). Unfortunately, the exterior gravitational field of Wahlquist's solution is unknown. Moreover, the surfaces of constant pressure are prolate, which suggests that the body is being deformed by the gravitational fields of external objects.

To construct numerical solutions for rotating stars should not be enormously more difficult in general relativity theory than in Newtonian theory, because the number of independent variables is the same—two. Already numerical solutions have been constructed for the case of the slow, rigid rotation of a fully relativistic star (Subsect. **3**.3 below), for the case of the rapid rotation of a nearly Newtonian star (Subsect. **3**.2 below), and for the case of a rapidly rotating, fully relativistic disk (Subsect. **3**.4 below).

The general case of a rapidly rotating, fully relativistic star is made particularly difficult by the fact that no analytic solution is known for the general external gravitational field. There are two different numerical methods which could be used to construct fully relativistic models: 1) numerical integration of the coupled, nonlinear, elliptic partial differential equations for N, ω, K, γ, Ω, ϱ, p. (Newtonian integrations of this sort have been performed by JAMES [19], STOECKLY [20], OSTRIKER and BODENHEIMER [21], OSTRIKER and MARK [22], and OSTRIKER and HARTWICK [23]). 2) Numerical application of a variational principle for rotating, relativistic configurations, which has been constructed by Hartle and Sharp [13]—see also Boyer and Lindquist [12].

Let us discuss briefly the Hartle-Sharp variational principle. HARTLE and SHARP have found functionals $\mathscr{M}[N, \omega, K, \gamma, \Omega, \varrho, p]$, $\mathscr{J}[N, \omega, K, \gamma, \Omega, \varrho, p]$ and $\mathscr{A}[N, \omega, K, \gamma, \Omega, \varrho, p]$ which have the following property: *extremize the quantity* $\mathscr{M}$ *with respect to variations of the functions* $N(r, \theta)$, $\omega(r, \theta)$, $K(r, \theta)$, $\gamma(r, \theta)$, $\Omega(r, \theta)$, $\varrho(r, \theta)$, $p(r, \theta)$, *in which variations the following constraints are imposed*: 1) $\mathscr{A}$ *is held constant*; 2) $\mathscr{J}$ *is held constant*; 3) p *and* ϱ *are constrained to obey a particular equation of state*, $p = p(\varrho, s)$ *where* s *is the entropy per baryon*; 4) s *is constrained to be constant throughout the configuration being perturbed* (*isentropic configuration*; *adiabatic temperature gradient*); 5) *the perturbation itself is adiabatic* ($\delta s = 0$ *everywhere*). *Those configurations which extremize* $\mathscr{M}$ *subject to these contraints are isentropic equilibrium configurations in rigid rotation. Moreover, every isentropic, rigidly rotating configuration for the given equation of state* $p(\varrho, s)$ *can be obtained by this variational procedure. Finally, the extremal value* *of $\mathscr{M}$ is the total mass-energy of the rotating configuration, M; the constant value of $\mathscr{J}$ is its total angular momentum, J, and the constant value of $\mathscr{A}$ is its total number of baryons, A.*

This variational principle for rotating configurations bears a close resemblance to the theorem of extremal mass-energy (Subsect. 3.5.3 of RSSD-65) for nonrotating configurations. However, there are these important differences: 1) The theorem of extremal mass-energy for nonrotating configurations is valid whether or not the configuration under study is isentropic, whereas the Hartle-Sharp variational principle is valid only when isentropy is imposed from the start. 2) A nonrotating equilibrium configuration has extremal mass-energy only when compared with *physically realizable*, momentarily-static configurations, obeying the same equation of state, and obtained from the equilibrium configuration by an adiabatic perturbation. However, equilibrium rotating configurations in the Hartle-Sharp variational principle have extremal values of $\mathscr{M}$ when compared with *any* set of functions (N, ω, K, γ, Ω, ϱ, p)—*physically realizable or not*— which obey the imposed contraints. Put in more mathematical terms: all comparison configurations in the nonrotating case are forced to obey the initial value equations of general relativity; whereas comparison configurations in the Hartle-Sharp variational principle are not constrained to obey the initial value equations. (In the case of rotation, the initial value equations have proved too difficult to be solved analytically before the variation is performed.)

Since these lectures were written BARDEEN [63] has developed a new variational principle for rotating stars, which is a complete analogue of the nonrotating theorem of external mass-energy, and which places no constraints on the distribution of entropy.

3.1.9. The external gravitational field. Only one physically acceptable exact solution to the vacuum field equations outside a rotating star is known. This is the Kerr [24] solution, which has the form

$$ds^2 = dt^2 - (r^2 + a^2)\sin^2\theta\, d\varphi^2 - \frac{2Mr}{r^2 + a^2\cos^2\theta}(dt + a\sin^2\theta\, d\varphi)^2 - (r^2 + a^2\cos^2\theta)\left(d\theta^2 + \frac{dr^2}{r^2 + a^2 - 2Mr}\right) \tag{3.33a}$$

when expressed in the co-ordinates of BOYER and LINDQUIST [12]. This solution involves two constants: the total mass-energy, M, and the angular momentum per unit mass,

$$-a = J/M\,. \tag{3.33b}$$

For zero angular momentum, the Kerr solution reduces to the Schwarzschild solution.

Any rotating star must be deformed; and as a result, the Newtonian potential, U, far away from it ($r \gg M$) must have nonzero multipole moments. HERNANDEZ [25] has shown that for the Kerr solution the multipole moments in the Newtonian realm are given by

$$(3.33c) \qquad \begin{cases} \left(2^{2n}_{\text{moment}}\right) \equiv \begin{pmatrix}\text{coefficient of } -r^{-(2n+1)}P_{2n}(\cos\theta) \\ \text{in distant Newtonian potential}\end{pmatrix} = (-1)^{n+1} M a^{2n}, \\ \left(2^{2n+1}_{\text{moment}}\right) \equiv 0 . \end{cases}$$

Because of this very special relationship between the multipole moments and the angular momentum, the Kerr solution cannot represent correctly the external field of *any* realistic stars (except for a « set of measure zero »). However, it seems quite likely that, if a rotating star undergoes relativistic gravitational collapse, it will necessarily radiate away precisely enough of each of its multipole moments so as to leave behind a « black hole » with Kerr's spacetime geometry (3.33*a*). The evidence for this is that, so far as we can tell, the Kerr solution is the only stationary, axially symmetric solution to the vacuum field equations which has a nonsingular event horizon with the topology of a 2-sphere; small, stationary perturbations of the Kerr geometry are singular at the horizon (see Subsect. 3˙3.4 below).

The problem of finding the most general solution for the external field of a rotating star can be reduced to the problem of solving the following coupled equations for two axially symmetric functions, α and β, in flat, Euclidean space:

$$(3.34) \qquad \nabla\cdot(\cosh^2\beta\nabla\alpha) = 0 , \qquad \nabla^2\beta + \tfrac{1}{2}\sinh 2\beta(\nabla\alpha)^2 = 0 .$$

This reduction of the problem has been performed and discussed by MATZNER and MISNER [25] (see also ERNST [26]). However, neither they nor anyone else has been able to find a physically acceptable exact solution more general than that of KERR. Several other solutions are known (see, *e.g.*, KINNERSLY [27]; KERR and DEBNY [28]; but each of these is either physically pathological or else so complicated that its physical meaning is as yet a mystery.

Approximate solutions for the external field are known in two cases: the nearly Newtonian case, and the slowly rotating case.

In any nearly Newtonian regime of the external field—*i.e.* far from *any* rotating star, and everywhere outside a nearly Newtonian, rotating star—the geometry of spacetime is

$$(3.35) \qquad ds^2 = [(1 - 2U + O(U^2)]\,dt^2 +$$
$$+ [4J/r^2 + O(J^2/r^4) + O(UJ/r^2)]r\sin^2\theta\,dt\,d\varphi - [1 + O(U)][dr^2 + r^2 d\theta^2 + r^2\sin^2\theta\,d\varphi^2] .$$

Here U is the Newtonian potential ($U \ll 1$), J is the star's angular momentum, and $O(x)$ means terms of the order of the dimensionless number x. (One uses group-theoretic arguments to identify the constant J in this line element with the star's angular momentum; see HARTLE and SHARP [29]; also PAPAPETROU [30]).

For a slowly rotating, fully relativistic star, the external field takes a more complicated form which will be presented and discussed in Subsect. 3'3.4.

3'1.10. Angular momentum and rotational energy. In proving theorems about rotating stars the following expression for the total angular momentum is sometimes useful:

$$(3.36a)\qquad J = -\int T^t_{\varphi}(-g)^{\frac{1}{2}}\,dx^1\,dx^2\,d\varphi = \int (\varrho + p)\,u^t |g_{\varphi\varphi}|(\Omega - \omega)(-g)^{\frac{1}{2}}\,dx^1\,dx^2\,d\varphi\,.$$

(Here g is the determinant of the metric tensor.) For example, this expression played a fundamental role in the derivation of the Hartle-Sharp [29] variational principle (Subsect. 3'1.8).

Another quantity of interest for theorems and practical applications is the rotational energy, E_{rot}. We define E_{rot} to be the total mass-energy of the rotating star minus the total mass-energy of a nonrotating star which contains the same number of baryons and has the same distribution of entropy per baryon. For any rigidly rotating, Newtonian star (even a rapidly rotating, highly deformed one), if the angular momentum is changed by an amount ΔJ while holding the entropy per baryon fixed, the rotational energy must change by

$$(3.3b)\qquad \Delta E_{\text{rot}} = \Omega\,\Delta J$$

(see, *e.g.*, OSTRIKER and GUNN [31]). ZEL'DOVICH [32] has argued that this must be true also in general relativity, since one can imagine increasing the angular momentum of a star by means of a long lever, with one end in the star and the other out in the Newtonian region far from the star. The angular momentum and energy are inserted into the lever—and thence transmitted to the star—in the Newtonian region, so the relation $\Delta E_{\text{rot}} = \Omega\,\Delta J$ must be valid. A similar argument which avoids the use of the lever can be made by letting electromagnetic or gravitational waves carry the rotational energy and angular momentum into the star (cf. OSTRIKER and GUNN [31]). The result again is $\Delta E_{\text{rot}} = \Omega\,\Delta J$.

A third derivation, which relies on manipulations of the Einstein field equations rather than on physical arguments, has been given by HARTLE [33]. He points out that, according to the Hartle-Sharp [29] variational principle for rigidly rotating, *isentropic* stars, any changes ΔA in the number of baryons and ΔJ in the angular momentum produce a change ΔM in the total mass-energy given by

$$\Delta M = \Omega \Delta J + [(\varrho + p)/nu^t]\Delta A \,. \tag{3.37}$$

Here $(\varrho + p)/nu^t$ is the radially constant injection energy per baryon (cf. Subsect. **3**˙1.3 and **3**˙1.4); and it is assumed that the star's entropy is held constant while the changes are made. For the special case $\Delta J = 0$ this equation gives back the injection-energy result of Subsect. **3**˙1.4; and for the special case $\Delta A = 0$ it gives the relation (3.36*b*) between rotational energy and angular momentum. For the generalization of expression (3.37) to the nonisotropic case see Subsect. **10**˙7 of [6].

In the limit of slow rotation, where the moment of inertia $I = J/\Omega$ can be considered independent of angular velocity, eq. (3.37) tells us that

$$J = I\Omega \,, \qquad E_{\text{rot}} = \tfrac{1}{2} I\Omega^2 \,. \tag{3.36c}$$

in general relativity as in Newtonian theory.

3˙2. *Rapidly rotating, nearly Newtonian stars.* – Every rotating, relativistic stellar model constructed to date, except the Bardeen-Wagoner disk (Subsect. **3**˙4), is either nearly Newtonian ($M/R \ll 1$) or slowly rotating ($\Omega^2 R^3/M \ll 1$). In this Subsection we describe the nearly Newtonian case; and in Subsect. **3**˙3 we present the slowly rotating theory.

3˙2.1. The post-Newtonian formalism. Rotating, nearly Newtonian stars are modeled using post-Newtonian expansions of general relativity theory. The ideal post-Newtonian formalism for studying stars made of perfect fluid is that of CHANDRASEKHAR [34]. Let us recall (cf. Subsect. **2**˙3.5 of RSSD-65) that the fractional errors in Newtonian theory due to post-Newtonian corrections are of order $2M/R$ ($\sim 10^{-6}$ for the sun; ~ 0.5 for massive neutron stars). Chandrasekhar's post-Newtonian approximation takes into account the dominant sources of those errors, but leaves errors of order $(2M/R)^2$ ($\sim 10^{-11}$ for the sun). The post-post-Newtonian approximation of CHANDRASEKHAR and NUTKU [64] takes account of the dominant remaining errors, but still has errors of order $(2M/R)^3$ ($\sim 10^{-17}$ for the sun).

3˙2.2. MacLaurin spheroids, Jacobi ellipsoids, and Roche model. CHANDRASEKHAR has used his post-Newtonian formalism to study the effects of general relativity on those rapidly but rigidly rotating fluid bodies which have played the key role in the development of our Newtonian intuition about rapid rotation—the MacLaurin spheroids, the Jacobi ellissoids, and the Roche model. To understand fully the results of these analyses, one must be familiar with the Newtonian theories of the MacLaurin spheroids, the Jacobi ellipsoids, and the Roche model—and I assume that most readers are not, However, several interesting effects can be described easily.

The MacLaurin spheroids are axially symmetric configurations of uniformly rotating fluid. In Newtonian theory their surfaces of constant pressure are oblate spheroids. The overstrong, post-Newtonian gravitational forces deform these « level surfaces »; roughly speaking, the surfaces are pulled inward at the equator, so that they are less eccentric.

CHANDRASEKHAR [35] describes the deformations in terms of the equations for the surfaces relative to a particular co-ordinate system. KREFETZ [36] describes them in terms of what an observer far away would see with his eyes. BECK (1966 unpublished work) has described them in terms of embedding diagrams (cf. Subsect. 3.5.1 of RSSD-65) for the intrinsic geometries of the level surfaces.

The Jacobi ellipsoids are uniform-density, rigidly rotating configurations which are *not* axially symmetric. In Newtonian theory their level surfaces are ellipsoids with all three axes unequal. The post-Newtonian deformations have been studied by CHANDRASEKHAR [37].

In both the MacLaurin spheroids and the Jacobi ellipsoids, the post-Newtonian deformations are huge (formally infinite!) for those special ratios of axes where the Newtonian configurations are in neutral equilibrium (*i.e.*, possess zero-frequency modes of oscillation). In effect, the post-Newtonian gravitational forces drive the Newtonian zero-frequency displacements until they become infinitely large—or, more precisely, until the assumptions of the perturbation theory break down. This phenomenon is a static analogue of the famous dynamical « general-relativistic instability » in spherical stars with adiabatic indices Γ_1 near $\frac{4}{3}$. Whenever a Newtonian configuration is in near-neutral equilibrium, one can expect such large post-Newtonian effects.

In the MacLaurin spheroids there is a particular eccentricity, $e = 0.812\,670$, at which a slight change in angular velocity can move the configuration either along the MacLaurin sequence, or into the nonaxially-symmetric Jacobi family of ellipsoids. This is called the « bifurcation point » between the MacLaurin spheroids and Jacobi ellipsoids. Here, as in configurations with zero-frequency modes, the post-Newtonian forces produce a marked effect—but this time the effect is an « indeterminacy » rather than an « infinity »: there is an infinite family of different post-Newtonian configurations, all with the same mass and angular velocity, at the bifurcation point; whereas, in Newtonian theory there is but one such configuration.

The Roche model for a rigidly rotating star consists of a tenuous, centrifugally deformed envelope in the gravitational field of a massive, undeformed core. CHANDRASEKHAR [65] has studied post-Newtonian effects on the Roche model, particularly for the case of maximum angular velocity, where mass is being shed at the equator. This analysis yields, as usual, post-Newtonian deformations; but nothing particularly exciting comes out of it.

3.2.3. The limit of slow rotation. Post-Newtonian effects on slowly rotating, nearly Newtonian stars are of considerable interest in connection with supermassive-star models for quasars. Since a fairly complete discussion of this subject was given by FOWLER [38] in the proceedings of Course XXXV of this summer school, we shall not review if here—except to remark that rotation is very effective in suppressing the general-relativistic instability when $M/R \ll 1$. For post-1966 work on the interaction between rotational effects and relativistic effects in supermassive stars see DURNEY and ROXBURG [39], and BISNOVATYI-KOGAN, ZEL'DOVICH, and NOVIKOV [40]. For a systematic formulation of the theory of the structures of slowly rotating stars in Chandrasekhar's post-Newtonian formalism see KREFETZ [41].

3.3. *Slowly rotating, fully relativistic stars.*

3.3.1. Relevance to pulsars. Just as the theory of rotating stars becomes tractable in the nearly Newtonian case, so also it is tractable in the fully relativistic, but slowly rotating case. By slow rotation we mean that the angular velocity $\Omega(x^1, x^2)$ is much smaller than the value

$$\Omega_{\text{crit}} \equiv (M/R^3)^{\frac{1}{2}}, \tag{3.38}$$

which would deform the star so greatly that it would come close to shedding mass at its equator.

The rotating neutron stars, which are thought to be in pulsars, are slowly rotating in this sense. For a «typical» neutron star the mass is $M \approx 1 M_\odot \approx 1.5$ km, the radius is $R \approx 10$ km; and, consequently, the critical angular velocity is

$$\Omega_{\text{crit}} \approx 1/10 \text{ km} \approx 3 \cdot 10^4/\text{s}. \tag{3.39a}$$

For comparison, the shortest period for any pulsar (as of June 1969) is 0.033 s, which corresponds to a neutron-star angular velocity at the star's surface of

$$\Omega_{\text{Crab pulsar}} \approx 2\pi/0.033 \text{ s} \approx 2 \cdot 10^2/\text{s} \approx 10^{-2}\, \Omega_{\text{crit}}. \tag{3.39b}$$

When first formed, the pulsar neutron stars are probably rapidly rotating ($\Omega \sim \Omega_{\text{crit}}$); but, apparently, the rapid rotation does not last long (much less than 100 years in the case of the Crab).

Since $\Omega/\Omega_{crit} \ll 1$, but $2M/R \sim 0.04$ to 0.5 for all the neutron stars in observed pulsars, those neutron stars can be studied by expanding the Einstein field equations for a fully relativistic, rotating star in powers of the dimensionless ratio Ω/Ω_{crit}. Such a slow-rotation expansion can be performed fairly easily for differentially rotating stars as well as for rigidly rotating stars, but only the rigid case has been treated thus far. In the following Sections we shall describe that treatment, which is due to HARTLE and his colleagues, and which is being presented in a series of papers in the Astrophysical Journal.

3'3.2. Equations of structure. HARTLE [42] (paper I of the series) has derived the equations of structure for a slowly and rigidly rotating, fully relativistic stellar model, accurate to second order in the ratio Ω/Ω_{crit}. (An analogous set of equations has been derived in the Russian-language literature by SEDRAKYAN and CHUBARYAN [43]). At first order in Ω/Ω_{crit}, the only effect of the rotation is to drag the inertial frames; but at second order it also deforms the star.

The conclusions of Hartle's work are summarized by the following prescription for calculating the stellar structure, which is taken with only a few changes from Sect. **2** of HARTLE and THORNE [44].

a) *An equation of state is assumed.* As the first step in the calculation of a slowly rotating stellar model, a one-parameter equation of state

$$\varrho = \varrho(p)\,, \qquad n = n(p) \tag{3.40}$$

is specified. Here p is the pressure, ϱ is the total density of mass-energy, and n is the number density of baryons. For neutron stars and white dwarfs this relation will be one of the equations of state for cold degenerate matter, while for supermassive stars, it will be the polytropic equation of state of index 3. (In order to avoid using such a one-parameter equation of state, one would have to derive the equations of thermal equilibrium and energy transfer for rotating, relativistic stars. This Hartle has not done. He has restricted himself to the theory of hydrostatic structure.)

b) *Values for the central density and angular velocity are chosen.* For slow rotation, once the equation of state is specified, there is a unique equilibrium configuration for each choice of the central density and angular velocity. The small perturbations away from a nonrotating equilibrium configuration are all proportional to the angular velocity or to its square. Consequently, for a given central density, all the models of different angular velocities can be obtained from a single model by scaling.

Having chosen a value, $\Omega \ll \Omega_{\text{crit}}$, of the angular velocity for each value of the central density, one constructs a sequence of equilibrium models by integrating the general-relativistic equations of structure for a sequence of central densities. The integration procedure is the following.

c) A nonrotating stellar model is computed. For a given density, the nonrotating equilibrium configuration is determined by integrating the Tolman-Oppenheimer-Volkoff equation of hydrostatic equilibrium for the pressure $p(r)$, and the mass, $m(r)$, interior to a given radius:

$$\frac{\mathrm{d}p}{\mathrm{d}r} = -\frac{(\varrho + p)(m + 4\pi r^3 p)}{r(r - 2m)}, \tag{3.41a}$$

$$\mathrm{d}m/\mathrm{d}r = 4\pi r^2 \varrho . \tag{3.41b}$$

The integration is performed outward, starting at the star's center, $r = 0$. At the star's center m is 0; ϱ is the given central density, ϱ_c; and p is $p(\varrho_c)$ as given by the equation of state (eq. (3.40)). The radius of the surface of the star, R, is that value of r at which $p(r)$ drops to zero; and the value of $m(r)$ there is the star's total mass, M.

The metric that describes the spherically symmetric geometry of the nonrotating star has the Schwarzschild form

$$\mathrm{d}s^2 = \exp[2\Phi(r)]\mathrm{d}t^2 - [1 - 2m(r)/r]^{-1}\mathrm{d}r^2 - r^2(\mathrm{d}\theta^2 + \sin^2\theta\,\mathrm{d}\varphi^2) . \tag{3.14c}$$

The remaining function in the metric, $\Phi(r)$, is determined by integrating outward from the center the equation

$$\mathrm{d}\Phi/\mathrm{d}r = -(\varrho + p)^{-1}(\mathrm{d}p/\mathrm{d}r) \tag{3.41d}$$

with the boundary condition $\Phi(\infty) = 0$. The total number of baryons in the nonrotating star can be found from the integral

$$A = \int_0^R n(r)[1 - 2m(r)/r]^{-\frac{1}{2}} 4\pi r^2 \,\mathrm{d}r . \tag{3.41e}$$

For further discussion of the construction of nonrotating stellar models see RSSD-65; Subsect. **3.1-3.4**, **5.2**, **5.3**, and **6.1**.

d) The rotational perturbations in the metric and stress-energy tensor are speciefied. When the equilibrium configuration described above is set into slow rotation, the geometry of spacetime around it and its interior distribution of stress-energy are changed. With an appropriate choice of co-ordinates, the perturbed geometry is described by

$$(3.42)\qquad ds^2 = \varepsilon^{2\Phi}[1+2(h_0+h_2P_2)]\,dt^2 - \frac{[1+2(m_0+m_2P_2)/(r-2m)]}{1-2m/r}\,dr^2 - \\ - r^2[1+2(v_2-h_2)P_2][d\theta^2+\sin^2\theta(d\varphi-\omega\,dt)^2] + O([\Omega/\Omega_{\text{crit}}]^3).$$

Here, $P_2 = P_2(\cos\theta) = (3\cos^2\theta - 1)/2$ is the Legendre polynomial of order 2; ω, which is the angular velocity of cumulative dragging as defined in equation (3.17a) (*), is a function of r and is proportional to the star's angular velocity, Ω; and h_0, h_2, m_0, m_2, v_2 are all functions of r that are proportional to Ω^2.

In the above co-ordinate system the fluid inside the star moves with the 4-velocity appropriate to rigid rotation, of which the contravariant components are

$$(3.43)\qquad \begin{cases} u^t = (g_{tt} + 2g_{t\varphi}\Omega + g_{\varphi\varphi}\Omega^2)^{-\frac{1}{2}} = \\ \qquad = \exp[-\Phi][1+\frac{1}{2}r^2\sin^2\theta(\Omega-\omega)^2\exp[-2\Phi] - h_0 - h_2P_2], \\ u^\varphi = \Omega u^t, \qquad u^r = u^\theta = 0. \end{cases}$$

The quantity

$$(3.44)\qquad \bar{\omega} \equiv \Omega - \omega,$$

which appears in the expression for u^t, is the angular velocity of the fluid relative to particles with zero angular momentum (cf. the beginning of Subsect. 3·1.7). It plays a fundamental role in the equations of structure below, as one might expect from eq. (3.7c) for the rotation tensor of the fluid.

The baryon number density, the density of mass-energy, and the pressure of the fluid are affected by the rotation because the rotation deforms the star. In the interior of the star at given (r, θ), in a reference frame that is momentarily moving with the fluid, the pressure is

$$(3.45a)\qquad p + (\varrho + p)(p_0^* + p_2^* P_2) \equiv p + \Delta p,$$

the density of mass-energy is

$$(3.45b)\qquad \varrho + (\varrho + p)(d\varrho/dp)(p_0^* + p_2^* P_2) \equiv \varrho + \Delta\varrho,$$

(*) In their series of papers, Hartle *et al.* give the quantity ω the misnomer « angular velocity of the local inertial frame ».

and the number density of baryons is

$$n + (\varrho + p)(\mathrm{d}n/\mathrm{d}p)(p_0^* + p_2^* P_2) \equiv n + \Delta n \,. \tag{3.45c}$$

Here, p_0^* and p_2^* are dimensionless functions of r, proportional to Ω^2, which describe the pressure perturbation; all other parameters were defined above. The stress-energy tensor for the fluid in the rotating star is, of course,

$$T_\mu{}^\nu = (\varrho + \Delta\varrho + p + \Delta p)\, u_\mu u^\nu - (p + \Delta p)\, \delta_\mu{}^\nu \,. \tag{3.46}$$

The rotational perturbations of the star's structure are described by the functions $\bar\omega$, h_0, m_0, p_0^*, h_2, m_2, v_2, p_2^*. These functions are calculated from Einstein's field equations as described below.

e) The angular velocity of cumulative dragging and the moment of inertia are determined. In equilibrium, a rotating star attains a balance between pressure forces, gravitational forces, and centrifugal forces. The magnitude of the centrifugal force, in the corotating reference frames relevant to Hartle's equations, is determined not by the angular velocity Ω of the fluid relative to a distant observer, but its angular velocity relative to particles with zero angular momentum, $\bar\omega(r)$. This quantity is of first order in Ω and is found by integrating the differential equation

$$\frac{1}{r^4}\frac{\mathrm{d}}{\mathrm{d}r}\left(r^4 j \frac{\mathrm{d}\bar\omega}{\mathrm{d}r}\right) + \frac{4}{r}\frac{\mathrm{d}j}{\mathrm{d}r}\,\bar\omega = 0 \,, \tag{3.47}$$

where

$$j(r) = \exp\left[-\Phi(r)\right]\left[1 - 2m(r)/r\right]^{\frac{1}{2}} \,. \tag{3.48}$$

The solution must be regular at the origin; outside the star it takes the form

$$\bar\omega(r) = \Omega - 2J/r^3 \,, \tag{3.49}$$

where J is the total angular momentum of the star. The moment of inertia of the star is defined as the ratio J/Ω and is conveniently expressed in terms of the radius of gyration, R_g:

$$R_g = (\text{moment of inertia}/M)^{\frac{1}{2}} = (J/\Omega M)^{\frac{1}{2}} \,. \tag{3.50}$$

In practice, one integrates eq. (3.47) outward from the center of the star, where the boundary conditions $\bar\omega = \bar\omega_c$ and $\mathrm{d}\bar\omega/\mathrm{d}r = 0$ are imposed. One chooses the constant $\bar\omega_c$ arbitrarily. When one reaches the surface—and only then—one can determine the angular velocity, Ω, and angular momentum, J, corresponding to $\bar{\omega}_c$:

$$J = \frac{1}{6} R^4 \left(\frac{\mathrm{d}\bar{\omega}}{\mathrm{d}r}\right)_{r=R}, \qquad \Omega = \bar{\omega}(R) + \frac{2J}{R^3} \tag{3.51}$$

(cf. eq. (3.49)). If a different value of Ω is desired, one rescales the function $\bar{\omega}(r)$ to obtain it:

$$\bar{\omega}(r)_{\text{new}} = \bar{\omega}(r)_{\text{old}} (\Omega_{\text{new}}/\Omega_{\text{old}}) . \tag{3.52}$$

HARTLE has shown from eq. (3.47) that the angular velocity of cumulative dragging, ω, decreases monotonically outward from the center of a rigidly rotating star.

An expression for the total angular momentum, equivalent to (3.51), which is sometimes useful is (cf. eq. (3.36a))

$$J = (8\pi/3) \int_0^R (\varrho + p) r^4 \exp[-\Phi] (1 - 2m/r)^{-\frac{1}{2}} \bar{\omega} \, \mathrm{d}r . \tag{3.51$'$}$$

The above results on frame dragging and angular momentum, in a different but equivalent form, predate Hartle's work; see BRILL and COHEN [51]; also HARTLE and SHARP [13].

f) The spherical deformations of the star are calculated. – The spherical part of the rotational deformation is calculated by integrating the $l = 0$ equations of hydrostatic equilibrium for the « mass perturbation factor » m_0 and the « pressure perturbation factor » p_0:

$$\frac{\mathrm{d}m_0}{\mathrm{d}r} = 4\pi r^2 \frac{\mathrm{d}\varrho}{\mathrm{d}p} (\varrho + p) p_0^* + \frac{1}{12} j^2 r^4 \left(\frac{\mathrm{d}\bar{\omega}}{\mathrm{d}r}\right)^2 - \frac{1}{3} r^3 \frac{\mathrm{d}j^2}{\mathrm{d}r} \bar{\omega}^2 , \tag{3.53a}$$

$$\frac{\mathrm{d}p_0^*}{\mathrm{d}r} = -\frac{m_0(1 + 8\pi r^2 p)}{(r - 2m)^2} - \frac{4\pi(\varrho + p) r^2}{(r - 2m)} p_0^* + \frac{1}{12} \frac{r^4 j^2}{r - 2m} \left(\frac{\mathrm{d}\bar{\omega}}{\mathrm{d}r}\right)^2 + \frac{1}{3} \frac{\mathrm{d}}{\mathrm{d}r} \left(\frac{r^3 j^2 \bar{\omega}^2}{r - 2m}\right) . \tag{3.53b}$$

These equations are integrated outward, with the boundary conditions that both m_0 and p_0 vanish at the origin. With these boundary conditions, the rotating star will have the same central density as the nonrotating one (cf. eq. (3.45b)). Outside the star, m_0, which is one of the perturbations in the metric tensor, is given by

$$m_0 = \delta M - J^2/r^3 , \tag{3.53c}$$

where δM is a constant. Consequently, the total mass-energy of the star with central density ϱ_c and angular velocity Ω is

$$M + \delta M = m(R) + m_0(R) + J^2/R^3 \,, \tag{3.54}$$

where R is the radius of the star's surface.

Once p_0^*, δM, and J have been calculated, one obtains the function $h_0(r)$ from the algebraic relations

$$h_0 = -\frac{\delta M}{r - 2M} + \frac{J^2}{r^3(r-2M)} \qquad \text{outside the star,} \tag{3.55a}$$

$$h_0 = -p_0^* + \tfrac{1}{3} r^2 \exp[-2\Phi]\bar{\omega}^2 + h_{0c} \qquad \text{inside the star.} \tag{3.55b}$$

Here h_{0c} is a constant determined by the demand that h_0 be continuous across the star's surface.

The binding energy of a relativistic star is the difference between its rest mass and its total mass-energy:

$$E_B = \mu_B A - M \,. \tag{3.56a}$$

Here the rest mass has been expressed as the product of the total baryon number, A, and the rest mass per baryon, μ_B. The change in binding energy, δE_B, of the rotating over the nonrotating star is calculated from m_0, p_0, and the density of internal energy,

$$\varepsilon = \varrho - \mu_B n \,, \tag{3.56b}$$

through the formulae

$$\left\{\begin{aligned} \delta E_B &= -J^2/R^3 + \int_0^R 4\pi r^2 B(r)\,\mathrm{d}r \,, \\ B(r) &= (\varrho + p) p_0^* \left\{\frac{\mathrm{d}\varrho}{\mathrm{d}p}\left[\left(1 - \frac{2m}{r}\right)^{-\frac{1}{2}} - 1\right] - \frac{\mathrm{d}\varepsilon}{\mathrm{d}p}\left(1 - \frac{2m}{r}\right)^{-\frac{1}{2}}\right\} + \\ &\quad + (\varrho - \varepsilon)\left(1 - \frac{2m}{r}\right)^{-\frac{3}{2}}\left[\frac{m_0}{r} + \frac{1}{3} j^2 r^2 \bar{\omega}^2\right] - \\ &\quad - \frac{1}{4\pi r^2}\left[\frac{1}{12} j^2 r^4 \left(\frac{\mathrm{d}\bar{\omega}}{\mathrm{d}r}\right)^2 - \frac{1}{3}\frac{\mathrm{d}j^2}{\mathrm{d}r} r^3 \bar{\omega}^2\right] . \end{aligned}\right. \tag{3.57}$$

The change in the total baryon number is then obtained from eq. (3.56). Notice that the change in binding energy δE_B (eq. (3.57)), the change in total mass-energy δM (eq. (3.54)), the rotational energy E_{rot} (eq. (3.36c)), and the angular momentum J (eq. (3.51)) must be related by

(3.58) $$E_{\text{rot}} = \tfrac{1}{2} J\Omega = \delta M - (\mathrm{d}M/\mathrm{d}A)\,\delta A = \delta M - (1 - 2M/R)^{\frac{1}{2}}(\delta M + \delta E_B)\,.$$

g) The quadrupole deformations of the star are calculated. One calculates the quadrupole part of the deformations by integrating the $l = 2$ equations

(3.59a) $$\frac{\mathrm{d}v_2}{\mathrm{d}r} = -2\frac{\mathrm{d}\Phi}{\mathrm{d}r}h_2 + \left(\frac{1}{r} + \frac{\mathrm{d}\Phi}{\mathrm{d}r}\right)\left[-\frac{1}{3}r^3\frac{\mathrm{d}j^2}{\mathrm{d}r}\bar{\omega}^2 + \frac{1}{6}j^2r^4\left(\frac{\mathrm{d}\bar{\omega}}{\mathrm{d}r}\right)^2\right],$$

(3.59b) $$\frac{\mathrm{d}h_2}{\mathrm{d}r} = \left\{-2\frac{\mathrm{d}\Phi}{\mathrm{d}r} + \frac{r}{r-2m}\left(2\frac{\mathrm{d}\Phi}{\mathrm{d}r}\right)^{-1}\left[8\pi(\varrho + p) - \frac{4m}{r^3}\right]\right\}h_2 -$$
$$- \frac{4v_2}{r(r-2m)}\left(2\frac{\mathrm{d}\Phi}{\mathrm{d}r}\right)^{-1} + \frac{1}{6}\left[\frac{\mathrm{d}\Phi}{\mathrm{d}r}r - \frac{1}{r-2m}\left(2\frac{\mathrm{d}\Phi}{\mathrm{d}r}\right)^{-1}\right]r^3j^2\left(\frac{\mathrm{d}\bar{\omega}}{\mathrm{d}r}\right)^2 -$$
$$-\frac{1}{3}\left[\frac{\mathrm{d}\Phi}{\mathrm{d}r}r + \frac{1}{r-2M}\left(2\frac{\mathrm{d}\Phi}{\mathrm{d}r}\right)^{-1}\right]r^2\frac{\mathrm{d}j^2}{\mathrm{d}r}\bar{\omega}^2\,,$$

subject to the four boundary conditions.

(3.59c) $$h_2 = v_2 = 0 \quad \text{at} \quad r = 0 \quad \text{and at} \quad r = \infty\,.$$

Outside the star h_2 and v_2 have the analytic forms

(3.60a) $$h_2 = J^2\left(\frac{1}{Mr^3} + \frac{1}{r^4}\right) + KQ_2^2\left(\frac{r}{M} - 1\right),$$

(3.60b) $$v_2 = -\frac{J^2}{r^4} + K\frac{2M}{[r(r-2M)]^{\frac{1}{2}}}Q_2^1\left(\frac{r}{M} - 1\right),$$

Where K is a constant and Q_n^m is the associated Legendre function of the second kind.

Once h_2 and v_2 have been calculated from eq. (3.59), the nonradial mass and pressure perturbation factors, m_2 and p_2^*, are determined from the algebraic relations

(3.61a) $$m_2 = (r - 2m)\left[-h_2 - \frac{1}{3}r^3(\mathrm{d}j^2/\mathrm{d}r)\,\bar{\omega}^2 + \frac{1}{6}r^4j^2\left(\frac{\mathrm{d}\bar{\omega}}{\mathrm{d}r}\right)^2\right],$$

(3.61b) $$p_2^* = -h_2 - \tfrac{1}{3}r^2\exp[-2\Phi]\bar{\omega}^2\,.$$

3.3.3. The level surfaces. The rotational deformation of the star is most clearly understood as follows: The surface of constant given density (level surface) that lies at radius r in the nonrotating configuration is displaced in the rotating configuration to radius

$$r + \xi_0(r) + \xi_2(r)\, P_2(\cos\theta)\,, \tag{3.62a}$$

where

$$\xi_0 = \delta r = -\, p_0^*(\varrho + p)/(\mathrm{d}p/\mathrm{d}r)\,, \qquad \xi_2 = -\, p_2^*(\varrho + p)/(\mathrm{d}p/\mathrm{d}r)\,. \tag{3.62b}$$

Equation (3.62*a*) describes the level surfaces in a particular co-ordinate system. An invariant description of a level surface can be obtained by embedding it in a three-dimensional flat space. To do this, we ask for the surface in a three-dimensional flat space with polar co-ordinates r^*, θ^*, φ^*, which has the same intrinsic geometry as the level surface in our star. Accurate to order Ω^2, the desired 3-surface in flat space is the spheroid

$$r^*(\theta^*) = r + \xi_0(r) + \{\xi_2(r) + r[v_2(r) - h_2(r)]\}\, P_2(\cos\theta^*)\,. \tag{3.63a}$$

The mean radius of this spheroid is

$$\bar{r}^* = r + \xi_0(r)\,, \tag{3.63b}$$

and its eccentricity is

$$e = [(\text{radius at equator})^2/(\text{radius at pole})^2 - 1]^{\frac{1}{2}} = [-3(v_2 - h_2 + \xi_2/r)]^{\frac{1}{2}}\,. \tag{3.63c}$$

The mean radius of the star and the eccentricity of its surface we denote by $\bar{R} = R + \delta R$ and e_s, respectively; and we calculate them by setting $r = R$ in eq. (3.63).

Recall from Subsect. **3˙1.5** that in a rigidly rotating star the level surfaces have not only constant density and pressure, ϱ and p, but also constant redshift factor, $1/u^t$. By combining eqs. (3.43) for u^t with eq. (3.62) for the level surface, and with the algebraic relations (3.41*a*), (3.55*b*), and (3.61*b*), one can prove the following: *The value of the redshift factor $1/u^t$ on a given level surface (surface of given ϱ and p) changes, in going from the nonrotating configuration to the rotating one, by a factor which is identically the same for all level surfaces:*

$$(1/u^t)_{\text{surface of given } \varrho \text{ in rotating configuration}} = (1 - h_{0c}) \times (e^{\Phi})_{\text{radius of same } \varrho \text{ in nonrotating configuration}}\,. \tag{3.64}$$

Notice that the factor involved is h_{0c}, which is the fractional change in $(g_{tt})^{\frac{1}{2}} = e^{\Phi}$ at the center of the star (cf. eqs. (3.42) and (3.55*b*)).

3·3.4. External gravitational field. (Based on HARTLE and THORNE [44].) The deformation of the star's external gravitational field is determined by the metric perturbation factors h_2, v_2 and m_2 of eqs. (3.60) and (3.61). Far from the star, $-h_2$ becomes the nonspherical perturbation in the Newtonian potential and thereby determines the star's quadrupole moment to be

$$Q = (\text{coefficient of } -r^{-3}P_2(\cos\theta) \text{ term in Newtonian potential}) = \tfrac{8}{5}KM^3 + J^2/M. \tag{3.65}$$

Here K is the constant in eq. (3.60), which is determined by numerical integration of eq. (3.59).

By combining eqs. (3.42), (3.49), (3.56a), (3.60), (3.61), and (3.65), we obtain for the line element outside the rotating star, accurate to second order in the angular velocity,

$$\begin{aligned} \mathrm{d}s^2 = {} & \left(1-2\frac{\mathscr{M}}{r}+2\frac{J^2}{r^4}\right)\left\{1+2\left[\frac{J^2}{\mathscr{M}r^3}\left(1+\frac{\mathscr{M}}{r}\right)+\right.\right. \\ & \left.\left.+\frac{5}{8}\frac{Q-J^2/\mathscr{M}}{\mathscr{M}^3}Q_2^2\left(\frac{r}{\mathscr{M}}-1\right)\right]P_2(\cos\theta)\right\}\mathrm{d}t^2-\left(1-2\frac{\mathscr{M}}{r}+2\frac{J^2}{r^4}\right)^{-1}\cdot \\ & \cdot\left\{1-2\left[\frac{J^2}{\mathscr{M}r^3}\left(1-\frac{5\mathscr{M}}{r}\right)+\frac{5}{8}\frac{Q-J^2/\mathscr{M}}{\mathscr{M}^3}Q_2^2\left(\frac{r}{\mathscr{M}}-1\right)\right]P_2(\cos\theta)\right\}\mathrm{d}r^2- \\ & -r^2\left[\!\!\left[1+2\left\langle-\frac{J^2}{\mathscr{M}r^3}\left(1+2\frac{\mathscr{M}}{r}\right)+\frac{5}{8}\frac{Q-J^2/\mathscr{M}}{\mathscr{M}^3}\cdot\right.\right.\right. \\ & \left.\left.\left.\cdot\left\{\frac{2\mathscr{M}}{[r(r-2\mathscr{M})]^{\frac{1}{2}}}Q_2^1\left(\frac{r}{\mathscr{M}}-1\right)-Q_2^2\left(\frac{r}{\mathscr{M}}-1\right)\right\}\right\rangle P_2(\cos\theta)\right]\!\!\right]\cdot \\ & \cdot\left\{\mathrm{d}\theta^2+\sin^2\theta\left[\mathrm{d}\varphi-\left(\frac{2J}{r^3}\right)\mathrm{d}t\right]^2\right\}. \end{aligned} \tag{3.66}$$

(Recall that Q_n^m is the associated Legendre function of the second kind.) The only constants that enter into this line element are the total mass of the rotating star, $M + \delta M \equiv \mathscr{M}$; the star's total angular momentum, J; and the star's mass quadrupole moment, Q. (Whenever M appears in a term of second order in Ω^2, it can be replaced by $\mathscr{M}$ without affecting the line element to order Ω^2).

For stars such as the sun, which are not very relativistic (*i.e.* $\mathscr{M}/r \ll 1$), the line element (3.66) can be expanded in powers of $\mathscr{M}/r$. For the sun, the relative magnitudes of the various quantities in the external line element are

$$\mathscr{M}/r < \mathscr{M}/R \approx 2\cdot 10^{-6}\,, \qquad J/r^2 < J/R^2 \sim 10^{-12}\,, \qquad Q/r^3 < Q/R^3 \lesssim 10^{-10}\,. \tag{3.67}$$

Consequently, to an accuracy of one part in 10^{15} for the sun, we can drop terms of order $(J/r^2)^2$, $(\mathscr{M}/r)(J/r^2)$, and $(\mathscr{M}/r)(Q/r^3)$ in the line element (3.62), obtaining

$$(3.68)\qquad ds^2=\left[1-\frac{2\mathscr{M}}{r}+\frac{2Q}{r^3}P_2(\cos\theta)\right]dt^2-\left[1-2\frac{\mathscr{M}}{r}+\frac{2Q}{r^3}P_2(\cos\theta)\right]^{-1}dr^2-$$
$$-\left[1-\frac{2Q}{r^3}P_2(\cos\theta)\right]r^2\left\{d\theta^2+\sin^2\theta\left[d\varphi-\frac{2J}{r^3}dt\right]^2\right\}.$$

The terms in $2\mathscr{M}/r$ account for the Newtonian gravitational attraction and the relativistic perihelion shift of mercury's orbit, while the terms in $2Q/r^3$ produce a perihelion shift associated with the sun's oblateness (DICKE and GOLDENBERG [45]). The term in $2J/r^3$ produces the dragging of inertial frames that also leads to a precession of mercury's perihelion with respect to the distant stars, but a precession that is negligibly small compared to mercury's total relativistic perihelion shift. From the line element (3.68), one can readily verify that the perihelion shift of mercury due to a slightly oblate, rotating sun is very accurately given by adding the relativistic shift for the spherical sun to the oblateness shift for the deformed Newtonian sun.

The line element given in eq. (3.66) describes correctly the geometry outside any slowly rotating configuration. It is interesting to compare it with the only known exact solution exterior to a rotating object—the Kerr metric of eq. (3.33). To second order in the angular velocity, the Kerr line element (3.33) can be transformed to the form of eq. (3.66) by the transformation

$$(3.69)\qquad r\to r\left\{1-\frac{a^2}{2r^2}\left[\left(1+\frac{2\mathscr{M}}{r}\right)\left(1-\frac{\mathscr{M}}{r}\right)+\cos^2\theta\left(1-\frac{2\mathscr{M}}{r}\right)\left(1+\frac{3\mathscr{M}}{r}\right)\right]\right\},$$
$$\theta\to\theta-a^2\cos\theta\sin\theta\frac{1}{2r^2}\left(1+2\frac{\mathscr{M}}{r}\right).$$

It is then seen easily that, for the Kerr solution,

$$J=\mathscr{M}a\quad\text{and}\quad Q=J^2/\mathscr{M}\qquad(\textit{i.e.},\ K=0;\ \text{cf. eq. (3.65)})\ .$$

These results are the same as those derived by HERNANDEZ [28] by a different method. The special relationship $Q=J^2/\mathscr{M}$ between the quadrupole moment and the angular momentum shows the very special nature of the Kerr solution (see Subsect. **3**·1.9 for additional discussion).

Notice another significance of the special relationship $Q=J^2/\mathscr{M}$: Because of the behavior of the Legendre functions

$$(3.70)\qquad Q_2^1(r/\mathscr{M}-1)\approx[2(r/\mathscr{M}-2)]^{-\frac{1}{2}}\,,\qquad Q_2^2(r/\mathscr{M}-1)\approx(r/\mathscr{M}-2)^{-1}\,,$$

near the gravitational radius, $r = 2\mathcal{M}$, the rotational and quadrupole deformations (3.66) of the Schwarzschild geometry are large there (infinite, in fact, at this order in the expansion in powers of J and Q), unless the Kerr condition, $Q = J^2/\mathcal{M}$ is satisfied.

This fact leads one to speculate that, just as the Schwarzschild geometry is the only static (*i.e.*, nonrotating and time-independent) solution to Einstein's vacuum equations with a nonsingular event horizon that has the topology of a 2-sphere (see ISRAEL [46]), so also the Kerr geometry may be the only stationary solution with a nonsingular horizon of 2-sphere topology. The significance of this speculation for the gravitational collapse of a rotating star was stated in Subsect. 3˙1.9.

3˙3.5. Numerical models for the slowly rotating neutron stars of pulsar theory. In paper II of the series on slow rotation (HARTLE and THORNE [44]) Hartle's equations are integrated numerically to yield models for slowly rotating neutron stars, white dwarfs, and supermassive stars. Some of the results for neutron stars, which are relevant to pulsars, are the following.

So long as the rotation is rigid, it does not change significantly the curve of mass *vs.* radius for neutron stars or white dwarfs. (See Fig. 2): For example, it can increase the maximum mass of a neutron star by 20% at most. Differential rotation, on the other hand, with $\Omega \gg (M/R^3)^{\frac{1}{2}}$ at the center and $\Omega \sim (M/R^3)^{\frac{1}{2}}$ at the surface, *might* produce very marked changes in neutron-star masses and radii; it is known to do so for Newtonian white dwarfs (OSTRIKER and BODENHEIMER [21]; OSTRIKER and HARWITT [23]; OSTRIKER and MARK [22].) For this reason it is very important to know whether neutron stars can rotate differentially—*i.e.*, whether long-range ordering of nucleon spins above nuclear densities, or agglomeration of nucleon into nuclei and quasi-nuclei below nuclear densities can produce sufficient crystalization to destroy differential rotation (see Salpeter's lectures in this volume); or whether the neutron star's magnetic field will extend sufficiently deep into the star to force it to rotate rigidly.

The moment of inertia of a massive neutron star exhibits an interesting anomaly due to relativistic effects: For a given mass M and radius R the moment of inertia is

$$I \approx (0.1 \text{ to } 0.4)\, MR^2\,. \tag{3.71}$$

This is somewhat larger than one would expect from Newtonian intuition, given the fairly rapid decrease of density outward in neutron-star models. The anomalously large moment of inertia seems to be associated with the anomalously

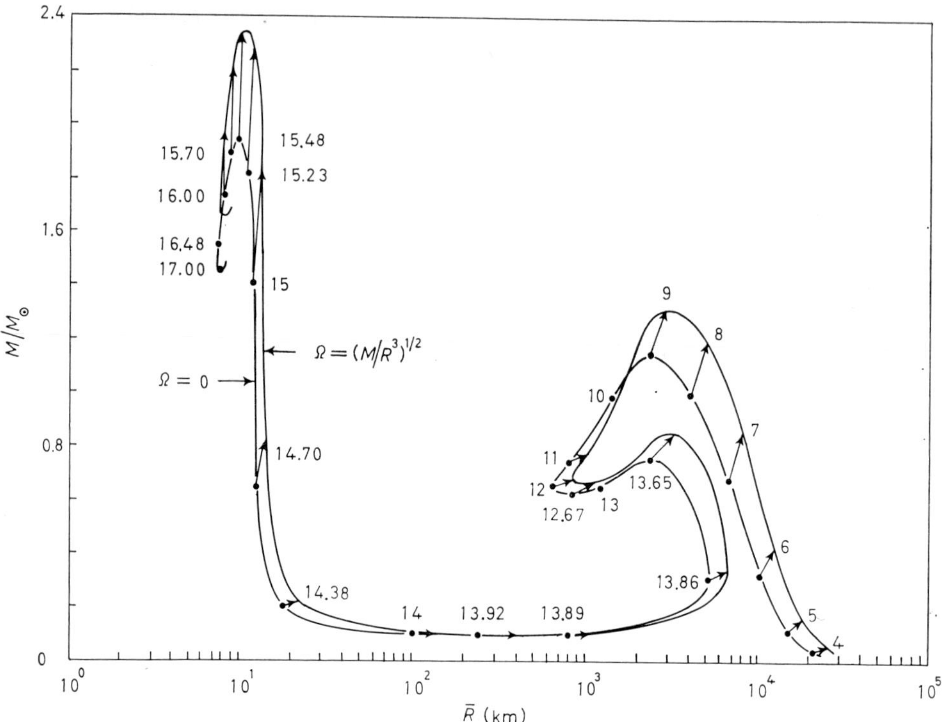

Fig. 2. – Effects of rotation on the masses and mean radii of Tsuruta-Cameron V_γ models for stars at the endpoint of thermonuclear evolution. (See Fig. 4 in RSSD-65 for discussion of the V_γ equation of state and the nonrotating V_γ models.) The curve with dots is a plot of mass, M, *vs.* radius, R, parametrized by the logarithm of central density in g cm^{-3}, for nonrotating V_γ configurations. Recall that for $\varrho_c < 10^9$ g/cm^3 the stars are stable white dwarfs; for $10^9 < \varrho_c < 10^{13.92}$ they are unstable; for $10^{13.92} < \varrho_c < 10^{15.48}$ they are stable neutron stars; and for $\varrho_c > 10^{15.48}$ they are unstable. The other curve is mass, $M + \delta M$, *vs.* mean radius $\bar{R} = R + \delta R$ for V_γ configurations rotating with uniform angular velocity, $\Omega = (M/R^3)^{\frac{1}{2}}$. This angular velocity is approximately the amount needed to produce shedding of mass at the star's equator, so that the slow-rotation method of computation is not actually valid for so large a value of Ω. For smaller angular velocities, where the method is valid, the deformation of the mass-radius curve is smaller by the dimensionless factor $\Omega^2 R^3/M$. The small arrows indicate the displacement, with increasing angular velocity, of configurations with the given central densities. To find the mass and mean radius of a configuration of given central density and given angular velocity, one moves out along the appropriate arrow by the fraction $\Omega^2 R^3/M$ of the total length of the arrow. (This figure is taken from HARTLE and THORNE [44]).

large volume

$$V \approx 1.5 \cdot \tfrac{4}{3}\pi R^3 \tag{3.72}$$

of a massive neutron star.

The dragging of inertial frames by a rotating neutron star looks very large when one characterizes it by

$$\frac{\omega(r)}{\Omega} = \frac{\text{angular velocity of comulative dragging}}{\text{angular velocity of star}}\,. \tag{3.73}$$

This quantity is shown for the V_γ family of neutron-star models in Fig. 3. Notice that near the center of a massive, stable neutron star the dragging angular velocity is about 75% of the star's angular velocity; and at the surface it is about 25%. Recall that outside the star

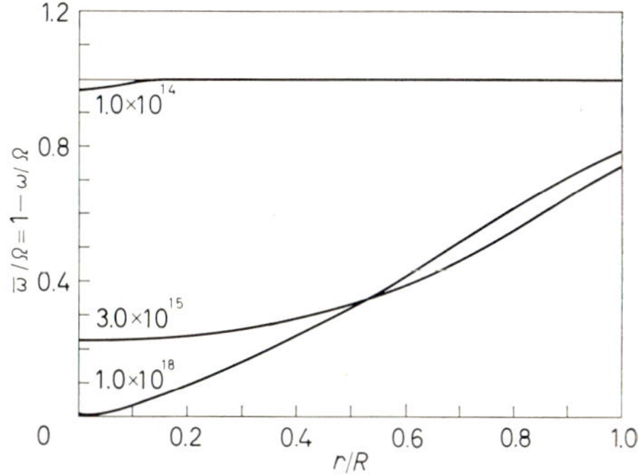

Fig. 3. – The fractional dragging of inertial frames as a function of radius in three V_γ models for stars at the endpoint of thermonuclear evolution (cf. Fig. 2). The model with $\varrho_c = 1.0\cdot10^{14}$ is a stable neutron star with $M = 0.101\,M_\odot$; that with $\varrho_c = 3.0\cdot10^{15}$ is a stable neutron star with $M = 1.95\,M_\odot$; that with $\varrho_c = 1.0\cdot10^{18}$ is an unstable star with $M = 1.49\,M_\odot$. Plotted against radius is $\bar{\omega}/\Omega = 1-\omega/\Omega$; recall (eq. (3.72)) that ω/Ω is the ratio of angular velocity of cumulative dragging to the angular velocity of the star.

$$\frac{\omega}{\Omega} = \frac{2J/r^3}{\Omega} \propto \frac{1}{r^3}\,. \tag{3.74}$$

For the neutron star in the Crab pulsar, $\Omega \approx 200/\text{s}$. Consequently, assuming the star has $M \sim (1.5 \text{ to } 2.0)\,M_\odot$, a gyroscope in a cavity at the star's center would precess with an angular velocity of

$$\omega_{IA} \sim (\omega/\Omega)_c\,\Omega \sim 100/\text{s}\,; \tag{3.75a}$$

a gyroscope just above the star's surface would precess with

$$\omega_{IA} \sim (\omega/\Omega)_s\,\Omega \sim 50/\text{s}\,; \tag{3.75b}$$

and a gyroscope at the « velocity of light circle » ($r = 1/\Omega$) would precess with

$$\omega_{IA} \sim (50/\text{s})(1/\Omega R)^3 \sim 10^{-5}/\text{s}\,. \tag{3.75c}$$

Although the above numbers for ω/Ω and ω_{IA} may look impressive, the effects of dragging of inertial frames on neutron-star models of pulsars seem to be negligible. This is because the rotation of the pulsar neutron stars is so slow ($\Omega/\Omega_{crit} \leqslant 10^{-2}$). A static observer near the surface of a massive rotating neutron star sees matter moving past him with velocity $\boldsymbol{v}$ to be accelerated, due to the dragging of inertial frames, by an amount

$$\boldsymbol{a}_{drag} = 2\boldsymbol{\omega}_{IA} \times \boldsymbol{v}$$

(cf. eq. (3.26)). The magnitude of this acceleration for a pulsar neutron star is

$$a_{drag} \sim \omega v \sim 0.2(R/r^2)^3 \Omega v \leqslant 40(R/r)^3 v \text{ s}^{-1} . \tag{3.76a}$$

For comparison, the acceleration due to the star's mass is

$$a_{mass} \sim M/r^2 \sim 0.3(R/r)^2 R^{-1} \sim 10^4(R/r)^2 \text{ s}^{-1} > 100\, a_{drag} . \tag{3.76b}$$

Consequently, the effects of inertial-frame dragging on the motion of matter near and above the star's surface are negligible.

A crucial feature of pulsar models not treated here is the strong magnetic field anchored in the rotating neutron star (see the lecture by PACINI in this volume). Because spacetime outside the star is rather strongly curved, the shape of the the magnetic field lines may differ somewhat (tens of percent near the star) from that predicted in Newtonian-Maxwellian theory (see GINZBURG and OZERNOY [47]). The differences might have an observable influence on the pulse shapes for pulsars.

3·3.6. Pulsations of slowly rotating stars. The theory of the effects of rigid rotation on the radial pulsations of relativistic stars is being developed in three of the papers by HARTLE and his colleagues (papers III, V, and VI). Paper III (HARTLE and THORNE [48]) develops « static criteria » for stability, analogous to the static criteria of the nonrotating case (Subsect. **4**·2.2-**4**·2.4 of RSSD-65). In these static criteria one compares the masses or binding energies of the rigidly rotating stars being tested, with the masses or binding energies of a certain family of differentially rotating stars; and from this comparison one diagnoses whether the rigid rotators are stable. In the post-Newtonian limit, for stars with adiabatic index Γ_1 very near $\frac{4}{3}$, these static criteria reduce to the simple analysis of FOWLER [38]. Fowler's analysis has revealed quantitatively the interplay between the stabilizing effects of rotation and the destabilizing effects of general relativity.

Paper V (CHITRE, HARTLE, and THORNE [49]), which is not yet finished, will derive the equations governing the quasi-radial pulsations of a relativistic star, accurate to second order in the angular velocity and first order in the pulsation amplitude. Those equations will be reduced to straightforward prescriptions for calculating i) the effects of rotation on the frequencies of the normal modes of pulsation, and ii) the rate at which the normal modes radiate gravitational waves because of the rotational deformation of the star. Paper VI (HARTLE and THORNE [50]) will present numerical calculations of these effects for realistic neutron-star models.

3·3.7. Mach's principle and slowly rotating configurations. In the last few years BRILL and COHEN (BRILL and COHEN [51]; COHEN and BRILL [16]; COHEN [52]) have derived analytic solutions to the differential equation for the angular velocity of cumulative dragging, ω, in several special cases; and using these solutions they have given a beautiful discussion of the extent to which Mach's principle is contained in general relativity theory. Two of their most interesting results are these:

First, consider a family of momentarily stationary, spherical clouds of dust with uniform density and uniform angular velocity Ω. Let all the dust clouds have the same rest mass, but give each successive one a smaller surface area, $4\pi R^2$. Examine the ratio ω_c/Ω, where ω_c is the angular velocity of cumulative dragging at the center and is also equal to the angular velocity of the inertial axes there. As R decreases, ω_c/Ω increases. Finally, at $R=0$, when the ball of dust has closed itself off from the rest of the universe and has become a closed Friedmann cosmological model, ω_c/Ω reaches unity. In fact, in this limit ω is constant and equals Ω throughout the interior of the dust ball. This means that the rotation has become completely undetectable; the inertial frames are now tied to the matter (cf. eq. (3.7*c*)).

Second, consider a family of concentric, rotating shells of masses M_i, radii R_i, and angular velocities Ω_i, in the nearly Newtonian limit $(R_i \gg M_i)$. Let all the shells rotate about the same axis. Then at the center—indeed, throughout the interior of the innermost shell—the angular velocity of cumulative dragging and the angular velocity of the local inertial axes are equal and are given by

$$\omega_c = \sum_i \tfrac{4}{3}(M_i/R_i)\,\Omega_i\,. \tag{3.77}$$

This confirms a conjecture by WHEELER [53] that inertial effects should go as $1/R_i$. It also tells us that, since the masses of shells with identical densities and thicknesses vary as $M_i \propto R_i^2$, it is the very distant, cosmological masses which should have the greatest effect on the inertial axes of an Earth-bound laboratory.

3·4. *Rapidly rotating, fully relativistic disk.* – BARDEEN and WAGONER [54, 55] are currently studying the general-relativistic theory of a body which rotates so rapidly that i) it is deformed into a thin disk, and ii) the centrifugal forces on each particle of matter dominate over the pressure forces so that the pressure can be neglected. They analyse such uniformly rotating disks by means of a post-Newtonian expansion in powers of

$$\gamma \equiv z_c/(1+z_c)\,, \tag{3.78}$$

where z_c is the redshift from the center of the disk to infinity. They have carried the expansion to five orders beyond Newtonian theory. The expansion seems to converge even in the ultrarelativistic regime ($z_c \to \infty, \gamma \to 1$). For example, the fractional binding energy has the form

$$E_b/M_0c^2 = (\gamma/5)[1 + (\tfrac{2}{5})\gamma + 0.1879\gamma^2 + 0.1002\gamma^3 + \\ + 0.0587\gamma^4 + 0.0367\gamma^5 + \ldots]\,, \tag{3.79}$$

which increases monotonically from 0 to about 0.4 as z_c increases from 0 to ∞.

This behavior of the binding energy is very different from the oscillatory behavior which occurs for nonrotating stars (RSSD-65, chapter 4), and which there signals the onset of one new relativistic instability after another with increasing z_c. From this contrast BARDEEN and WAGONER conjecture that rapidly rotating disks might not experience a relativistic instability, even in the limit of arbitrarily large redshift! They might be stable against gravitational collapse!

Another intriguing aspect of these disks is that their structure seems to remain nonsingular in the limit as $z_c \to \infty$. Also, for large z_c the angular momentum J is related to the total mass-energy M by $J \approx M^2$, and the quadrupole moment satisfies the corresponding Kerr relation $Q \approx J^2/M \approx M^3$.

These and other preliminary results of the analysis are summarized in WAGONER [54], and in BARDEEN and WAGONER [55].

4. – Gravitational collapse and black holes.

The theory of relativistic gravitational collapse, and of the « black holes » (*i.e.*, collapsed or « frozen » stars) which result from it, dates back to the pioneering work of OPPENHEIMER and SNYDER [56]; but most of the work on the subject was done in a big burst of effort following the discovery of quasars (1963 to about 1966). The review of this subject given in the oral version of these lectures was a shortened version of chapter 7 of RSSD-65 and of chapter 8 of RSSD-66, plus a description of some of the material reviewed in chapter 13 of ZEL'DOVICH and NOVIKOV [6].

The theory of gravitational collapse and black holes can be broken up into three subtopics: spherical collapse, nonspherical collapse, and possible astrophysical roles of collapse and black holes.

Spherical collapse was researched to death in the period 1963 to 1966. In my opinion it is fully understood, and it deserves very little further research. For the most detailed review of it to date see MISNER [57].

Nonspherical collapse is a difficult topic which is just beginning to crack under the vigorous efforts of a number of workers. Because I expect extensive developments in this subject in the next year or two, I shall not waste my time by giving a review of it now. However, I will say that it is now « ninety-nine and eighty-eight one-hundredths percent » certain that collapse with *small* nonspherical perturbations is characterized by the same type of event horizon as spherical collapse.

Interest in possible astrophysical roles of collapse and black holes is mounting rapidly right now. The only review now in existence of the wide variety of possibilities is that by ZEL'DOVICH and NOVIKOV [6].

5. – Gravitational waves.

My oral lectures on this subject consisted of a review of three recent developments in the theory of gravitational waves: i) the beautiful solution, by ISAACSON [58], to the problem of how to define and describe the energy carried by a gravitational wave; ii) the powerful methods devised by BURKE [59] for incorporating radiation reaction into slow-motion expansions of general relativity; and iii) the solution by Campolattaro, Price, and Thorne (see Thorne [60]) to the problem of the emission of gravitational waves by *fully relativistic*, pulsating stars.

My recent « Stony-Brook lectures » (Thorne [61]) contain a qualitative review of these three topics; and the paper by BURKE and THORNE [62] contains a more detailed review of topics ii) and iii).

REFERENCES

[1] K. S. THORNE: *High-Energy Astrophysics*, edited by L. GRATTON (New York, 1966). Cited in text as « RSSD-65 ».

[2] K. S. THORNE: *High-Energy Astrophysics*, vol. **3**, edited by C. DE WITT, P. VÉRON and E. SCHATZMAN (New York, 1967). Cited in text as « RSSD-66 ».

[3] J. R. OPPENHEIMER and G. VOLKOFF: *Phys. Rev.*, **55**, 374 (1939).

[4] B. K. HARRISON, K. S. THORNE, M. WAKANO and J. A. WHEELER: *Gravitation Theory and Gravitational Collapse* (Chicago, 1965).

[5] J. A. Wheeler: *Annual Reviews Astron. Astrophys.*, vol. **4** edited by L. Goldberg (Palo Alto, Cal., 1966).
[6] Ya. B. Zel'dovich and I. D. Novikov: *Stars and relativity*, Vol. **1** of *Relativistic Astrophysics* (Chicago, 1971).
[7] B. Carter: in preparation (1970).
[8] B. Carter: *Journ. Math. Phys.*, **10**, 70 (1969).
[9] A. Papapetrou: *Ann. Inst. H. Poincaré*, A-**4**, 83 (1966).
[10] R. H. Boyer: *Proc. Camb. Phil. Soc.*, **62**, 495 (1966).
[11] R. H. Boyer: *Proc. Camb. Phil. Soc.*, **61**, 527, 531 (1965).
[12] R. H. Boyer and R. W. Lindquist: *Phys. Lett.*, **20**, 504 (1966).
[13] J. B. Hartle and D. H. Sharp: *Phys. Rev. Lett.*, **15**, 909 (1965).
[14] Ya. B. Zel'dovich: *Pis'ma Žurn. Ėksp. Teor. Fiz.*, **1**, 40 (1965).
[15] H. Thirring: *Phys. Zeits.*, **19**, 33 (1918).
[16] J. M. Cohen and D. R. Brill: *Nuovo Cimento*, **56** B, 209 (1968).
[17] L. D. Landau and E. M. Lifshitz: *Classical Theory of Fields*, second edition (Reading, Mass., 1962).
[18] H. D. Wahlquist: *Phys. Rev.*, **172**, 1291 (1968).
[19] R. A. James: *Astrophys. Journ.*, **140**, 552 (1964).
[20] R. Stoeckly: *Astrophys. Journ.*, **142**, 208 (1965).
21] J. P. Ostriker and P. Bodenheimer: *Astrophys. Journ.*, **154**, 1089 (1968).
[22] J. P. Ostriker and J. W.-K. Mark: *Astrophys. Journ.*, **151**, 1075 (1968).
[23] J. P. Ostriker and F. D. A. Hartwick: *Astrophys. Journ.*, **153**, 797 (1968).
[24] R. P. Kerr: *Phys. Rev. Lett.*, **11**, 522 (1963); *Quasistellar Sources and Gravitational Collapse*, edited by I. Robinson, A. Schild and E. Schucking (Chicago, 1965).
[25] R. A. Matzner and C. W. Misner: *Phys. Rev.*, **154**, 1229 (1967).
[25*a*] W. Hernandez: *Phys. Rev.*, **159**, 1070 (1967).
[26] F. J. Ernst: *Phys. Rev.*, **167**, 1175 (1968).
[27] W. Kinnersly: *Journ. Math. Phys.*, in press (1969).
[28] R. P. Kerr and G. C. Debney: in press (1969).
[29] J. B. Hartle: *Astrophys. Journ.*, **147**, 317 (1967).
[30] A. Papapetrou: *Proc. Roy. Irish Acad.*, **52**, 11 (1948).
[31] J. P. Ostriker and J. E. Gunn: *Astrophys. Journ.*, **157**, 1395 (1969).
[32] Ya. B. Zel'dovich: private communication (1969).
[33] J. B. Hartle: *Astrophys. Journ.*, **161**, 111 (1970).
[34] S. Chandrasekhar: *Astrophys. Journ.*, **142**, 1488 (1965).
[35] S. Chandrasekhar: *Astrophys. Journ.*, **142**, 1513 (1965); **147**, 334 (1967).
[36] E. Krefetz: *Astrophys. Journ.*, **148**, 613 (1967).
[37] S. Chandrasekhar: *Astrophys. Journ.*, **148**, 621 (1967).
[38] W. A. Fowler: chapter in *High-Energy Astrophys*, edited by L. Gratton (New York, 1966).
[39] B. Durney and I. Roxburgh: *Proc. Roy. Soc.*, A **296**, 189 (1967).
[40] G. Bisnovaty-Kogan, Ya. B. Zel'dovich and I. D. Novikov: *Astr. Žur.*, **44**, 525 (1967).
[41] E. Krefetz: *Astrophys. Journ.*, **143**, 1004 (1966).
[41*a*] E. Krefetz: *Astrophys. Journ.*, **148**, 589 (1967).
[42] J. B. Hartle: *Astrophys. Journ.*, **150**, 1005 (1967).
[43] D. M. Sedrakyan and E. B. Chubaryan: *Astrofizika*, **4**, 551 (1968).
[44] J. B. Hartle and K. S. Thorne: *Astrophys. Journ.*, **153**, 807 (1968).
[45] R. H. Dicke and H. M. Goldenberg: *Phys. Rev. Lett.*, **18**, 313 (1967).

[46] W. ISRAEL: *Phys. Rev.*, **164**, 1776 (1967).

[47] V. L. GINZBURG and L. M. OZERNOY: *Žurn. Ėksp. Teor. Fiz.*, **47**, 1030 (1964) (English translation in *Sov. Phys. JETP*, **20**, 689 (1965)).

[48] J. B. HARTLE: *Astrophys. Journ.*, in press (1969).

[49] S. M. CHITRE, J. B. HARTLE and K. S. THORNE: *Astrophys. Journ.*, in preparation (1971).

[50] J. B. HARTLE and K. S. THORNE: *Astrophys. Journ.*, in preparation (1971).

[51] D. R. BRILL and J. M. COHEN: *Phys. Rev.*, **143**, 1011 (1966).

[52] J. M. COHEN: *Phys. Rev.*, **173**, 1258 (1968).

[53] J. A. WHEELER: *Gravitation and Relativity*, edited by H.-Y. CHIU and W. HOFFMAN (New York, 1964), p. 303.

[54] R. V. WAGONER: *Ann. Rev. Astron. Astrophys.*, **7**, 553 (1969).

[55] J. M. BARDEEN and R. V. WAGONER: *Astrophys. Journ. Lett.*, in press.

[56] J. R. OPPENHEIMER and H. SNYDER: *Phys. Rev.*, **56**, 455 (1939).

[57] C. W. MISNER: *Proceedings of 1968 Brandeis Summer Institute in Theoretical Physics* (1968).

[58] R. A. ISAACSON: *Phys. Rev.*, **166**, 1263, 1272 (1968).

[59] W. L. BURKE: Ph. D. Thesis, Caltech (available from University Microfilms, Ann Arbor, Mich.); paper in preparation (1970).

[60] K. S. THORNE: *Phys. Rev. Lett.*, **21**, 320 (1968).

[61] K. S. THORNE: *Proceedings of Third Summer Institute for Astronomy and Astrophysics*, State University of New York at Stony Brook, edited by H.-Y. CHIU and R. SEARS (in press) (1970).

[62] W. L. BURKE and K. S. THORNE: *Proceedings of the Relativity Conference in the Midwest* (in press) (1970).

[63] J. M. BARDEEN: *Astrophys. Journ.*, **162**, 71 (1970).

[64] S. CHANDRASEKHAR and Y. NUTKU: *Astrophys. Journ.*, **158**, 55 (1969).

[65] S. CHANDRASEKHAR: *Astrophys. Journ.*, **148**, 621 (1967).

Quasi-Stellar Objects: Their Importance for Cosmology and General Relativity.

G. R. Burbidge and E. M. Burbidge
University of California - San Diego, Cal.

The discovery of quasi-stellar objects (QSOs) with large red-shifts was thought to open up a new era in observational cosmology. The confusion that reigns at present is associated with the fact that the nature of the QSOs remains in doubt and there is still a serious question as to their distances and the interpretation of the red-shifts. Despite this, they are bound to have a large impact on cosmology. Their impact will depend on which of a number of hypotheses turns out be correct. The situation is summarized in Table I.

Table I.

Some tests of observational cosmology currently of importance	Relevance of QSOs
1) Hubble relation, determination of q_0	Large red-shifts (assumed cosmological)
2) Mean density of mass-energy in Universe (detection of intergalactic matter)	Absorption in spectra (if intergalactic in origin)
3) Radio source counts (interpretation of the log N-log S curve)	They comprise a significant fraction of identified radio sources
4) Nature of microwave background radiation	Discrete source models would involve QSOs or related objects

If the red-shifts are cosmological in origin, then they are likely to have an impact on observational cosmology in all of the areas shown in Table I. If they are comparatively close by and the red-shifts are intrinsic, then they are irrelevant as far as items (1) and (3) are concerned, but they are still important for (2) and (4).

If they are local, then the redshifts must in large part be due to their moving at very high speeds, or exceedingly strong gravitational fields must be present, or something else which no one has yet considered must be responsible. In any case we shall show that, whatever the origin of the red-shift is, there is strong evidence that very strong gravitational fields are present in QSOs and nuclei of galaxies. Thus, these are probably the laboratories where we can study general relativity in the strong-field approximation.

We now turn to a discussion of some observational properties of QSOs which bear on these questions.

1. – The red-shift apparent magnitude relation.

The red-shift apparent magnitude relation for 136 QSOs is shown in Fig. 1 (*). This shows the large scatter in the relation between red-shift and apparent brightness and the sharp cut-off at red-shifts slightly greater than 2. The

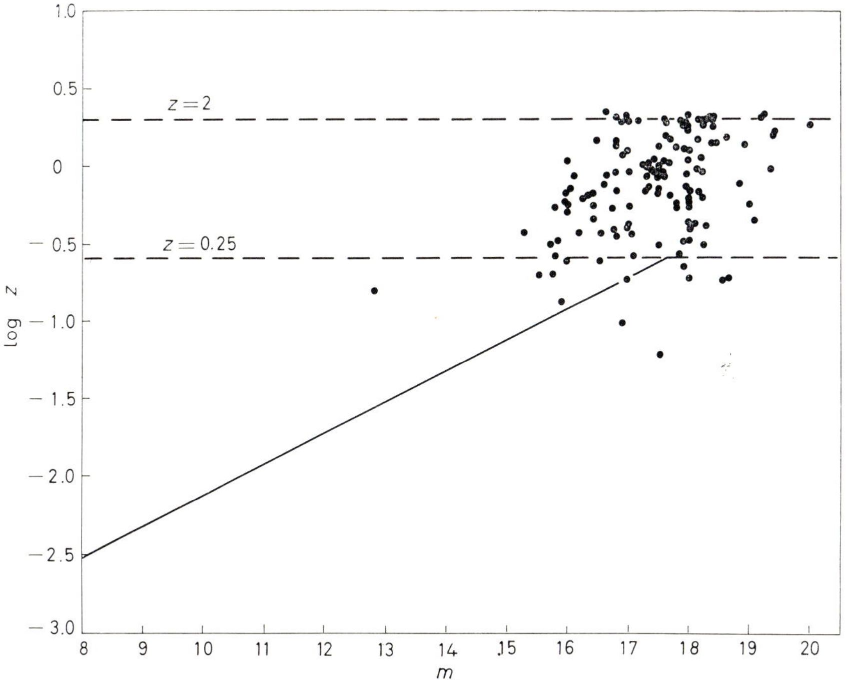

Fig. 1. – Red-shift-apparent magnitude relation for 136 QSOs. Full line is the Hubble relation for normal galaxies.

(*) At present (May 1970) red-shifts for about 200 QSOs have been measured. Accurate magnitudes are not available for all of these. However, the objects which can now be added to Fig. 1 do not materially change the effect seen in the Figure.

apparent cut-off for apparent magnitudes fainter than about 19 is due to the difficulty of observing objects much fainter than this, but the red-shift cut-off is real. It is not an absolutely sharp cut-off, but there is a rather sharp decline in objects as the value of z increases beyond two. This is one of the most significant results to come out of nearly a decade of observations of QSOs. Its interpretation depends on whether one accepts the local or cosmological interpretation for QSOs. If they are local, then the cut-off must be due to the fact that intrinsic red-shifts greater than a critical value close to 2.3 are rarely possible. If the red-shifts are cosmological in origin, then this is strong evidence that, at an epoch corresponding to a red-shift of about 2, some rather remarkable events occurred. One possibility is that we live in a universe of the Lemaitre type. This has been explored recently by SALPETER and PETROSIAN, by SHKLOVSKY and KARDASHEV, and by others. If one takes the view that the Lemaitre model is not plausible, either because models with a nonzero cosmological constant are taboo (for historical or philosophical reasons), or because it is frequently stated that the value of q_0 lies between $+\frac{1}{2}$ and $+1$ (based on rather uncertain data), then it is necessary to explain the cut-off in terms of an evolutionary effect either in a hypothetical intergalactic medium or in the QSOs themselves. In the latter case one can make the rather *ad hoc* proposal that, prior to the epoch corresponding to a red-shift near 2, the rate formation of QSOs was very low. For z in the range 0 to 2 the rate of formation may be a function of $(1+z)^n$. Alternatively, it has been proposed by REES that the intergalactic gas was more highly ionized at epochs $z<2$ than at earlier epochs. This would mean that there was progressively more obscuration and reddening as one attempted to detect QSOs with z greater than about 2.3, because of absorption by neutral hydrogen. QSOs with greater red-shifts should be much redder and fainter than those so far observed (*).

2. – The distribution of red-shifts.

There is a sharp peak in the redshifts close to 1.95. That such an effect was present in the absorption red-shifts was first indicated in 1967. It can be seen from Table II and Fig. 2 that it is strongly present. In addition, an even sharper effect in emission red-shifts at 1.955 has now appeared. It has been argued that this is not a chance effect and implies either that the red-shifts

(*) In 1970 LYNDS and WILLS found a QSO, 4C 05.34, with an emission-line red-shift of 2.88. This object has a magnitude of about 18 and normal colors. This is not to be expected on Rees's hypothesis unless there are irregularities in the intergalactic medium as far as its state of ionization is concerned. LYNDS and WILLS have reiterated, following this discovery, that there is a real paucity of objects with red-shifts greater than about 2.3.

TABLE II. – *Extragalactic emission-line objects with* $0.02 < z < 2.40$ ([a]).

Object	Type	α_{1950}	δ_{1950}	z_{em}	z_{abs}
PKS 1358-11	RG	13 58 58	−11 22.0	0.025	
Markarian 10 ([b])	Seyfert			0.029	
3C 120	Seyfert	04 30 31	+05 15.0	0.032	
Markarian 9	Seyfert			0.039	
3C 371 ([c])	N	18 07 19	+69 49.0	0.0508	
PKS 1514 + 00 = 4C 0.56	RG	15 14 07	+00 29.5	0.053	
PKS 0634 — 20	RG	06 34 23	−20 00	0.056	
Cygnus *A* ([b])	RG	19 57 44	+40 35.8	0.0561	
3C 445	N	22 21 16	−20 21.9	0.0568	
3C 390.3	N	18 45 39	+79 43.2	0.0569	
B 234	QSO	13 00 43	+36 07.8	0.060	
3C 33 ([b])	RG	01 06 14	+13 03.7	0.060	
PKS 0521 — 36	N	05 21 14	−36 30.2	0.061 or 0.055	
I Zw 0051 + 12	Compact	00 51 00	+12 25	0.061	
II Zw 2130 + 09	Compact	21 30 00	+09 56.0	0.061	
Ton 1542	QSO	12 29 48	+20 25.0	0.0636	
PKS 0349 — 27 ([b])	RG	03 49 33	−27 53.1	0.066	
PKS 1131 + 21	RG	11 31 21	+21 22.2	0.066	
3C 227	N	09 45 08	+07 39.3	0.0855	
III Zw 008 + 10	Compact	00 08 00	+10 42.0	0.089	
B 264	QSO	12 59 31	+32 22.0	0.095	
PKS 1417 — 19	N	14 17 02	−19 15.0	0.1192	
Object in VV 172 ([c])	Compact	11 29 12	+71 05.0	0.123	
3C 135	N	05 11 32	+00 53.1	0.127	
PKS 2300 — 18 ([b])	N	23 00 23	−18 57.8	0.129	
Ton 256	QSO	16 12 00	+26 13	0.131	
3C 223	RG	09 36 50	+36 07.6	0.137	
3C 273	QSO	12 26 33	+02 19.7	0.158	
PKS 2349 — 01	N	23 49 22	−01 26.1	0.174	
B 154	QSO	12 55 02	+35 21.4	0.183	
B 340 ([d])	QSO	13 04 47	+34 40.6	0.184	
3C 234	N	09 58 57	+29 01.6	0.1846	
BSO 2	QSO	12 48 18	+33 47.1	0.186	
PKS 0736 + 01	QSO	07 36 43	+01 44.0	0.191	
PKS 0837 — 12	QSO	08 37 28	−12 03.8	0.200	
PKS 2135 — 14	QSO	21 35 00	−14 46.5	0.200	
3C 287.1	N	13 30 21	+02 16.1	0.2156	
3C 17	N	00 35 46	−02 24.2	0.2201	
3C 459	N	23 14 01	+03 49.1	0.2205	

(*a*) The references for the red-shifts or the original red-shifts can be found in 1, 2, 18, 20, 26, 27, 28, 29, 30.
(*b*) Emission-line extension visible on spectrograms.
(*c*) Possible stellar component.
(*d*) Image nonstellar; possibly *N*.

TABLE II (*continued*).

Object	Type	α_{1950}	δ_{1950}	z_{em}	z_{abs}
B 114	QSO	12 52 58	+35 55.4	0.221	
3C 171	N	06 51 11	+54 12.8	0.239	
PKS 1217 + 02	QSO	12 17 38	+02 20.4	0.240	
PKS 1004 + 13	QSO	10 04 44	+13 05.3	0.240	
4C 55.27	QSO	13 32 17	+55 15.8	0.249	
3C 79	N	03 07 11	+16 54.7	0.256	
PHL 1093	QSO	01 37 23	+01 16.3	0.260	
3C 323.1	QSO	15 45 31	+21 01.6	0.264	
Ton 616 = 4C 25.40	QSO	12 23 12	+25 14.5	0.267	
PHL 1186	QSO	01 47 36	+09 01	0.270	
B 46 ([d])	QSO	12 46 29	+34 40.8	0.271	
PHL 1194	QSO	01 48 42	+09 02	0.299	0.283 (one line assumed to be Mg II λ2798)
3C 109	N	04 10 55	+11 04.9	0.306	
PHL 1078	QSO	01 35 36	−05 44	0.308	
3C 249.1	QSO	11 00 26	+77 15.0	0.311	
3C 277.1	QSO	12 50 15	+56 50.6	0.320	
PKS 0231 + 022	QSO	02 31 15	+02 16.3	0.322	
PKS 2251 + 11	QSO	22 51 40	+11 20.9	0.323	
LB 2136 = 4C 49.22	QSO	11 50 48	+49 50.0	0.334	
PKS 1049 − 09	QSO	10 48 59	−09 02.2	0.344	
PKS 1233 − 24	QSO	12 32 59	−24 57.7	0.355	
PKS 1510 − 08	QSO	15 10 09	−08 54.8	0.361	0.351 (one line assumed to be Mg II λ2798)
PHL 1027	QSO	01 30 30	+03 22	0.363	
3C 48	QSO	01 34 50	+32 54.3	0.367	
4C 37.43	QSO	15 12 46	+37 02.5	0.370	
3C 351	QSO	17 04 03	+60 48.6	0.371	
4C 48.28	QSO	10 12 49	+48 53.2	0.385	
PKS 1229 − 02	QSO	12 29 27	−02 07.6	0.388	0.395 (one line assumed to be Mg II λ2798)
PHL 3375	QSO	01 28 24	+07 28	0.390	
PKS 0812 + 02	QSO	08 12 48	+02 04.3	0.402	0.384, 0.344 (each based on one line assumed to be Mg II λ2798)
PHL 1226	QSO	01 51 48	+04 34	0.404	
PKS 0214 + 10 = 4C 10.6	QSO	02 14 27	+10 50.4	0.408	
3C 215	QSO	09 03 44	+16 58.2	0.411	
3C 47	QSO	01 33 40	+20 42.3	0.425	
4C 21.35	QSO	12 22 24	+21 39.4	0.434	
PHL 658	QSO	00 03 20	+15 52	0.450	

TABLE II (*continued*).

Object	Type	α_{1950}	δ_{1950}	z_{em}	z_{abs}
B 312	QSO	13 04 53	+37 29.6	0.450	
5C 2.10	QSO	10 49 41	+48 55.9	0.478	
PKS 2128 — 12	QSO	21 28 52	—12 20.1	0.501	
PKS 1327 — 21	QSO	13 27 23	—21 26.7	0.528	
Ton 469 = 3C 232 = 4C 32.33	QSO	09 55 26	+32 34.8	0.534	
3C 279	QSO	12 53 35	—05 31.2	0.538	
3C 147	QSO	05 38 44	+49 49.7	0.545	
PKS 1136 — 13	QSO	11 36 38	—13 34.1	0.554	
3C 334	QSO	16 18 07	+17 43.8	0.555	
3C 275.1	QSO	12 41 27	+16 39.2	0.557	
PKS 1634 + 26 = 3C 342 = = 4C 26.49 = NRAO 510	QSO	16 34 21	+26 54.3	0.561	
MSH 04 — 12	QSO	04 05 27	—12 19.5	0.567	
PKS 0403 — 13	QSO	04 03 14	—13 16.5	0.574	
3C 345	QSO	16 41 18	+39 54.2	0.595	
PKS 1335 + 023	QSO	13 35 07	+02 22.1	0.61	
MSH 03 — 19	QSO	03 49 10	—14 38.3	0.614	
3C 261	QSO	11 32 16	+30 22.0	0.614	
PKS 0155 — 10	QSO	01 55 15	—10 58.5	0.616	
MSH 13 — 011	QSO	13 35 31	—06 11.9	0.625	
3C 263	QSO	11 37 09	+66 04.4	0.652	
PKS 0932 + 02 = 4C 2.27	QSO	09 32 42	+02 16.3	0.659	
3C 57 = PKS 0159 — 11	QSO	01 59 30	—11 47.4	0.669	
4C 2.4 + PKS 0115 + 02	QSO	01 15 43	+02 40.2	0.672	
PKS 2344 + 09	QSO	23 44 04	+09 14.0	0.677	
3C 207	QSO	08 38 03	+13 23.2	0.684	
PKS 0225 — 014	QSO	02 25 35	—01 29.2	0.685	
3C 380	QSO	18 28 13	+48 42.7	0.692	
4C 39.25	QSO	09 23 55	+39 15.4	0.699	
PHL 923	QSO	00 56 32	—00 09.4	0.717	
PKS 1354 + 19	QSO	13 54 42	+19 33.5	0.720	
3C 254	QSO	11 11 53	+40 53.7	0.734	
3C 138	QSO	05 18 17	+16 35.4	0.760	
3C 175	QSO	07 10 15	+11 51.3	0.768	
3C 286	QSO	13 28 50	+30 46.0	0.849	
3C 454.3	QSO	22 51 29	+15 52.9	0.859	
PKS 1252 + 11	QSO	12 52 08	+11 57.2	0.871	
4C 20.33	QSO	14 22 38	+20 13.8	0.871	
3C 196	QSO	08 09 59	+48 22.1	0.871	
PKS 1317 — 00	QSO	13 17 04	—00 34.2	0.89	
PKS 0922 + 14	QSO	09 22 21	+14 57.4	0.895	
PKS 2216 — 03	QSO	22 16 16	—03 50.6	0.901	
3C 309.1	QSO	14 58 58	+71 52.3	0.904	
PKS 0957 + 00	QSO	09 57 44	+00 19.6	0.906	
4C 37.24	QSO	08 27 55	+37 52.3	0.914	

TABLE II (*continued*).

Object	Type	α_{1950}	δ_{1950}	z_{em}	z_{abs}
3C 336	QSO	16 22 32	+23 52.0	0.927	
MSH 14-121	QSO	14 53 12	—10 56.8	0.940	
3C 288.1	QSO	13 40 30	+60 36.9	0.961	
3C 94	QSO	03 50 05	—07 20.1	0.962	
PKS 2115—30	QSO	21 15 11	—30 31.9	0.98	
4C 29.68	QSO	23 25 41	+29 20.6	1.012	
3C 245	QSO	10 40 06	+12 19.3	1.029	
CTA 102	QSO	22 30 07	+11 28.5	1.037	
3C 2	QSO	00 03 49	—00 21.0	1.037	
3C 287	QSO	13 28 16	+25 24.6	1.055	
3C 186	QSO	07 40 57	+38 00.5	1.063	
PKS 0122—00	QSO	01 22 55	—01 21.4	1.070	
3C 208	QSO	08 50 23	+14 04.1	1.110	
PKS 1055+20 = 4C 20.24	QSO	10 55 37	+20 08.3	1.11	
3C 204	QSO	08 33 18	+65 24.1	1.112	
PKS 1127—14	QSO	11 27 36	—14 32.9	1.187	
BSO 1	QSO	12 46 29	+37 47.0	1.241	1.241 (one line assumed to be C IV λ1550)
PKS 1454—06	QSO	14 54 04	—06 05.5	1.249	
RS 13	QSO	13 31 30	+27 45.6	1.287	
PKS 2153—204	QSO	21 53	—20 30	1.31	
PKS 0859—14	QSO	08 59 55	—14 04.0	1.327	
B 201	QSO	12 57 27	+34 39.6	1.375	
3C 181	QSO	07 25 20	+14 44.4	1.382	
3C 268.4	QSO	12 06 42	+43 56.1	1.400	
3C 446	QSO	22 23 11	—05 12.4	1.403	
PHL 1377	QSO	02 32 36	—04 15.2	1.436	
3C 298	QSO	14 16 38	+06 42.4	1.439	1.419 (one line assumed to be C IV λ1550)
4C 17.46	QSO	08 56 03	+17 03.2	1.444	1.439 (one line assumed to be C IV λ1550)
AO 0952+17	QSO	09 52 11	+17 57.8	1.472	
PHL 3632	QSO	01 39 54	+06 10	1.479	
3C 270.1	QSO	12 18 04	+33 59.8	1.519	1.498 (one line assumed to be C IV λ1550)
3C 205	QSO	08 35 10	+58 04.7	1.534	
4C 31.38	QSO	11 53 44	+31 44.8	1.557	
4C 42.01	QSO	00 32 23	+42 21.8	1.588	
3C 280.1	QSO	12 58 14	+40 25.2	1.659	
4C 19.31	QSO	08 36 15	+19 33.0	1.689	
PKS 0922+005	QSO	09 22 36	+00 32.1	1.72	

TABLE II (*continued*).

Object	Type	α_{1950}	δ_{1950}	z_{em}	z_{abs}
3C 454	QSO	22 49 09	+18 32.8	1.757	
PKS 2146 − 13	QSO	21 46 46	−13 18.7	1.800	1.775 (one line assumed to be C IV λ1550)
3C 432	QSO	21 20 24	+16 51.7	1.805	
PKS 2354 + 14	QSO	23 54 44	+14 29.5	1.810	
PHL 3424	QSO	01 31 12	+05 32	1.847	
B 194	QSO	12 56 08	+35 44.9	1.864	1.8366, 1.8946
PHL 1222	QSO	01 51 12	+04 48	1.910	1.934 (one line assumed to be Ly α)
4C 29.50	QSO	17 02 11	+29 51.0	1.927	
PHL 61 = PKS 2134 + 004	QSO	21 34 04	+00 28.2	1.94	
PKS 0119 − 04	QSO	01 19 56	−04 37.3	1.955	1.965
PHL 938	QSO	00 58 12	+01 56	1.955	1.9064 (one line assumed to be Ly α), 0.6128
3C 191	QSO	08 02 04	+10 24.1	1.956	1.947
BSO 6	QSO	12 59 30	+34 27.2	1.956	
PHL 5200	QSO	22 25 54	−05 34	1.98	1.90 ÷ 1.98, 1.9502, 1.8910 variable?
PKS 1148 − 00	QSO	11 48 10	−00 07.2	1.982	
PHL 1127	QSO	01 41 30	+05 14	1.990	1.95 (one line assumed to be Ly α)
LB 8755	QSO	08 48 05	+15 33.5	2.010	
3C 9	QSO	00 17 50	+15 24.3	2.012	
Ton 1530	QSO	12 22 57	+22 53	2.046	1.9362, 1.9798, 2.0553
PHL 1305	QSO	02 26 24	−03 52	2.064	
PKS 0229 + 13	QSO	02 29 02	+13 09.7	2.065	
B 189	QSO	12 56 51	+36 48.2	2.075	
BSO 11	QSO	13 11 22	+36 16.5	2.084	
PKS 2254 + 024	QSO	22 54 44	+02 27.3	2.09	
PKS 0106 + 01	QSO	01 06 02	+01 19.1	2.107	
PKS 1116 + 12	QSO	11 16 20	+12 51.0	2.118	1.947
QS 1108 + 285	QSO	11 08 26	+28 57.9	2.192	
PKS 0237 − 23	QSO	02 37 53	−23 22.1	2.228	2.2017, 1.9556?, 1.6744, 1.6715, 1.6564, 1.5958, 1.5132, 1.3646
4C 25.5	QSO	01 23 56	+25 43.9	2.358	2.3683
5C 2.56	QSO	10 55 18	+49 55.6	2.380	

have a large intrinsic component or that a cosmological model of Lemaitre type is indicated.

There is a very sharp peak in red-shifts close to 0.06. This was originally pointed out when five objects were found to have red-shifts of 0.060 or 0.061. Most of the objects with red-shifts of 0.06 are not QSOs because they are not quite stellar in appearance, but in all other ways—continuum, line emission, etc.—they are similar but not identical. It has also been argued that

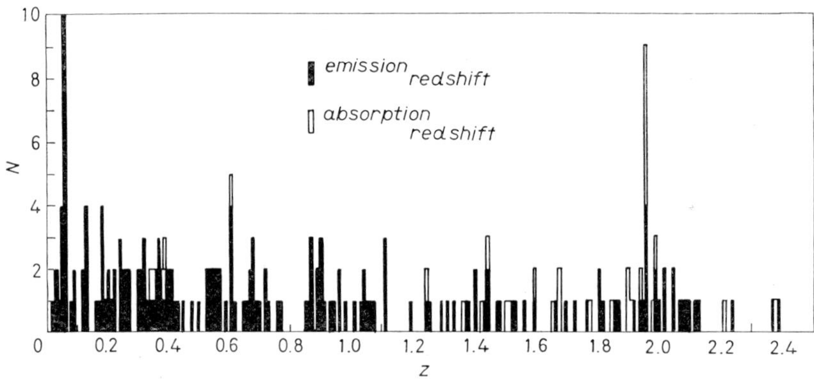

Fig. 2. – Histogram of the distribution of z among all 186 objects listed in Table II, plotted at intervals of 0.01. Note large peak near 0.061 and tendency for peaks at multiples of this. Black columns indicate emission redshifts; white columns indicate absorption redshifts.

the distribution of red-shifts suggests that there are periodical peaks at multiples of about 0.06. A number of statistical analyses of these effects have been carried out. The conclusion from these is that the peak at 0.06 and the combined emission and absorption peak at 1.95 are statistically significant, while there is still argument about the significance of the absorption and emission peaks at 1.95, treated separately. On the other hand, there is much more doubt about the peaks at multiples of 0.06, and most workers do not find them convincing.

It should be stressed that to see any *sharp* peaks at all in the red-shift is very remarkable, as the following argument shows.

If this (or any other) class of objects in the universe contains a component consisting of subgroups i which have an intrinsic red-shift component z_i, where z_i can take individual discrete values, then the total observed red-shift, z_0, is given by

$$1 + z_0 = (1 + z_c)(1 + z_r)(1 + z_i) ,$$

where z_c and z_r result from the expansion of the universe and random motions respectively. Thus to see any sharp effect at all requires that z_c and z_r must be small; it is very easy to mask an effect of this type if it is present.

3. – Attempts to measure the distances of the QSOs.

If the QSOs could be shown to be physically associated with galaxies whose distances are known, the question of whether the QSOs lie at cosmological distances or not would be answered. Various attempts to investigate the relation between galaxies and QSOs have been made. Deep plates have been taken of the regions around some of the QSOs with small red-shifts and, in particular, 3C 273 and 3C 48, and no clusters of galaxies have been found. On the other hand, there has recently been an investigation of B 246, which was originally classified as a QSO and which has a redshift of only 0.09. It has been shown by BAHCALL, GUNN, and SCHMIDT that this object does lie in a cluster of galaxies whose red-shifts are similar to that of B 264. However, a more detailed examination of B 264 by OKE, LYNDS, ARP, and others has shown that it was incorrectly classified as a QSO, and it should properly be classified as an N galaxy. This is because it is not stellar and its line spectrum and continuum are characteristic of other N systems, also because several of the galaxies in the cluster are brighter than B 264.

Thus at present we have no direct evidence that QSOs are associated with distant galaxies.

On the other hand, ARP has attempted to show that some QSOs with large red-shifts are physically associated with comparatively nearby galaxies with $z \leqslant 0.01$. His method of attack has been to look at alignments of objects across bright peculiar galaxies and evidence of geometrical forms which suggest that QSOs and radio galaxies have been ejected from such systems. Because his statistical arguments are open to criticism, his work has been disregarded by many, but one cannot fail to be impressed by some of the examples of alignments between galaxies and QSOs shown by ARP, for example, the case involving the peculiar galaxy NGC 520 and four QSOs which appear to be aligned with it. If alignments such as this are not due to chance, there must be a genetic connection between extragalactic objects with very different red-shifts, and despite the difficulties the most plausible idea is then that the QSOs are being thrown out of the nuclei of galaxies. Now undoubtedly evidence is accumulating that in the phenomenon of explosion on the galactic scale coherent plasma clouds are ejected. If we exclude the QSO phenomenon for the moment, we have evidence of the following kinds.

1) In the famous radio galaxy with the jet, M 87, it seems that the radio and optical structures in the jet are made up of several very small condensed

objects which presumably have been ejected from the center, and WADE pointed out many years ago that the neighboring small radio source, 3C 272.1, which is associated with the galaxy M 87, is exactly aligned with the jet in M 87.

2) Detailed studies of the structure of the radio source generated by NGC 1275 in the Perseus cluster suggest that small clouds of plasma have been ejected and have travelled through the cluster forming a fan-shaped distribution. In fact, the distribution of the galaxies in the Perseus cluster follows a highly elongated contour, and, though most of them are normal galaxies, the distribution is somewhat reminiscent of some of the configurations around nearby galaxies shown by Arp.

3) Frequently, high-resolution studies have shown that very small radio components (diameter $\leqslant 0''.01$) are associated with radio galaxies and the fact that they can sometimes be found at great distances from the parent galaxies suggests that coherent objects have been thrown out.

4) In a chain of galaxies known as VV 172 one compact galaxy has a red-shift more than twice as large as the other members. The chance that this is a background object is small. If it is physically associated with the other systems, either it is being ejected with a velocity 0.07 c (21 000 km/s), or it must have a red-shift component that is not a Doppler shift.

Thus there is growing evidence that coherent objects are ejected from galaxies. However, there is no definite evidence so far which shows conclusively that QSOs have been ejected from comparatively nearby galaxies.

Consequently, all attempts to measure the distances of QSOs by showing that they are physically associated with galaxies have so far failed.

4. – Absorption in the spectra of QSOs.

It can also be seen from Table II and the histogram that the absorption lines in QSOs tend to appear more frequently as the emission red-shifts increase. Is this a real effect, and, more important, do the absorption lines in the QSOs arise in the object itself or in the intervening intergalactic medium? The latter conclusion is favored by many, because it would show that the QSOs are at cosmological distances, would allow investigation of the properties of the intergalactic medium, and would mean that the QSOs and related systems could be used for cosmology. Some investigations have been made in which it is assumed that the absorption is intergalactic in origin. However, it appears that this approach may be premature.

Of the 155 QSOs with known red-shifts, listed in Table II, twenty have been found to show absorption lines in their spectra. In most of these twenty objects the absorption lines have a red-shift close to the emission-line red-shift, and therefore arise in gas associated with the QSO. Does the observational evidence point to any of the absorption lines arising in intervening galaxies or intergalactic gas clouds?

On the assumption that the QSOs are at cosmological distances, BAHCALL and SALPETER and REES and SCIAMA discussed the probability of the line of sight intercepting one or more individual galaxies or clouds of intergalactic gas lying in clusters of galaxies. WAGONER calculated equivalent widths of the stronger absorption lines to be expected if a QSO were seen through an interstellar cloud of normal chemical composition in an intervening galaxy; reddening by dust in the galaxy and a detectable amount of Faraday rotation might also be produced. For QSOs with $z > 1$, the probability of intersection of the light path with a galaxy is significant; indeed, if absorption lines resulting from such intersections are not found, an argument against the cosmological interpretation of the red-shifts would be provided.

We consider the observational evidence in the twenty SQOs in Table II that show absorption lines. In many of these only a single absorption line is seen, in the wing or close to one of the strong resonance emission lines. The three lines or doublets in which absorption of this kind is found are Mg II $\lambda 2798$ ($\lambda 2796 + \lambda 2803$), C IV $\lambda 1549$ ($\lambda 1548 + \lambda 1551$), and Ly-$\alpha$ $\lambda 1216$. Of the twenty QSOs with absorption lines, four with $0.28 < z_{em} < 0.40$ show absorption resulting from Mg II; in these $|z_{em} < z_{abs}| < 0.02$, while one of these, PKS 0812 + 02, has a possible second component with $z_{em} - z_{abs} \approx 0.06$. In one, PKS 1229.02, $z_{abs} > z_{em}$. In the range $1.2 < z_{em} < 1.6$, there are another four QSOs having a single absorption line of CIV with $z_{em} - z_{abs} \leqslant 0.02$; in these objects z_{abs} is always less than z_{em}.

The remaining twelve QSOs have $z_{em} \geqslant 1.8$. Some show absorption only in Ly-α and some have many absorption lines. It is in this group that the multiple absorption red-shifts have been found; outside this group, PKS 0812 + 02 is the only object that has a possible second absorption-line red-shift. Five of the twelve have an absorption-line red-shift greater than the emission-line red-shift. Counting the individual multiple red-shifts found in five of these twelve QSOs, a total of twenty-four separate redshifts have been identified in the literature, although one or two of these have been questioned. 4C 25.5, with $z_{abs} > z_{em}$, has many additional absorption lines which may give multiple red-shifts, ten of them (in nine objects) have $|z_{em} - z_{abs}| \leqslant 0.03$. In these cases, the absorptions must arise in gas associated with the QSO, and not in separate intervening galaxies. A histogram showing the distribution of values of $(z_{em} - z_{abs})$ is shown in Fig. 3.

Of the remaining fourteen separate absorption-line red-shifts which differ from the respective values of z_{em} by more than 0.03, nine occur in three objects having in addition one absorption red-shift with $|z_{em} - z_{abs}| \leqslant 0.3$, and, most crucially, in all these cases there is no difference in the profiles of the lines in the systems with z close to or widely different from z_{em}. It can be concluded therefore that, in these cases, all red-shift systems are produced in gas in rather similar conditions. Certainly it is not reasonable to suppose that one system arises in gas closely associated with the QSO and others in intervening galaxies. These three objects, containing most absorption red-shifts considerably different from z_{em}, are PHL 5200, Ton 1530 and PKS 0237-23.

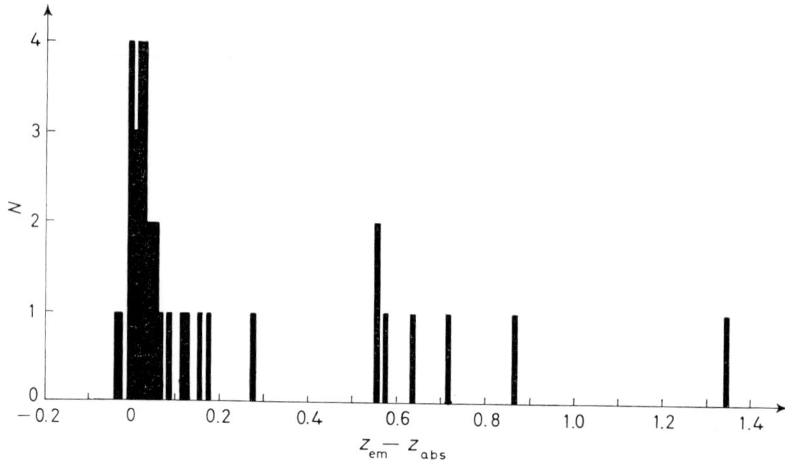

Fig. 3. – Histogram showing the distribution of the differences, $z_{em} - z_{abs}$, for QSOs which show absorption lines in their spectra as well as emission lines. Note that most values cluster around zero, and some have $z_{abs} > z_{em}$.

PHL 5200 has very broad absorptions in several lines, with the absorption-band red-shift extending from $z = 1.88 \div 1.98$ and adjoining with a fairly sharp edge the emission lines at $z = 1.98$. The gas producing the absorption could lie in a thick shell expanding outward at velocities up to 10 000 km s^{-1}. The two discrete red-shifts at 1.9502, 1.8910 belong to narrow absorption maxima occurring within the broad absorptions. There is no doubt that all the absorption arises in gas associated with the QSO.

Ton 1530 is one of the three QSOs in which multiple absorption red-shifts have been found. Ton 1530 has one absorption red-shift close to, but slightly greater than, z_{em}; the other absorptions have $z_{abs} < z_{em}$ and the strongest system has $z_{em} - z_{abs} = 0.11$. In addition, there are many unidentified absorption lines, mostly at wavelengths less than the emission line of Ly-α, which it has been suggested are individual Ly-α absorption lines. Two further red-shifts have been suggested recently, both with z_{abs} much less than z_{em}. Again all lines have similar profiles and because one set has to be associated with the QSO we conclude that all are.

PKS 0237-23 has the largest number of discrete red-shifts, and in addition still has many unidentified absorption lines, mostly shortward of Ly-α, which might be individual Ly-α absorptions. The differences from z_{em} for the most conclusively identified systems are 0.557, 0.572 and 0.863. Again, all absorption-line profiles are similar (deep and narrow), and all seem to have arisen in gas in similar conditions. The large red-shift differences here make it tempting to attribute the lines to intervening galaxies or clouds, but the probability of finding one object with six or seven such clouds or galaxies lined up along the line of sight, when twenty-two in the same redshift range $z_{em} > 1.8$ have none, is negligibly small.

What of the remaining three objects that have $z_{em} - z_{abs} > 0.3$? These are PHL 938, PHL 1127 and PKS 1116 + 12. PHL 938 contains two red-shifts, at 1.906 and 0.613. The 1.906 system is most likely to be associated with the QSO, because the difference from $z_{em} = 1.955$ is not much larger than our chosen dividing line of 0.03. The 0.613 system contains nine lines, all Mg II resonance lines or zero-volt lines of Fe II. The lines are extremely sharp and heavily saturated. The absence of lines arising from excited fine-structure states of the ground level of Fe II means the radiation field is very dilute and the electron density $n_e < 10^3$. The line widths are less than those corresponding to a spread of velocity of 160 km/s. If therefore the absorbing gas were in an intervening cloud or galaxy, the differential rotation and/or turbulent motions in the line of sight would have to be very small. Absorption in an intervening, rather face-on, spiral galaxy cannot be ruled out, but is unlikely because the UBV colors of PHL 938 show no evidence for reddening by dust in such a galaxy. Faraday rotation or 21 cm absorption cannot, unfortunately, be searched for because PHL 938 is radio-quiet. Finally, the absorption-line red-shift of 0.613 occurs close to one of the peaks in the general red-shift distribution.

PHL 1127 has only one absorption line, Ly-α, at $z_{em} - z_{abs} = 0.4$, and PKS 1116 + 12 has one or possibly two lines at $z_{em} - z_{abs} = 0.17$. In both of these, $z_{abs} = 1.95$. On this basis alone, it is extremely improbable that the absorbing gas is in an intervening cloud or galaxy. If it were, one would be forced to look to the Lemaitre model universe to explain the $z = 1.95$ effect.

We conclude that, with one possible exception ($z_{abs} = 0.613$ in PHL 938), none of the absorption lines found in QSOs are likely to be intergalactic; all are probably associated with the objects, and it is necessary to devise a model of QSOs which will allow for them to be surrounded by gas at very different, very sharply defined, red-shifts.

Is the apparent tendency for absorption lines to occur preferentially in objects with large z_{em} real? If the absorption lines were intergalactic, and the red-shifts cosmological, this would be expected because as z increases the path length is greater, and in an evolving universe the density of galaxies or clouds was higher in the past. WEYMANN and WILCOX have argued that the apparent concentration of absorption in high red-shift objects may not be real, but may result from selection effects.

There are three resonance absorptions that can be expected to be strong —Mg II, C IV and Ly-α. WEYMANN and WILCOX added Si II, but these lines are very rarely found in emission, whereas Mg II emission is one of the strongest and most characteristic features of QSO spectra. Further, Mg II absorption, when found, is comparable in strength with C IV and, on the rare occasions when both Mg II and Si II are accessible in the same spectrum, Mg II is the stronger. Indeed, Mg II absorption lines are strong enough in PHL 938 to have been mistaken for C IV in the first observations of this QSO. We conclude that Mg II, C IV and Ly-α may be treated on an equal basis as indicators of the presence of absorbing gas. Ly-α is undoubtedly the strongest and most definitive indicator, but C IV and Mg II are comparable with one another and are considerably stronger than Si II. The Fe II zero-volt lines are weaker than Mg II, probably comparable with Si II, from the evidence found in PKS 0237-23 and PHL 938.

WEYMANN and WILCOX based their analysis chiefly on a search for correlation between red-shift and the appearance of C IV lines in a QSO. They found no correlation. If our argument is accepted, however, we can make a comparison between the number of QSOs having Mg II absorption and having $0.24 \leqslant z_{em} < 0.75$, in which Mg II occurs in the most easily observed spectral range, and the corresponding numbers of objects showing C IV and Ly-α in absorption. Only four out of a total of sixty-one objects with $0.24 \leqslant z_{em} < 0.75$ show Mg II absorption. By contrast, five of twenty-three with $1.2 \leqslant z_{em} \leqslant 1.8$, where only C IV is easily available, have absorption, and eleven out of twenty-eight with $1.8 < z_{em} < 2.4$, where Ly-α is easily visible, have absorption. Further, nine of the latter eleven have other lines besides Ly-α in absorption, either at the same red-shift as Ly-α or at a different red-shift. We conclude that absorption lines are more likely to be found in QSOs with $z_{em} > 1.8$, independently of spectroscopic effects.

A final point to consider is whether absorptions from intergalactic clouds could be present in the spectra of many of the QSOs with large red-shifts, but have remained so far undetected. It is sometimes suggested that high-resolution spectra might reveal such lines which would be invisible on the usual low-dispersion spectra obtained for determining red-shifts.

While the possibility of finding very weak narrow absorption lines on high-resolution spectra cannot be ruled out, it must be pointed out that all the interesting absorption-line QSOs discussed here revealed the absorption lines on the first low-dispersion spectra that were obtained, even though it required higher resolution to disentangle the multiple red-shifts and identify the lines. For example, the C IV absorption in Ton 1530, with a red-shift considerably different from z_{em}, was seen on the first spectra reported by HILTNER, COWLEY and SCHILD, although it was not identified. Similarly, KINAM detected the discrepant absorption lines in PHL 938, although he did not identify them, and the first low-dispersion spectrograms of PKS 0237-23 revealed an unexplained set of complex absorption lines.

There is indeed a gap in red-shift from about $z = 0.75$ to 1.2 between which none of the three strong resonance transitions of Mg II, C IV and Ly-α occur in the region most easily observed with conventional spectrographs. But the image-tube spectrograms of LYNDS extend the region of easy visibility of Mg II upward from $z = 0.75$. Thus we think it unlikely that large numbers of hitherto undetected intergalactic absorption lines exist in the well observed spectra of such QSOs as 3C 9.

To summarize: three lines of argument lead to the conclusion that the absorptions found in the QSOs are likely to be associated with the objects themselves and not with an intervening medium, namely, the actual values of the absorption red-shifts so far found, spectroscopic considerations, and the very small probability that the multiple absorption red-shift found in objects like PKS 0237-23 could be due to intergalactic clouds.

5. – Space density of QSOs.

The frequency of radio-quiet QSOs in now known to be very high, and it amounts to something like 10^7 objects down to magnitude 22. If these objects lie at cosmological distances they are still rare compared with ordinary galaxies. However, it must be remembered that a number of arguments suggest that the quasi-stellar phase in the overall lifetime of a galaxy (if this is the underlying object) or of a massive object is only $10^6 \div 10^7$ years so the number of underlying objects may be $\sim (10^{10} \div 10^{11})$ and this makes them comparable in numbers with galaxies. They may indeed be galaxies or the precursors to them in this picture.

If they are not at cosmological distances, but are closer, similar arguments suggest that the space density is higher, and they may be more frequent than galaxies of average mass and luminosity in the universe.

In either event, if QSOs are a manifestation of a new class of extragalactic objects and not simply exploding galaxies, they probably make a significant if not a dominant contribution to the mass energy in the universe.

6. – QSOs and the $\log N$-$\log S$ relation.

While radio emission is only a minor feature from an energetic standpoint in the power radiated from QSOs, they do comprise an important class of radio sources. Thus in the discussion of the $\log N$-$\log S$ curve for radio sources they are of great importance. For the 3C catalogue which is the only radio catalogue for which the identifications are fairly complete, about $\frac{2}{3}$ of the sources are thought to be galaxies with about $\frac{1}{3}$ QSOs, with $(5 \div 10)\%$ not identified with any optical object. If we treat this as prime data, since we have red-shifts for perhaps $(40 \div 45)\%$ of all of the optically identified sources, we find that we are not looking at a distance-volume effect which is required if we are to use it to test cosmological models, but a luminosity effect, and the shape of the $\log N$-$\log S$ curve is determined by that predominantly. The sources with measured red-shifts consist of radio galaxies with red-shifts no greater, except in one case, than 0.3, and QSOs. If the fainter galaxies for which we have as yet no red-shifts are indeed further away, then we may expect to see cosmological effects. Alternatively, these galaxies may simply be intrinsically faint objects so that we are only looking further down the luminosity function. It is interesting to point out that, as one makes identifications and gets red-shifts for radio galaxies associated with weaker radio sources in the 4C, Parkes and 5C catalogues, there is yet no evidence that these sources are systematically further away. For example, the largest red-shift known at present for a 4C radio galaxy is only 0.04.

If we wish to use the identified QSOs and make counts, we can only make cosmological investigations, if we *assume*, since we have not been able to prove, that the red-shifts have a cosmological origin.

Beyond these arguments there are serious questions that can be asked about how the observed data for the largely unidentified sources are handled. The common statement based on the Cambridge data and the Parkes catalogue is that the $\log N$-$\log S$ curve has a slope of about -1.8, indicating that there is an excess of faint (and therefore, it is asserted, more distant) sources. But, as one goes to even fainter sources, the slope flattens to values close to -1.4 and eventually to values $\lesssim -1$. Thus, in the commonly accepted picture this requires an evolutionary history of sources requiring an epoch when they were very plentiful and, before this, an epoch when the source production was low.

However, if one carries out source counts at higher frequencies than those used in the Cambridge surveys, different values for the $\log N$-$\log S$ slope are obtained, and reports of Australian work now suggest that through the whole range from the brightest to the faintest sources counted, the $\log N$-$\log S$ slope has always the Euclidean value or less. This, of course, is compatible with steady-state models, as is the flatter slope at the faint end in the Cambridge surveys. Moreover, the mere fact that one finds different slopes for counts at different frequencies should be a further argument to proceed very cautiously.

The fact that we don't know what we are looking at in the majority of cases, the fact that where we do know the luminosity function is dominating everything else, and the fact that only some of the data show a slope incompatible with the simple steady-state theory is reason enough to say that this cosmological test is in doubt.

If it is assumed that the QSOs are at cosmological distances, optical studies show that their space density is not uniform. The number is a steeply increasing function of $(1+z)$ with a sharp decrease corresponding to the cut-off in red-shifts beyond $z \simeq 2.3$. This is compatible with the evolutionary picture developed by the Cambridge radio astronomers to explain the $\log N$-$\log S$ curve.

7. – QSOs and the microwave background radiation.

The strongest positive evidence in favor of an evolving universe is the presence of the microwave background radiation. The reason why many people feel this way is simply because the existence of black-body radiation at this epoch, at a temperature of a few degrees Kelvin, was predicted if the universe started from a highly condensed state. When evidence began to appear showing that such radiation is indeed present, the argument became, to many, almost irresistible. Despite this, the observational evidence at present is quite unclear in the critical region near the peak of the black-body curve. We are either seeing black-body radiation at ~ 3 °K, and gross errors have been made in direct measurements from balloons and rockets, or we are seeing black-body radiation together with an excess flux which might be of galactic or extragalactic origin, or we are seeing radiation which does not have a big bang origin, so that perhaps the universe did not begin in this way at all. We simply do not know which is correct.

If part, or all, of the microwave radiation did not arise in a big bang, it must be non-thermal radiation generated by discrete sources. Discrete source models have been investigated by a number of people, and in some detail by WOLFE and BURBIDGE. Models can be made which will explain the observed flux, but to explain the observed limits on the small-scale fluctuations of the microwave background, the number density of discrete sources must be comparable with the space density of ordinary galaxies, and much greater in the steady-state theory. Such sources are likely to be types of galactic nuclei of QSOs which are microwave or infrared emitters. Surveys with infrared and millimeter wave telescopes may eventually show us whether such a population of discrete sources exists, though if they are all rather weak and present in large numbers, only their integrated flux may be detectable.

The main point is that, with the discovery of the QSOs, it became apparent that there is a large class of objects in the universe emitting energy of a non-thermal character. They are distinct from galaxies which emit thermal radiation of thermonuclear origin. Thus this discovery makes more plausible the possibility that large populations of objects exist which emit nonthermal radiation peaking in other parts of the electromagnetic spectrum.

8. – Strong gravitational fields in QSOs and galactic nuclei.

So far we have only discussed QSOs in so far as they have an impact on cosmology. We now discuss briefly the reasons why they are of importance for the study of strong gravitational fields.

It is clear by now that there are two very different processes which give rise to electromagnetic radiation in the universe. They are:

i) Thermonuclear processes which give rise to the thermal radiation from stars,

ii) nonthermal processes which take place in the nuclei of galaxies and QSOs where the energy released is of gravitational origin.

In addition to this, there may be radiation remaining which was generated in a big bang, if the universe began in that way.

It is ii) that we are particularly concerned with here. It is now clear that as well as the QSOs many, if not all, galaxies have active nuclei, though the scales of activity cover a very wide range. Average luminosities are shown in Table III. These values should be compared with the luminosity of a bright galaxy of stars which is about 10^{44} erg/s.

TABLE III.

QSOs (erg/s)		Nuclei of galaxies (erg/s)
Local	Cosmological	
L (radio) 10^{40}	$\leqslant 10^{45}$	$10^{40} \div 10^{45}$
L (optical) 10^{41}	10^{46}	$10^{41} \div 10^{44}$
L (infrared $(1 \div 20)$ μm) 10^{42}	10^{47}	$\leqslant 10^{42} \div 10^{46}$

There are several ways of calculating the total energy which is released by activity in the nuclei. If we accept that the radiation is due to the synchrotron process, or to Compton scattering, we can calculate the total energy instantaneously present in particles and magnetic or radiation fields as a function of the magnetic or radiation field if we know the size of the source. This type of calculation is well known, and for synchrotron radiation it is known that the total energy is a minimum when the energy in the particles is $\frac{4}{3}$ the energy in the magnetic field. There are many uncertainties associated with the fact that we only have information about the electron flux and we know nothing about the relative numbers of electrons and nucleons. However, these calculations have led to the conclusion that minimum total energies $\geqslant 10^{61}$ erg are required to explain some of the most extended radio sources as they are seen at present. The corresponding time scale obtained from the radiation lifetimes for the electrons are in the range $10^7 \div 10^9$ years for the extended sources.

On the other hand, calculations made for the much smaller sources which remain confined to the nuclear regions where they were generated require much smaller minimum total energies in particles and fields, but the corresponding lifetimes of the electrons are very much less, with time scales only of the order $0.1 \div 100$ years in many cases. If the activity therefore continues for much longer periods in such sources, the total energies required over the total times of activity may be comparable to the very large energies required to explain the extended radio sources.

Another way to estimate the total energy released through nonthermal processes is to determine Lt where L is the luminosity given in Table III and t is the total time over which activity takes place. Attempts can be made to estimate t by estimating the frequency of occurrence of strong radio galaxies, or of Seyfert nuclei, etc. These estimates lead to the conclusion that $t \sim \sim (10^6 \div 10^9)$ years in different classes of objects. Estimates of a dynamical time scale also lead to the conclusion that the time scales for activity are in this range. Thus, if we take into account the wide range of total luminosities and the fact that weak activity may well persist through the whole life of a galaxy,

$$Lt = (10^{54} \div 10^{63})\ \text{erg}\,, \qquad \text{or} \qquad (1 \div 10^9)\, M_\odot\, c^2.$$

The radiation of nonthermal flux from nuclei and QSOs is not the only evidence for large energy release. There is spectroscopic evidence that gas is being ejected from Seyfert nuclei, from QSOs, and from galaxies which are not known to be extremely powerful nonthermal emitters. In some cases the mass loss rates are very high. In the Seyfert galaxy and radio source NGC 1275 (Perseus A), about $10^8 M_\odot$ of gas is being ejected with a velocity of at least 300 km/s, so that the kinetic energy is at least 10^{58} erg. In another Seyfert galaxy, NGC 4151, the mass loss rates from the nuclear region have been estimated to lie in the range $(10 \div 1000)\, M_\odot/\text{y}$ (*). Even in the case of the nucleus of our own Galaxy evidence from 21 cm line studies indicates that perhaps $10^7 M_\odot$ has been ejected over the last few million years.

(*) Data obtained since these lectures were given suggest that the mass loss rates in NGC 4151 may be closer to about $1 M_\odot/\text{y}$.

There is also the possibility that the gravitational waves which WEBER may have detected have come from the galactic center. In this case the mass loss rates must be exceedingly large—from 10 to about 1000 $M_{\odot}$/y.

All of this evidence for very large energy releases must be combined with the fact that we now know that the active nuclear regions are exceedingly small. Size measurements are made either indirectly, by using the fact that the nuclei are often found to be rapidly variable, or directly. The estimates of size made from the flux variations depend on the fact that, provided the emitting surface is not moving relativistically, the upper limit to the size R of a region which shows a significant variation in a time τ is given by the relation

$$R \leqslant c\tau \,.$$

It is known that QSOs are frequently variable in their optical continua on a time scale of years, months or days, so that in many cases

$$R \leqslant (10^{15} \div 10^{18}) \text{ cm} \,.$$

Similar variations are seen in the nuclei of active galaxies. Also, rapid variations are seen at high frequencies in the radio spectra, suggesting similar upper limits.

Direct measurements of sizes are now being made by the radio astronomers using very long base-line (intercontinental) interferometers. With these they are able to measure sizes as small as $(10^{-2} \div 10^{-3})$ s of arc. In the case of NGC 1275 the nuclear radio source is variable at high frequencies, and its size can also be measured directly. The results lead to the conclusion that the size is only about a light year and that this component of the source is expanding nonrelativistically.

The estimates of the energy released, together with limits on the size, enable us to make some rough calculations of the possible range of values of

$$\alpha = \frac{2GM}{Rc^2} \,.$$

If the total energy released in the nucleus is E, we can put

$$M = \frac{E}{c^2}\beta \qquad (\beta > 1) \,,$$

where the efficiency of the process of energy release is given by β^{-1}.

If we now put

$$R = 10^{17}\gamma \text{ cm} \,,$$

we find that

$$\alpha \simeq 10^{-66} E \left(\frac{\beta}{\gamma}\right) .$$

We are unsure about the true values of β and γ in any given case. However, all of the attempts to understand the efficiency of conversion of mass into energy which is able to escape have led to the view that the efficiency is not likely to be much higher than about 5%, *i.e.* $\beta = 20$. Even this may be far too high an efficiency factor since it is not clear how energy is converted into relativistic particles, for example. At the same time we frequently find *upper limits* to the sizes of variable components corresponding to values of $\gamma \leqslant 1$, without knowing how small they really are. In Table IV we have therefore

TABLE IV.

E (erg)	10^{60}	10^{61}	10^{62}	10^{63}
α ($\beta = 20$, $\gamma = 1$)	$2 \cdot 10^{-5}$	$2 \cdot 10^{-4}$	$2 \cdot 10^{-3}$	$2 \cdot 10^{-2}$
α ($\beta = 20$, $\gamma = 10^{-2}$)	$2 \cdot 10^{-3}$	$2 \cdot 10^{-2}$	$2 \cdot 10^{-1}$	2

given a considerable range of possible values for γ. From this Table it can be seen that there is certainly reason to believe that it is at least possible that in some cases $\alpha \sim 10^{-2} - 1$, so that we are in a region of very strong gravitational fields where the general theory of relativity has never been tested.

The conclusion that we can draw from these results in that the physics of the nuclei of galaxies and QSOs is an excellent topic for investigation for general relativists. Not only is there a finite possibility that we are seeing here the behavior of matter in exceedingly strong gravitational fields, but we also can ask quite fundamental questions concerned with the way that such massive dense regions have reached the state in which they are now seen.

Is this a result of condensation of galaxies in an early stage of an evolving universe?

Are these delayed cores which are present, perhaps, in all galaxies?

Are these regions in which matter is being created?

BIBLIOGRAPHY

The subject matter treated in these lectures has been discussed in more detail in the following publications:

G. R. BURBIDGE and E. M. BURBIDGE: *Quasi-Stellar Objects* (San Francisco, 1967).

G. R. BURBIDGE and E. M. BURBIDGE: *Nature*, **222**, 735 (1969).

G. R. BURBIDGE and E. M. BURBIDGE: *Nature*, **224**, 21 (1969).

G. R. BURBIDGE: *The nuclei of galaxies*, in *Annual Reviews of Astronomy and Astrophysics*, Vol. **8** (1970).

Optical Observations Relevant to Cosmology: Hubble Diagram.

E. M. BURBIDGE

University of California, San Diego - La Jolla, Cal.

1. – Definitions and nature of problem.

A systematic relation between the luminosities of galaxies and their red-shifts was discovered by HUBBLE and HUMASON in 1929 and was followed up by HUBBLE and HUMASON and more recently by SANDAGE. The nature of this relation was that the most obvious interpretation was that the galaxies are moving apart—*i.e.* the Universe is expanding—and over the range of red-shift covered by HUBBLE and HUMASON the velocity of recession of any galaxy, v, is proportional to its distance D:

$$v \propto D\,, \tag{1}$$

where the red-shifts are small enough for the first-order Doppler formula to be applicable to the red-shifts, z:

$$z = \frac{\Delta\lambda}{\lambda} = \frac{v}{c} \tag{2}$$

($\Delta\lambda$ = wavelength shift in spectral line of rest wavelength λ). The luminosity l of a galaxy as observed follows the universe square law:

$$l \propto D^{-2}\,, \tag{3}$$

where D is the distance (« luminosity distance » if one is dealing with large distances where it is necessary to specify what particular definition of distance is to be used).

Astronomers measure luminosities on a logarithmic scale where « magnitude » m is defined by specifying that 5 magnitudes is equivalent to a factor of 100 in luminosity, so that two objects of observed magnitudes m_1, m_2 have luminosities l_1, l_2 such that:

$$\frac{l_1}{l_2} = 10^{0.4(m_2-m_1)} . \tag{4}$$

Combining eqs. (1)-(4), then, if it can be assumed that the *intrinsic* luminosity of galaxies at all distances is the same, one obtains a relation between magnitude, m, and red-shift, z:

$$m = \text{const} + 5 \log z . \tag{5}$$

This in fact was the relation empirically found by HUBBLE. We can see how well it fits the most recent data of Sandage's by looking at the diagram shown in the first Figure, where measured magnitude (corrected for effects I shall discuss later) is plotted as abscissa, for a number of galaxies, and $\log z$ or $\log cz$ (c in km/s) is the ordinate.

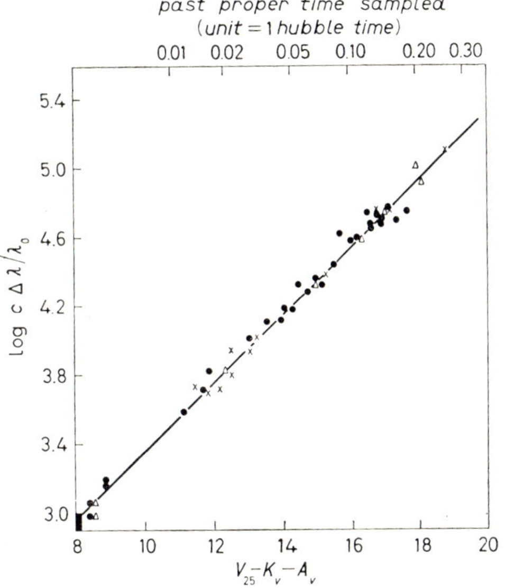

Fig. 1. – Hubble relation, from SANDAGE (1968). Black rectangle at lower left gives area over which Hubble's points extended in 1929. Brightest cluster galaxy: ▵ • non-radio, × radio.

Note that:

1) The straight line fits the points very well, and it was *not drawn to fit* them but is simply a line at slope $\frac{1}{5}$, as it shoul be from eq. (5).

2) There are hardly any points above $z = 0.2$.

3) The small black rectangle at the lower left is, as SANDAGE pointed out, the area over which HUBBLE plotted the points from which he made his original deduction that the universe is expanding!

For larger distances where the first-order equations do not fit and D takes different forms in different cosmological models, the relation between m and $\log z$ will in general depart from eq. (5) and the problem SANDAGE has been tackling is to find the model that best fits the observed relation.

In general a comparison between observations and theory has been made only between two classes of models—the exploding or big-bang models introduced by FRIEDMANN in 1922 and investigated by ROBERTSON and WALKER, and the simple steady-state model investigated by BONDI, GOLD, and HOYLE.

In the big-bang models, until quite recently it has been customary to set the cosmological constant $\Lambda = 0$ (models in which a nonzero Λ is taken were originally considered by LEMAITRE for the purposes of having a halt in the expansion during which it would be easier to form galaxies; it has become desirable to look again at these models because the QSO observations may require such interpretation; I shall not discuss them further in this lecture).

I shall now give some definitions, ahead of the lectures by ELLIS in which they will be discussed properly, because I need to define the observational quantities we are trying to determine.

Starting with a Friedmann singularity, a nonempty universe expands and the subsequent expansion velocity is *decelerated* in a nonempty universe by the self-gravitation of the matter. In the steady-state theory, the expansion *accelerates* as new matter is created to fill the void left by the expansion, so that a density constant in space and time is maintained (but there can be large-scale inhomogeneities). In McCrea's version of the steady-state theory, matter is created in regions of already-existing high matter-density, *i.e.* in the centers of galaxies.

The Hubble constant, H, has a present-day value H_0 where

$$H_0 \equiv \left(\frac{\dot{R}}{R}\right)_0 .$$

R is the radius of the universe; the subscript 0 denotes present-day values, and $H_0 D = cz$. The « deceleration parameter » q is given by

$$-q_0 H_0^2 \equiv \left(\frac{\ddot{R}}{R}\right)_0 ,$$

The present-day mean matter-density in the universe is related to q_0 by

$$\varrho_0 = \frac{3H_0^2 q_0}{4\pi G} ,$$

G being the gravitational constant.

For Friedmann universes with positive curvature the deceleration will eventually reduce the expansion to zero and reverse it. The cycle time for a complete oscillation of the universe is then

$$T = \frac{2\pi q_0}{H_0(2q_0 - 1)^{\frac{3}{2}}} .$$

The limiting q_0 for which this occurs is set by $q_0 > \frac{1}{2}$.

In the steady-state model, at time t,

$$R \propto \exp [Ht]$$

and

$$q_0 = -1 .$$

In Fig. 1 I showed the line

$$m = \text{const} + 5 \log z ,$$

which fitted the points quite well, and which is the solution *in all models* for small z, is also the solution for large z in the Friedmann model with

$$q_0 = +1 .$$

2. – Observational problems.

In plotting a diagram such as Fig. 1, there are various observational problems to tackle:

1) Calibration for nearby galaxies must be determined since this gives H (determine distance of nearest galaxies which do partake of the expansion of the universe).

2) Galaxies of the same intrinsic luminosity must be selected (« standard candles »).

3) The magnitude (luminosity) is measured through a filter of limited band-pass and must be corrected, ideally to a total or bolometric magnitude, but in any case must be corrected for red-shift of the galaxy.

4) Evolutionary effects in the luminosity of galaxies due to the aging of stars in an evolutionary universe must be evaluated—one must show that $dl/dt \approx 0$ or determine its value.

5) One must check whether the expansion velocity is isotropic or, if not, evaluate departures from isotropy.

The details of these various steps are as follows:

1) *a*) Cepheid variables (pulsating stars) are a good distance criterion (tight relation between intrinsic luminosity and period of light variation.) They must be calibrated *in our own galaxy* by using those that lie in star clu-

sters whose distances can be determined (tied ultimately to trigonometric parallaxes via the parallax of the Hyades star cluster, about which there has been some recent controversy).

b) Distances to galaxies like M33 and M31 (Andromeda Nebula) in the local group can then be determined by means of Cepheid variables in them. These are of course too close to be partaking of the general expansion.

c) Since Cepheid variables are not bright enough to be seen in more distant galaxies such as the Virgo Cluster galaxies, some more luminous criterion must be selected. SERSIC and SANDAGE use the diameters of HII regions (clouds of ionized gas). The diameters appear to have a well-defined distribution for any one morphological class of galaxy. Globular clusters can also be used. The brightest globular cluster in M31 has a blue absolute magnitude of $M_{\text{B}} = -9.83$.

d) The globular clusters around the galaxy M87 in the Virgo Cluster have recently been measured by RACINE (there are some 2000 of them). The brightest has $m_B = 21.3$, so that the distance modulus recently determined by SANDAGE for M87 is $m - M = 21.3 - (-9.8) = 31.1$.

e) Assume all galaxies in Virgo Cluster are at approximately the same distance and use $m - M = 31.1$ for the brightest galaxy in the Virgo Cluster, which is NGC 4472. Then M_{B} for NGC 4472 is -21.68. Use this as the « standard candle » for all distant clusters of galaxies.

2) SANDAGE has found that the brightest galaxy in clusters appears to be independent of the number of galaxies in a cluster. There is some controversy about this; ABELL thinks there may be a selection effect (richer clusters tending to have brighter galaxies for their most luminous members). Such a possible selection effect was first discussed by SCOTT and is often referred to as the « Scott effect ».

3) The corrections to the measured apparent magnitudes are first the « aperture effect », *i.e.* correcting for the fact that small distant galaxies may have all their luminosity included in the measuring diaphragm while the outer parts of nearby galaxies might have some luminosity excluded. Second, there is the « K correction. » The K correction consists of two parts:

a) the band-width correction—a fixed filter band-width used for measurement will pass *too few* angströms in a red-shifted galaxy because each wavelength unit is increased by red-shift (or *too many* cycles per second if frequency units are used).

b) A galaxy has some complicated distribution of flux with wavelength ($F(\lambda)$ or $F(\nu)$). The filter admits a different part of this in galaxies of different red shift. Let $S(\lambda)$ or $S(\nu)$ be the transmission functions of the filter, and $F(\lambda_0)$ or $F(\nu_0)$ be the intrinsic flux distributions emitted in the rest wavelength system. Then the total K corrections are

$$K = \underset{\substack{\uparrow \\ \text{band-width correction}}}{2.5 \log (1+z)} + 2.5 \log \left\{ \frac{\int_0^\infty F(\lambda_0) S(\lambda)\, d\lambda}{\int_0^\infty F[\lambda_0/(1+z)] S(\lambda)\, d\lambda} \right\}$$

in wavelength units, or

$$K = -2.5 \log (1+z) + 2.5 \log \left\{ \frac{\int_0^\infty F(\nu_0) S(\nu)\, d\nu}{\int_0^\infty F[\nu_0(1+z)] S(\nu)\, d\nu} \right\}$$

in frequency units.

K is in the sense

$$m_{\text{bol}} + \text{const} = m_{\text{meas}} - K\,.$$

The filter system usually used is the V filter—a width of about 900 Å centered on a wavelength 5500 Å. Corrections for $z = 0.28$ are quite large (0.83 mag). SANDAGE and SMITH have advocated a red-filter system (R_S), covering (5800÷6800) Å with maximum transmission at 6300 Å. The correction at $z = 0.28$ is less in this system, only 0.93 mag).

OKE and SANDAGE have calculated the intrinsic flux from the average of several nearby giant ellipticals and have calculated K corrections for red-shifts out to $z = 0.28$. They also calculated color differences $m_B - m_V$ to be expected between blue and yellow magnitudes, and have compared their calculated curve with observations of a number of more distant galaxies; the comparison is shown in the second Figure. The calculated curve goes very well through the observed points. This demonstrates the absence of any excess reddening of distant galaxies, which was originally thought by STEBBINS and WHITFORD to be present. This so-called Stebbins-Whitford effect was produced partly because STEBBINS and WHITFORD used a *dwarf* elliptical M32 as their standard, and it has intrinsic differences from giant ellipticals.

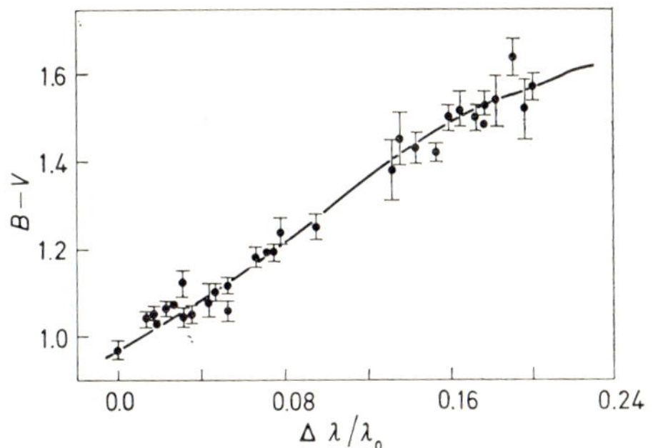

Fig. 2. – Comparison between measured color differences and those calculated using adopted standard flux, from OKE and SANDAGE (1968).

For still bigger red-shifts, we shall in the future need to know the intrinsic ultraviolet flux measured above the earth's atmosphere (which cuts off $\lambda < 3000$ Å). There may be a problem here; first results from the orbiting astronomical observatory OAO AII show that M31 has a turn-up in $F(\lambda)$ in the far ultra-violet, and this may be due to nonthermal radiation whose amount may be variable from one galaxy to another.

Ultimately we ought to use spectrum-scanning methods on all galaxies.

4) SPINRAD's studies of the stellar populations of ellipticals suggest they are mainly composed of low-mass elderly stars for which little change in luminosity can be feared out to $z \approx 0.46$ because this corresponds to a time-scale of only $\sim 3 \cdot 10^9$ y. Evolutionary times for such stars are likely to be $\sim 7 \cdot 10^9$ y or longer. However, TINSLEY has disagreed with this conclusion. The argument hinges on our interpretation of the large observed mass-to-light ratios of elliptical galaxies—are they due to a large population of very low-mass stars or to some other cause?

5) DE VAUCOULEURS has found evidence for a considerable anisotropy of expansion velocities among galaxies (*e.g.* northern-southern galactic hemisphere differences) which he attributed in 1958 to a rotation+expansion of the supercluster of which our local group and the Virgo Cluster are supposed to be members. Such a supercluster was originally suggested by SHAPLEY. KRISTIAN and SACHS have discussed such a velocity anisotropy as a « shear velocity field », and SCIAMA and STEWART have been interested in it from the point of view of finding our own velocity with respect to the distant background.

DE VAUCOULEURS believes the ratio of the Hubble constant outside the supercluster to that inside is about 1.35, self-gravitation within the cluster countering the general expansion to some extent. He originally related the suggested rotation of the supercluster to its flattening, but the time-scale for one rotation is about an order of magnitude greater than the Hubble time-scale. Both expansion and rotation velocities are suggested to be of order $(200 \div 300)$ km/s.

SANDAGE believes there is a velocity anisotropy, but somewhat smaller than deVaucouleurs' Figures. He believes

$$\frac{H_\infty}{H_{\text{local}}} \approx 1.17 \,.$$

The effect can be eliminated in fitting the Hubble diagram if one can go directly to distant galaxies outside the region of the anisotropy, *i.e.* to $v = cz > 4000$ km/s or so.

3. – Sandage's results.

His best fit to the observations is shown in Fig. 3 which plots the same points shown in Fig. 1 on a grid of curves for different values of q_0.

He gets $H_0 = 75.3^{+19}_{-15}$ km/s per megaparsec.

Then $H_0^{-1} = 12.9^{+3.7}_{-2.7} \cdot 10^9$ y.

He believes $q_0 = +1$ provides a reasonable fit to the data: if so then the time since the last Friedmann singularity is

$$t = 7.4 \cdot 10^9 \text{ y}$$

and the whole cycle time $T = 8.2 \cdot 10^{10}$ y.

The mean present-day density

$$\varrho_0 = 2 \cdot 10^{-29} \text{ g/cm}^3 ,$$

i.e. a factor $\sim 10^2$ greater than that given by counting up the luminous matter in the universe (using the recent Lick counts plus a «reasonable» estimate of average mass-to-light ratios of 35 or so); I believe the observed value is then about $3 \cdot 10^{31}$ g/cm³.

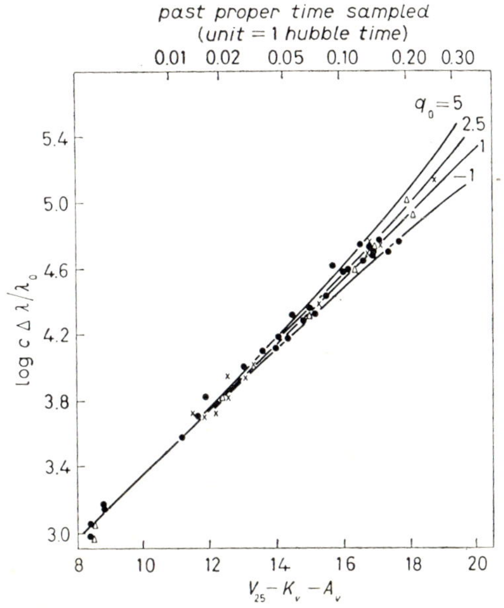

Fig. 3. – Points from Fig. 1 fitted to lines calculated for various values of q_0 (SANDAGE, 1968). Brightest cluster galaxy: ▵ • non-radio, × radio.

If the effect of the velocity anisotropy is such that all the nearby clusters (mainly northern hemisphere measures) have to be corrected *upward* in the Hubble diagram, then the whole grid of calculated lines should be moved to the left in the diagram and a fitting would be made toward the value $q_0 = 0$.

I believe the value of q_0 is indeterminate at present. Clearly we need to investigate the velocity anisotropy much more carefully. We need large telescopes in the Southern Hemisphere.

4. – Additional problems with red-shifts.

Quite apart from the local velocity anisotropy, there are some instances of red-shift measurements which make one wonder if one sometimes sees red-shifts that have a noncosmological origin. The QSO's are a particular case of this problem, but even among galaxies there are some puzzling instances, *e.g.*

1) 3C 390.3—a radio galaxy—has a mean red-shift giving $v = 17\,000$ km/s. The spectrum lines have two peaks separated by some 4600 km/s, so that we have that much intrinsic red-shift to explain anyway.

2) VV 172 is a chain of 5 galaxies; SARGENT has found that 4 have similar red-shifts giving $v \approx 15\,600$ km/s but one has $v = 38\,000$ km/s. Is it a background object or not? It looks like the rest of the chain and appears to belong to it.

We need more work on these problems.

BIBLIOGRAPHY

A. SANDAGE: *Observatory*, **88**, 91 (1968).
J. B. OKE and A. SANDAGE: *Astrophys. Journ.*, **154**, 21 (1968).
A. SANDAGE: *Astrophys. Journ.*, **152**, L149 (1968).
G. DE VAUCOULEURS and W. L. PETERS: *Nature*, **220**, 868 (1968).
M. L. HUMASON, N. U. MAYALL and A. SANDAGE: *Astron. Journ.*, **61**, 97 (1956).

Some Current Ideas on Galaxy Formation.

M. J. REES

Institute of Theoretical Astronomy, - Cambridge

1. - Introduction.

Most theoretical cosmology deals with the properties of homogeneous « perfect fluid » models. The evident nonuniformity displayed by the actual universe—as a consequence of the existence of galaxies, stars and observers—does not however render these simple models inapplicable. They in fact provide a valid overall description provided that the typical fluctuations $\Delta\varrho$ of the mean density over length scales r satisfy

$$G(\Delta\varrho)\, r^2 \ll c^2 \tag{1.1}$$

for all r. This ensures that local fluctuations in the gravitational potential should be $\ll c^2$, and that the associated peculiar velocities should be $\ll c$. Both for galaxies and for clusters of galaxies, (1.1) is fulfilled by a margin of at least $\sim 10^4$, and there are strong observational indications that it holds on all larger scales up to a Hubble radius.

Observations of the Hubble diagram (SANDAGE [1]), the radio source counts (RYLE [2]) and—much more convincingly—the spectrum and isotropy of the microwave background radiation, all indicate that the universe is isotropic, and has expanded from a much denser state. If the overall *mean* density was once much higher than the present density *in galaxies*, clearly galaxies (and, *a fortiori*, clusters of galaxies) could not then have existed in their present form. Protogalaxies may have been merely regions of slightly enhanced density in the expanding universe, whose expansion was retarded, and eventually halted, by their excess gravitational field. Although gravitational forces promote instability, pressure gradients (especially radiation pressure in the fireball phase) and viscous dissipation tend to counteract the growth in the « hot big bang » model. My main intention is to describe the interaction between these various processes, and to discuss to what extent one can understand the occurrence of structure with the characteristically observed scales without artificially choosing the initial conditions with this end in view. No very spectacular theoretical results have yet been achieved, but the fact that these investigations can even be contemplated illustrates how the discovery of the microwave background radiation has extended the range of phenomena amenable to quantitative discussion.

Most recent work has been predicated on the view that the present structural features evolved from small perturbations in an almost homogeneous universe, but it must be emphasized that this is no more than a working hypothesis. These studies have already led to some encouraging results, and it would in any case be premature to discard the hypothesis until its implications have been more fully investigated. As is well known, AMBARTZUMIAN [3] has long been voicing the opinion that matter emerges locally from a high-density singular state, but this idea cannot be developed quantitatively in our present state of knowledge. A further incentive for pursuing studies of the fate of perturbations in cosmological models is that only in this approach is it consistent to assume a homogeneous background model: if galaxies had always had the same contrast density $\Delta\varrho/\varrho$, then (1.1) would be violated when ϱ was large, so these mathematically tractable models would be irrelevant.

2. – Gravitational instability of Friedmann models.

The first full relativistic treatment of the behaviour of perturbed Friedmann models was given by LIFSHITZ [4]. Many authors have subsequently discussed various aspects of this problem, and excellent reviews have been given by, among others, ZELDOVICH [5], HARRISON [6] and FIELD [7]. The reader is referred to these and to the papers cited therein for fuller details; I shall give only a simplified discussion of the main points here.

The local behaviour of a homogeneous isotropic cosmology is described by the scale factor $R(t)$ (which may be conveniently thought of as the radius, in a Schwarzschild co-ordinate system, of an arbitrary co-moving sphere of mass m contained within the particle horizon), and the pressure $p(t)$. The number density of particles is proportional to R^{-3}, and the total density ϱ varies according to

$$\frac{\dot{\varrho}}{\varrho} = -3\left(1+\frac{p}{\varrho c^2}\right)\frac{\dot{R}}{R}\,. \tag{2.1}$$

If $p \ll c^2$, and we take the cosmical constant Λ to be zero (*), the overall

(*) The special features of Lemaître-type models are discussed in the Appendix. However, even when Λ is non zero it exerts a negligible effect during the early dense phases of the expansion.

dynamics is described by the standard Friedmann « dust » model whose expansion is given parametrically by

$$R = (Gmt^{*2})^{\frac{1}{3}} \times \begin{cases} 1 - \cos\eta & (K = 1)\,, \\ \frac{1}{2}\eta^2 & (K = 1)\,, \\ \cosh\eta - 1 & (K = -1)\,, \end{cases} \tag{2.2a}$$

$$t \;= t^* \times \begin{cases} \eta - \sin\eta & (K = 1)\,, \\ \frac{1}{6}\eta^3 & (K = 0)\,, \\ \sinh\eta - \eta & (K = -1)\,, \end{cases} \tag{2.2b}$$

where K/t^* measures the space curvature. When $t \ll t^*$, all three models approximate closely to the Einstein-de Sitter ($K = 0$) model, with $R \propto t^{\frac{2}{3}}$: when $t \gtrsim t^*$ (*i.e.* $\eta \gtrsim 1$), the $K = -1$ model starts to expand freely with no deceleration, and $R \propto t$; and the $K = 1$ model starts to recontract, returning to a state of infinite density when $\eta = 2\pi$. (In the $K = 0$ model, t^* has no physical significance and may be scaled arbitrarily.) The total energy of a spherical region containing a mass m is positive for $K = -1$ but negative for $K = +1$. For $K = 0$ this energy is zero; for other models the deviations from zero energy are greater when t^* is small.

Provided we can neglect pressure forces, the expansion of a spherical region would be unaffected by the removal of the rest of the universe. As a corollary, the behaviour of the remainder of the universe would be unaltered if the material within a comoving sphere were replaced by any spherically symmetrical distribution with the same gravitational mass (if we envisage many such spherical holes, we have the « swiss cheese » model of EINSTEIN and STRAUSS [8]). It is now easy to see that a Friedmann dust model is unstable to the growth of density perturbations. Suppose that a perturbed spherical region contains matter which is expanding in the manner appropriate to a different model with lower energy. As the expansion proceeds, it will lag more and more behind the rest of the universe. When $\eta \ll 1$ in (2), one can easily see that the values of R attained by different models at a given t differ by a term of order η^4. This means that, while the perturbation is still small,

$$\frac{\Delta R}{R} \propto \frac{\Delta\varrho}{\varrho} \propto R \propto t^{\frac{2}{3}}\,. \tag{2.3}$$

This shows that an expanding dust universe is unstable to the growth of those perturbations which can be regarded as fluctuations in the local total energy, or in the local space curvature. (One can also imagine initial irregularities in which the energy is unperturbed, but the expansion starts at different times.

This type of fluctuation also emerges from a more elaborate mathematical treatment of the problem but is of little interest because it *diminishes* in amplitude in proportion to $R^{-\frac{3}{2}}$ during the expansion.) Provided we interpret $\Delta\varrho/\varrho$ as the fractional deviations from the background density which different observers would measure *at the same proper time t*, (3) still holds even when the scale of the fluctuations is large compared to the particle horizon (*). Of course, this will be true for *any* scale at sufficiently early times, because in models which obey (2) the mass M_H within the particle horizon varies as

$$M_H \propto R^{\frac{3}{2}} \propto t\,. \tag{2.4}$$

Relation (3) obviously ceases to apply when the fluctuations attain order unity. In a $K=0$ background universe, the total energy of any perturbation will be negative. It will stop expanding when (to be precise) its density is $\frac{9}{16}\pi^2$ times that of the background, and will then start to collapse. Qualitatively similar conclusions hold in any model, provided that $\Delta\varrho/\varrho$ becomes ~ 1 when $t \ll t^*$. In a model with $K=-1$, any perturbation which still has small amplitude when $t \simeq t^*$ will retain almost *constant* amplitude thereafter. This is because the self-gravitation of the perturbed region would become unimportant when $t > t^*$, so that it would continue, like the background universe, in undecelerated expansion. In a closed model ($K=1$) a fluctuation whose amplitude is still small at t^* will not separate out until the whole background universe is in a state of collapse. Any fluctuations which start to re-contract while they are still larger than M_H would not really constitute a part of our observable universe.

If p is nonzero, fluctuations on sufficiently small scales will be stabilized by pressure gradients (**). They will not grow, but instead oscillate like sound waves. The minimum length scale over which gravitational forces overwhelm pressure is $\sim c_s/(G\varrho)^{\frac{1}{2}}$, where c_s is the sound speed. The corresponding « Jeans

---

(*) In the case when the perturbed sphere is small compared with the particle horizon, and its boundary expands with a speed $v \ll c$, we can derive (3) by the following elementary argument. A change in energy corresponds to a given change $\Delta(v^2)$ in v^2 which is independent of R. For an Einstein-de Sitter (« zero-energy ») model, $v \propto R^{-\frac{1}{2}}$, so $\Delta v/v \propto R$. This means that the fractional difference in the times taken by the perturbed and unperturbed spheres to attain a given radius (or, equivalently, the fractional density perturbation at a given time) increase in proportion to R.

(**) When the gas is capable of cooling by emission of radiation, the pressure and density are no longer related by a simple equation of state, and there is a possibility of nongravitational *thermal instabilities*. These will be mentioned in Sect. **3**, when we consider the conditions which arise after the reheating of intergalactic gas in a « hot big bang » cosmology.

mass » (JEANS [9]) is

$$M_J \simeq \frac{c_s^3 \varrho}{(G\varrho)^{\frac{3}{2}}} \tag{2.5}$$

in a $K = 0$ model, this mass is such that the differential expansion velocity across it is $\sim c_s$. The effects of pressure are negligible for all fluctuations with masses $\gg M_J$, so in such cases the above discussion of dust models is applicable.

Apart from the dust model, another perfect fluid cosmology of physical interest, and for which a simple expression for $R(t)$ can be derived, is the « radiation universe » for which $p = \frac{1}{3}\varrho c^2$. We then have $\varrho \propto R^{-4}$, and

$$R \propto \begin{cases} \sin\eta & (K = 1)\,, \\ \eta & (K = 0)\,, \\ \sinh\eta & (K = -1)\,, \end{cases} \tag{2.6a}$$

$$t \propto \begin{cases} 1 - \cos\eta & (K = 1)\,, \\ \frac{1}{2}\eta^2 & (K = 0)\,, \\ \cosh\eta - 1 & (K = -1)\,. \end{cases} \tag{2.6b}$$

For small t, $R \propto t^{\frac{1}{2}}$. In this model, $c_s = c/\sqrt{3}$, so that pressure stabilizes all masses substantially smaller than the particle horizon (in other words, $M_J \simeq M_H$). However this universe *is* unstable to perturbations with scales $\gg M_H$, and we find (by comparing $R(t)$ for different models, as we did earlier for the dust case) that

$$\frac{\delta\varrho}{\varrho} \propto R^2 \propto t\,. \tag{2.7}$$

Because M_H increases with t, a perturbation with any given scale will start off by growing according to (7), but as soon as it comes within the horizon pressure gradients it will become important and it will start oscillating.

The original studies of this subject were motivated by the hope that galaxies and clusters might have condensed from the random $\sqrt{N}/N$ fluctuations which would naturally be expected in a universe composed of discrete atoms. For a galactic mass of $\sim 10^{11} M_\odot$, however, the « statistical » fluctuations are only $\sim 10^{-34}$, and these would not have condensed out by the present epoch unless one assumes that the growth was initiate at a stage when the particle horizon encompassed only a few atoms. Although there is no reason to single out any special stage in the expansion, it is not clear that much physical significance attaches to this result. This has been further discussed by PEEBLES [10] (*). The problem is that the overall expansion transforms the growth rate of the instability from an exponential to a (much slower) power law, which means that one must either suppose that the fluctuations come into being at an exceedingly early epoch, or else assume larger initial amplitudes. For example, the existence of galaxies could be accounted for in a « dust » universe if there were perturbations with $\delta\varrho/\varrho \simeq 10^{-7}$ at the time when such a mass first comes within the horizon. Though much larger than the « statistical » amplitude, these are still « small » perturbations in the sense that they would cause the geometry of the universe to deviate only negligibly from the strictly homogeneous case.

However, this type of hypothesis has the unsatisfactory feature that it really explains nothing. The initial amplitude must be chosen in a purely *ad hoc* manner: if it were too small, galaxies would not yet have condensed, but if it were too large they would have stopped expanding at an early epoch, and would have higher mean densities than are observed. In addition, when $p = 0$, perturbations of all scales grow at the same rate, given by (23). This means that there is no preferred scale of instability, but that the postulated initial spectrum must be weighted in favour of whatever masses are required. It is true that nonzero pressure could stabilize small scales which were $\lesssim M_J$ during certain stages of the expansion, with the result that the growth of large scales would be favoured. However, this effect does not seem capable of impressing any prominent features on an initially smooth mass spectrum of fluctuations.

The detection of the thermal microwave background, and its interpretation as « primeval » radiation, stimulated renewed interest in this approach to the problem of galaxy formation. There are several reasons why this should have been so:

i) It constituted the first really compelling evidence for a hot, dense, opaque early phase in the evolution of the universe—the so-called « primordial fireball ».

ii) The isotropy of the background radiation has been established to within $\lesssim 0.1\%$—unprecedented precision for a cosmological measurement. This suggests (**) that a simple Friedmann-type model may provide an adequate dynamical description of the early universe.

(*) A more precise statement of the amplitude required at a given epoch cannot be made without specific assumptions regarding the equation of state and the multiplicity of particles which may exist in the high-density phases. KUNDT (this volume p. 368) shows that the situation is somewhat more favourable if Hagedorn's theory of high-energy thermodynamics is correct.

(**) Homogeneous anisotropic cosmologies have been considered in which the anisotropy dies away rapidly during the expansion (see G. F. R. Ellis' lectures) so this is certainly not a completely firm inference.

iii) The « hot big bang » model allows us to predict a definite equation of state throughout the expansion (except for $t \ll 1$ s): during the first $(10^3 \div 10^5)$ y the universe would have been radiation-dominated and would expand according to (2.6), whereas the expansion would obey a law of type (2.2) during the later stages.

iv) The fate of perturbations in the fireball is *not*, as in perfect fluid models, governed solely by competition between pressure gradients and gravitation: a variety of dissipative processes operate, as a consequence of the *imperfect* coupling between matter and radiation. The resulting damping acts preferentially on certain scales (and certain types) of perturbation. It would be gratifying if the irregularities most likely to survive and amplify bore some relation to the characteristic scales of observed structures in the universe.

3. – The fate of perturbations in the primordial fireball.

The main aspects of the « hot big bang » model are outlined by SCIAMA elsewhere in the volume (see also the review by FIELD [11]). The present mean density of matter can be taken as

$$\varrho_{m_0} = 3.6 \cdot 10^{-29} \sigma_0 \left(\frac{H_0}{100}\right)^2 \text{g cm}^{-3} , \tag{3.1}$$

when H_0 km s^{-1} Mpc^{-1} is the Hubble constant and $\sigma_0 = 4\pi G \varrho_{m_0}/3H_0^2$. When $\Lambda = 0$, σ_0 is equal to the deceleration parameter q_0. Taking the primeval radiation to have a black-body spectrum with temperature 2.7 °K, its energy density is equivalent to

$$\varrho_{\gamma_0} = 5 \cdot 10^{-34} \text{ g cm}^{-3} . \tag{3.2}$$

The « standard » model predicts that there should, in addition, be a background density of thermal neutrinos, with a density 7/11 times ϱ_γ which affects the overall expansion but does not (except when $t \ll 1$ s (MISNER [12])) interact with fluctuations smaller than the particle horizon. Although radiation is now dynamically negligible compared to the matter, it would dominate in the early stages, because $T \propto R^{-1}$ and $\varrho_\gamma \propto R^{-4}$, whereas $\varrho_m \propto R^{-3}$. The thermal pressure of the matter is always negligible compared to that of the radiation, so these relations fix the equation of state. (Throughout what follows we take $H_0 = 100$ km s^{-1} Mpc^{-1}: the results can all be straightforwardly scaled to fit any other value). In fact the changeover from radiation-dominated to matter-dominated expansion occurs when $R/R_0 \simeq 2 \cdot 10^{-5} \sigma_0^{-1}$, or when $T \simeq 1.3 \cdot 10^5 \sigma_0$ °K.

The corresponding time is

$$t_{\mathrm{eq}} \simeq 5 \cdot 10^2 \sigma_0^{-2}\ \mathrm{y}\,. \tag{3.3}$$

Before t_{eq}, the expansion obeys (2.6), with $\eta \ll 1$, so that $R \propto T^{-1} \propto t^{\frac{1}{2}}$; after t_{eq}, it approximates to a « dust » model (2.2), the appropriate value of K depending on whether $\sigma_0 \gtrless \frac{1}{2}$. In fact (FIELD and SHEPLEY [13]), the whole of the expansion can be represented by

$$\frac{R_1}{R} = \begin{cases} \sin^2\dfrac{\eta}{2} + r\sin\eta\,, & (K = \ \ 1)\,, \\[2ex] \dfrac{(\eta)^2}{2} + r\eta\,, & (K = \ \ 0)\,, \\[2ex] \sinh^2\dfrac{\eta}{2} + r\sinh\eta\,, & (K = -1)\,, \end{cases} \tag{3.4}$$

where

$$t = \frac{1}{c}\int R(\eta)\,\mathrm{d}\eta\,. \tag{3.5}$$

R_1 is the value of R at which the effects of the curvature become important —*i.e.* its value when, in our earlier notation, $t \simeq t^*$—and r the ratio ϱ_γ/ϱ_m at that stage (which, for all permissible models, is $\lesssim 10^{-1}$). (3.4) accurately describes the dynamics through the transition period when $\varrho_\gamma \simeq \varrho_m$.

Another key phenomenon in the expansion is the recombination of hydrogen, which occurs when the temperature falls to $\sim 4000\ ^\circ\mathrm{K}$. This happens at the time

$$t_{\mathrm{rec}} \simeq \begin{cases} 10^5\,\sigma_0^{-\frac{1}{2}}\ \mathrm{y} & (\sigma_0 \gtrsim 0.03)\,, \\[1ex] 6 \cdot 10^5\ \mathrm{y} & (\sigma_0 \lesssim 0.03)\,. \end{cases} \tag{3.6}$$

After t_{rec}, the universe becomes transparent, and motions of the matter are unaffected by the black-body radiation. When $t < t_{\mathrm{rec}}$, on the other hand, matter and radiation interact strongly, the main coupling being via Compton scattering of photons by electrons. The mass $M_{\tau=1}$ of matter within a sphere of optical depth unity would be $\sim 3 \cdot 10^6\,\sigma_0^{-2}\,M_\odot$ just before recombination, and diminishes rapidly ($\propto T^{-6}$) as we extrapolate further back into the past. Thus, at least for perturbations on a galactic scale, the matter and radiation behave approximately as a single fluid. The speed of sound in this composite fluid is

$$c_s = \frac{c}{\sqrt{3}}\left(1 + \frac{3}{4}\frac{\varrho_m}{\varrho_\gamma}\right)^{-\frac{1}{2}}. \tag{3.7}$$

Before t_{eq}, $c_s \simeq c/\sqrt{3}$, and the Jeans mass M_J is comparable with the mass M_H within the particle horizon. Between t_{eq} and t_{rec}, M_J remains constant, at a value

$$M_J \simeq 10^{15}\sigma_0^{-2} M_\odot ; \tag{3.8}$$

after t_{rec} it drops drastically owing to the fact that the thermal pressure of the particles is only $\sim 10^{-7}\sigma_0$ of the radiation pressure.

Nonuniformities in ϱ_γ on scales much less than M_J will give rise to oscillations. Small amplitude sinusoidal perturbations filling the universe would be like sound waves. If the coupling between matter and radiation were perfect, and there were no damping, the behaviour of the amplitude $(\delta\varrho/\varrho)$ would be governed by the usual theory of adiabatic invariants—the energy within one wavelength varies inversely as the period $\varkappa$ of the oscillation. Let us focus attention on perturbations involving a specific mass of matter. The energy within one wavelength varies during the expansion as

$$\left(\frac{\delta\varrho}{\varrho}\right)^2 c_s^2(\varrho_m + \varrho_\gamma)\, R^3 \, . \tag{3.9}$$

When $t < t_{eq}$, $\varkappa \propto R$; but when $t > t_{eq}$, $\varkappa \propto R^{\frac{3}{2}}$, because $c_s \propto R^{-\frac{1}{2}}$. Therefore

$$\frac{\delta\varrho}{\varrho} \simeq \text{constant} \qquad (t < t_{eq}) \, , \tag{3.10}$$

but

$$\frac{\delta\varrho}{\varrho} \propto R^{-\frac{1}{4}} \propto t^{-\frac{1}{6}} \qquad (t_{eq} < t < t_{rec}) \, . \tag{3.11}$$

The physical interpretation of (3.10) is that the inertial mass within a co-ordinate volume varies as R^{-1}, so that the reduction of energy ($\propto R^{-1}$) required by the adiabatic invariant can be absorbed by the decrease of mass, without necessitating any diminution of the amplitude.

However, the distances which photons are able to traverse between successive scatterings are not *completely* negligible, so they tend to leak out of regions where the pressure is above average. A more accurate treatment allowing for photon diffusion and viscosity reveals that the oscillations would, as a consequence, suffer significant damping. Detailed calculations have been carried out by SILK [14], PEEBLES [15] and MICHIE [16]. PEEBLES considered accoustic waves small compared with M_J, treating the particles and the photons as two separate fluids. The imperfect coupling leads to a phase difference in which the matter wave lags slightly behind the wave, in the photon gas, and to attenuation of the waves (*i.e.* an imaginary component of the sound speed.)

SILK and MICHIE have given rather more complicated discussions for spherically symmetric perturbations, utilizing Misner and Sharp's [17] relativistic hydrodynamic equations. When $\varrho_m > \varrho_\gamma$ (that is, $t > t_{eq}$), oscillations of wavelength λ are dissipated in a time comparable with the time it takes a photon to diffuse a distance λ. Scales much smaller than (3.12) would therefore have damped out well before t_{rec}. One can thus determine for a given epoch t, a *minimum undamped mass* M_D for which the damping time $\sim t$; one finds that $M_D \simeq (M_{\tau=1} \cdot M_H)^{\frac{1}{2}}$. In a radiation-dominated regime, when the photons contribute to the inertia of the waves as well as to the pressure, the efficacy of diffusive damping is reduced—the damping time then exceeds that given by the above naive criterion by a factor ϱ_γ/ϱ_m (SILK [14]). An additional process which is then of comparable importance is viscous damping arising from the slight anisotropy which develops in the local photon distribution during oscillations.

Dissipative processes are clearly more important on the smaller scales, and these studies lead to the important conclusion that all oscillations with masses up to

$$M_2 \simeq 10^{12} \sigma_0^{-\frac{5}{4}} M_\odot \qquad (\sigma_0 \gtrsim 0.03) \tag{3.12}$$

will have been severly attenuated by the time hydrogen starts to recombine. MICHIE in fact continued his numerical computations until the hydrogen was 99% neutral, finding that the dissipation can then affect scales somewhat larger than (3.12). It would be interesting to follow such calculations right through to the completion of recombinations, allowing for deviations from Saha equilibrium (PEEBLES [18], ZEL'DOVICH, KURT and SUNYAEV [19]). Although the decoupling is effectively accomplished within a period only $\sim 10\%$ of the expansion time scale, damping during this phase may be significant because the reduced electron density results in longer photon path lengths.

Figure 1 illustrates certain features of a standard « hot big bang » model with $\sigma_0 = \frac{1}{2}$. The mass M_H of the *matter* lying within the particle horizon at time t is plotted (we observe that a galactic mass first comes within the horizon when $t \simeq 1$ y). Also plotted are M_J, $M_{\tau=1}$ and M_D. The amplitude of oscillations of a given mass varies according to (3.9) and (3.10) until the mass falls below M_D, at which time damping becomes important. The maximum value of M_J is $4 \cdot 10^{15} M$, and pressure is never significant on scales larger than this.

Photons are the main dissipative agents throughout the era when galactic masses undergo oscillations (*i.e.* for $1\text{ y} \lesssim t \lesssim 10^5$ y), but various other processes may operate in the very hot early phases. Of particular interest is neutrino viscosity, which plays a role when $T \simeq 2 \cdot 10^{10}$ °K ($t \simeq 0.5$ s) and has an important influence in diminishing large-scale anisotropies in the universe (MISNER [20], STEWART [21], CARSWELL [22]). Neutrinos can also damp out

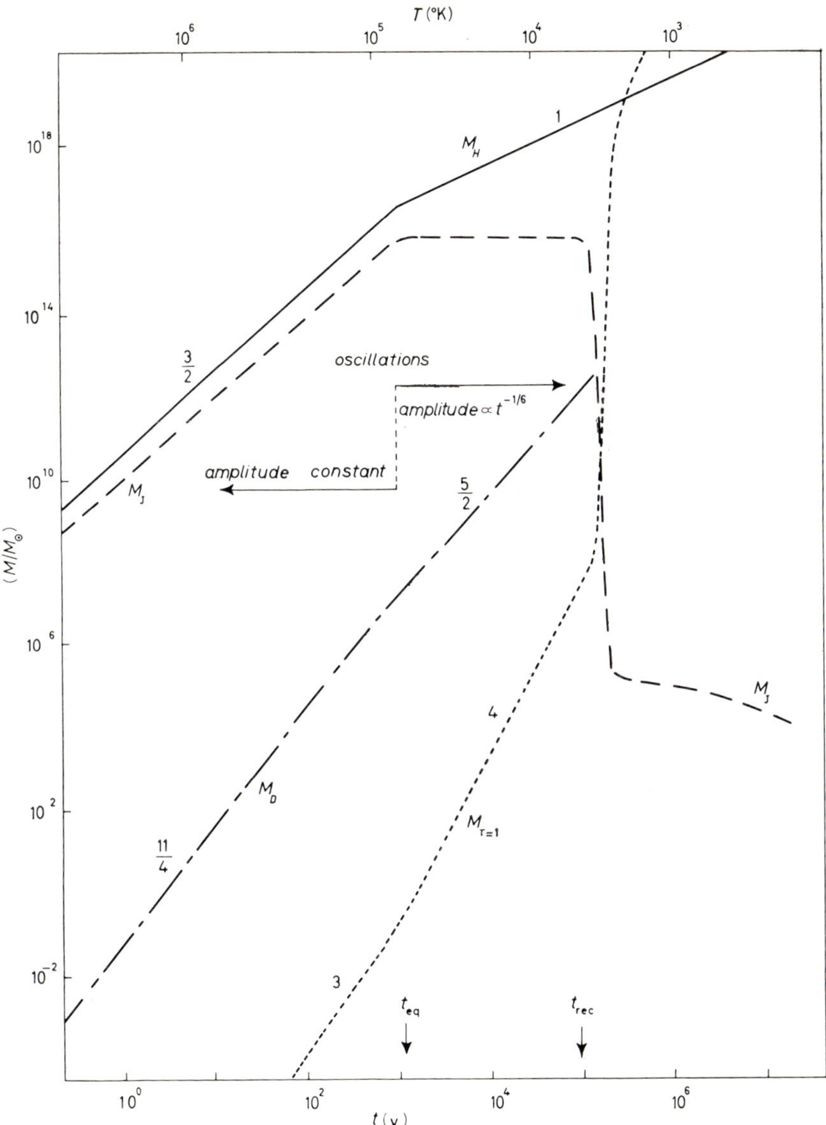

Fig. 1. – Some characteristics of an Einstein-de Sitter ($\sigma_0 = \frac{1}{2}$) model with present black-body temperature $T \simeq 3$ °K and present density $\sim 2 \cdot 10^{-29}$ g cm^{-3}. The expansion is matter-dominated for $t > t_{eq}$ but radiation-dominated at earlier times, M_H denotes the mass of baryons within the particle horizon (note that a galactic mass comes within the horizon when $t \simeq 1$ year). $M_{\tau=1}$ is the mass of a sphere whose optical depth is unity and M_J is the Jeans mass. The plasma recombines at t_{rec}, when $T \simeq 4000$ °K. In calculating M_J, we include radiation pressure before t_{rec}, but not afterwards (assuming the gas is not reheated). Perturbations in the radiation density on scales $\lesssim M_D$ are severely damped. During the expansion, the quantities vary as powers of t which are given alongside the curves.

oscillations within the horizon (MISNER [12]), in an analogous manner to photons at later stages in the expansion, but this is of no direct interest for galaxy formation, because $M_H \simeq 10^{-4} M_\odot$ at the relevant epoch. (Inhomogeneities on scales much larger than M_H would cause the local expansion to simulate a slightly anisotropic universe, so it is conceivable that neutrinos may also affect these).

Two other processes—one a damping process, one an instability—should be mentioned briefly. The first arises from the bulk viscosity which is a general property of partially relativistic gases (STEWART and ANDERSON [23]). This may be especially important when $T \simeq 5 \cdot 10^9$ °K—just before the electron-positron pairs annihilate, but after they have ceased to be ultra-relativistic, and could damp out oscillations. Secondly, FIELD [7] has discussed a *non*-gravitational instability, first pointed out by GAMOW [24], which arises because each particle tends to shield its neighbours from the intense ambient radiation. The resulting « mock gravity »—an inverse-square-law attraction—dominates ordinary gravity when $t \leqslant 80$ y, but is effective only for scales $\leqslant M_{\tau=1}$. This mechanism, like the operation of bulk viscosity and (probably) neutrino viscosity, is thus restricted to uninterestingly small scales. There are no doubt other such processes which remain to be discovered (for example, neutrinos may interact gravitationally with adiabatic fluctuations even when they have become collisionless).

The thermal coupling between matter and radiation before t_{rec} is so strong as to prevent the development of any substantial temperature difference between matter and radiation (except possibly as a result of large amplitude small scale oscillations). This fact, together with the negligible heat capacity and pressure of the matter in comparison with the radiation, inhibits the development of *thermal* instabilities. As mentioned later, however, the latter may be important during later stages.

Although all oscillations involving masses $\leqslant 10^{12} M_\odot$ would have been damped before the plasma starts to recombine, only in the special case of *adiabatic* perturbations which preserve the photon/baryon ratio (*i.e.* for which $\delta\varrho_\gamma/\varrho_\gamma = \frac{4}{3}(\delta\varrho_m/\varrho_m)$ does the radiative damping also eliminate inhomogeneities in the matter density (*). In general, so-called *isothermal* perturbations in the particle density remain even after the radiation has smoothed out. The amplitude of isothermal perturbations below M_J stays almost constant until t_{rec} because radiation drag and radiation pressure prevent the self-gravitation of the matter from being effective. A perturbation which was isothermal when it first came within the horizon would of course *never* oscillate (though scales $\gg M_J$ start to grow as R after t_{eq}).

There is at present no special motive for postulating one type of initial perturbation rather than another. Precisely adiabatic perturbations would occur only if, for some reason, the entropy per baryon were strictly uniform.

One process which would lend to almost purely isothermal fluctuations has been discussed by HARRISON [25]. He shows that, in a universe which possesses overall symmetry between matter and antimatter, small nonuniformities in the matter-antimatter ratio when $T \gtrsim 10^{13}$ °K (before baryon pairs have annihilated) could lead to isothermal fluctuations of amplitude ~ 1 at later times.

Nor is there any *a priori* reason for postulating a particular form for the mass spectrum of the perturbations. Since no damping mechanism has been suggested for the largest scales—and indeed all masses $\gtrsim M_3$ (3.8) grow continuously in amplitude—the apparent large-scale uniformity of the universe forces us to assume that the corresponding initial amplitudes were very small. The least artificial assumption seems to be to postulate a smooth initial spectrum in which the « power » is concentrated in smaller masses (or higher wave numbers). The necessary amplitude must be fixed *a posteriori*, but the possibility that some perturbations be selectively damped or amplified raises the hope that certain preferred scales of irregularity may be especially prominent when they emerge from the « fireball ».

4. – Processes occurring after t_{rec}.

The work discussed in the last Section leads to the conclusion that all purely adiabatic fluctuations with masses up to those of the largest galaxies, or even somewhat larger, would have been attenuated by radiative viscosity. However, isothermal fluctuations would survive with their initial amplitude, as would the isothermal component of any more general initial perturbation. Any fluctuations whose amplitude at t_{rec} was already ~ 1 would form condensations with density $\gtrsim 10^4$ atoms cm^{-3}. Therefore any agglomeration of galactic mass which is destined to form an ordinary galaxy must have had amplitude $\ll 1$ just after decoupling. This provides some reassurance that a linear treatment may be appropriate before t_{rec}.

After the recombination, the free electron density drops so low that the universe becomes transparent to electron scattering. Detailed computations of the recombination (PEEBLES [18]; ZEL'DOVICH, KURT and SUNYNEV [19]) show that when the radiation temperature has fallen below ~ 1000 °K, n_e/n_H is only $\sim 10^{-5}$. This means that, even though the radiation may still affect the overall dynamics of the universe, there is no significant dynamical coupling

(*) This is actually strictly true *only* when the damping occurs gradually over many oscillations. However a mass of $\sim 10^{12}\, M_\odot$ may not oscillate more than ~ 10 times before t_{rec}. If the damping is spread over, say, y oscillations, the residual isothermal fluctuation may have up to y^{-1} times the amplitude of the undamped oscillation, even if the latter were strictly adiabatic.

between radiation and matter as far as fluctuations with $M \ll M_H$ are concerned. (One can easily check that both the radiation drag acting on transparent fluctuations (PEEBLES [26]), and the pressure due to the high-frequency tail of the radiation spectrum in the Lyman lines or Lyman continuum, are indeed negligible.) The gravitational instability of density fluctuations after t_{rec} is thus inhibited by gas pressure alone, which is $\sim 10^{-7}\sigma_0$ times the radiation pressure. During the recombination, the Jeans mass therefore drops by the three-halves power of this factor, to

$$M_1 \simeq 2\cdot 10^5 \sigma_0^{-\frac{1}{2}} M_\odot . \tag{4.1}$$

Until condensations have formed which are able to reheat the gas, its temperature falls according to some function of R between R^{-2} and R^{-1}—the former law applies if the gas cools adiabatically with $\gamma = \frac{5}{3}$, and the latter if there is still thermal coupling between gas and radiation (PEEBLES [18]) shows that, in a low density model, inverse Compton heating of the residual electrons provides efficient coupling until T falls to ~ 200 °K) (*). M_J therefore tends to decrease below (4.1) as the temperature drops.

Masses in the range which interests us can therefore condense unimpeded by pressure. In order to discuss the growth rates for low density models as well as for the Einstein-de Sitter ($\sigma_0 = \frac{1}{2}$) universe, it is convenient to remember that the time t^* defined earlier is related to σ_0 according to

$$t^* \simeq 10^{10}\sigma_0\,\text{y} . \tag{4.2}$$

The significance of t^* (which is only meaningful when $\sigma_0 < \frac{1}{2}$ (**)) is that for $t \leqslant t^*$, the universe approximates closely to an Einstein-de Sitter model, whereas for $t > t^*$ the expansion is undecelerated as in the Milne model. t^* is not defined when $\sigma_0 = \frac{1}{2}$, but it will abbreviate the following discussion if we then take it to be $\sim 10^{10}$ y (*i.e.* the present epoch).

For $\sigma_0 \geqslant 0.03$, density perturbations whose masses exceed (4.1) grow according to (2.3) from $t_{rec}\,(R/R_0 \simeq 10^{-3})$ until $t^*\,(R/R_0 \simeq 2\sigma_0)$—in other words from $t \simeq 10^5 \sigma_0^{-\frac{1}{2}}$ y to $t \simeq 10^{10}\sigma_0$ y. If $\sigma_0 \leqslant 0.03$, corresponding to a present mean density of $\leqslant 10^{-30}$ g cm^{-3}, the universe is still radiation-dominated at

(*) Some authors (SASLAW and ZIPOY [27], PEEBLES and DICKE [28]) have considered the possibility that H_2 may form via H^-, the residual electrons acting as a catalyst. It does not appear that more than $\sim 10^{-5}$ of the material can be molecular, but if this fraction *were* significant, thermal contact between gas and radiation could be maintained until an even later stage.

(**) The behaviour of a closed model with $\sigma_0 > \frac{1}{2}$ does not exhibit any marked qualitative differences from the Einstein-de Sitter case (at least until the stage when the overall expansion of the background has been halted).

recombination *i.e.* $t_{eq} > t_{rec}$). Growth according to (2.3) does not then commence until t_{eq}, because until this epoch the time scale of the overall expansion is governed by the radiation, and the mass of the matter (and therefore the gravitational effect of a perturbation which involves the matter alone) are not significant. In summary, density irregularities grow according to $\delta\varrho/\varrho \propto R$ when

$$\max\,[t_{rec}, t_{eq}] < t < t^* \,. \tag{4.3}$$

Conditions for the growth of perturbations after decoupling are thus most propitious in the Einstein-de Sitter model (or in models with $\sigma_0 > \frac{1}{2}$): any region with $\delta\varrho/\varrho \gtrsim 10^{-3}$ would by now have stopped expanding. Larger amplitudes would be necessary if the bound systems in fact separated out at large redshifts, and also, from (4.3), in low-density models.

Since all perturbations larger than the Jeans mass grow at essentially the same rate, the first condensations to separate out would correspond to the scale for which, at t_{rec}, $\delta\varrho/\varrho$ was greatest. If the initial fluctuations were strictly *adiabatic* the smallest surviving scales would have masses $\sim M_2$ (3.12). If the amplitudes decreased for still larger masses, the first condensations to form would be of this size, and it is tempting to identify them with large galaxies. The time at which they formed clearly depends on the maximum dimensions attained by protogalaxies. Studies of old halo stars in our Galaxy with highly eccentric orbits (EGGEN, LYNDEN-BELL and SANDAGE [29]) suggest that these formed during a stage when our galaxy was undergoing a free-fall collapse from a radius of, perhaps, ~ 50 kpc. The mean density corresponding to this maximum radius is $\sim 10^{-2}$ particles cm^{-3}, and one easily calculates that, if $\sigma_0 = \frac{1}{2}$, the expansion must have halted at a redshift $z \simeq 5$, and the collapse been completed by $z \simeq 2.5$. The required value of $\delta\varrho/\varrho$ at t_{rec} is $\sim 0.5\%$. (For $\sigma_0 < \frac{1}{2}$ these redshifts, and the inferred amplitude of the perturbation at t_{rec}, would be bigger).

If the perturbations were *not* precisely adiabatic, and the initial spectrum favoured the smaller masses, we would expect condensations of $\sim 10^6\,M_\odot$ to develop first. PEEBLES [30] has considered how the mean mass depends on various assumptions regarding the slope of the mass spectrum of the perturbations, but in all cases the typical condensations would be within an order of magnitude of M_J. PEEBLES and DICKE [28] have suggested that some of these might form globular clusters. This hypothesis accounts for the remarkably « standardized » properties of these objects, and would predict that there should be large numbers of them in intergalactic space. The discovery of two « intergalactic » clusters at distances $\sim 10^5$ pc by E. M. BURBIDGE and SANDAGE [31] is helpful to this view but does not constitute strong confirmation, since they could have broken away from a galaxy in the Local Group. Galaxies would, on this picture, result from inelastic collisions between clouds of $\sim 10^6\,M_\odot$

occurring before they had condensed into stars. This theory does not predict any preferred mass for galaxies (*).

DOROSHKEVICH, ZEL'DOVICH and NOVIKOV [32] have proposed a more complex chain of events to explain the existence of galaxies. They suggest that «primary» condensations of $\sim 10^6\, M_\odot$ do not develop in the manner envisaged by PEEBLES and DICKE, but form unstable «superstars» (HOYLE and FOWLER [33]) which explode and heat up their surroundings. If even ~ 1 part in 10^4 of the material in the universe exploded in this fashion, enough heat would be generated to raise the temperature of all the remaining gas to $\sim 10^6$ °K. It this happened at a redshift $z \leqslant 20$—as would be the case if $\delta\varrho/\varrho$ were $\leqslant 2\%$ at t_{rec}—the gas would not have time to cool down again. The consequent increase in the Jeans mass acts like a thermostat and inhibits the formation of further superstars. The irregular heating would, however, create inhomogeneities of $(10^9 \div 10^{11})\, M_\odot$, even if such scales were completely absent from the initial spectrum of perturbations. These could then condense to form quasars or galaxies.

After the gas has been reheated, the so-called thermal instability can play a role, as well as gravitational instabilities. FIELD [34] has given a detailed discussion of thermal instabilities in a specifically cosmological context (see also KATO, NARIAI and TOMITA [35]). Radiative losses *per unit mass* go as ϱ, and a gas is «isobarically» unstable—so that a region which is slightly compressed may cool so much more efficiently that its pressure drops below that of its surroundings, and it is squeezed further—if the cooling rate varies as T^α when $\alpha < 1$. For a high temperature gas where free-free cooling dominates, $\alpha \simeq \frac{1}{2}$, and over the range $(1.5 \cdot 10^4 \div 4 \cdot 10^5)$ °K, α is generally negative. The criterion must be modified when there is a heat input, but provided that (as is usually the case) this is independent of ϱ, it remains true that a gas hotter than $\sim 15\,000$ °K is thermally unstable, even in an expanding cosmological environment. The characteristic growth time would be comparable with the radiative cooling time.

During the fireball phase before t_{rec}, the strong thermal coupling between matter and radiation, and the negligible heat capacity of the matter, obviously precludes thermal instabilities. After recombination, the cooling time for the gas is too long (and in any case the radiation would be hotter than the matter, so that there would be a tendency for thermal *overstability*, rather than instability). However if the gas was heated to $\sim 10^6$ °K at a red shift ~ 20, as DOROSHKOVICH, ZEL'DOVICH and NOVIKOV [32] suggest, conditions would be propitious for the subsequent formation of condensations through thermal

(*) The presence of density irregularities on scales $\gg 10^6\, M_\odot$ would perhaps lead to the formation of bound systems of globular clusters (or proto-globular clusters) whose members would stand a better chance of colliding than would «field» clusters.

instability: not only would the cooling time be $\sim t$, but the heat input would be sufficiently irregular to provide substantial irregularities to initiate the instability. An upper limit to the scale of condensations that can form in this way is set by the fact that pressure forces are only effective when the differential expansion velocity across the condensing region is $\lesssim c_s$. Thermal conductivity sets a lower limit, which is generally too small to be significant. The end result of a thermal instability would be a cool cloud in pressure equilibrium with a hotter background. However if the cool cloud were gravitationally unstable, pressure could not halt its further collapse.

(The re-ionization of the medium would restore coupling between photons and electrons, and large scale irregularities may suffer further radiative damping (BARDEEN (36]) which could be especially important in low-density models.)

5. – Scope for further work.

Most theoretical discussions have so far dealt exclusively with spherical or sinusoidal linear perturbations in an isotropic Friedmann model. The only achievement has been to show that if the initial fluctuation spectrum is « smooth », but decreases in amplitude as the scale increases, there are three characteristic scales which are favoured:

M_1: the minimum mass able to condense freely after t_{rec} $((10^5 \div 10^6)\, M_\odot$, depending on $\sigma_0)$

M_2: the mass of the smallest adiabatic perturbation which escapes severe damping during its oscillatory phase before t_{rec} $((10^{12} \div 10^{14})\, M_\odot)$

M_3: the minimum mass which enjoys uninterrupted growth throughout the expansion $((10^{15} \div 10^{18})\, M_\odot)$.

At present, attempts to associate these scales with observed agglomerations in the universe are highly conjectural, and any firmer inferences must await further progress on both the observational and theoretical fronts.

Observations. Three categories of observation seem to be especially relevant.

a) The most obvious prerequisite for any serious comparison between theory and observation is fuller information on the actual mass spectrum of galaxies and clusters (and superclusters, if they exist). The most prominent galaxies have masses $(10^{10} \div 10^{12})\, M_\odot$. However this certainly does not imply that most of the material in the universe is necessarily gathered into such systems, since intrinsically weak objects would be invisibly faint, unless they lay comparatively close to us, and so would be under-represented in surveys down to a given limiting magnitude. A further effect militating against the discovery of low-luminosity systems is that either their angular size is so small that they are indistinguishable from stars, or their surface brightness is too low for them to be seen at all. ARP [37] has presented an instructive diagram which clearly shows that the known types of galaxy range over all combinations of radius and luminosity which permit their recognition. Some « compact » galaxies are less than 100 pc across. Galaxies of $\lesssim 10^8 M_\odot$ exist within the Local Group, and it seems difficult to exclude the existence of a continuous spectrum of galactic masses stretching right down to $\sim 10^6 M_\odot$ (though we may recall that both HUBBLE and BAADE firmly believed that there was a definite gap separating the luminosity functions of galaxies and globular clusters (ZWICKY [38]). The *absence* of any characteristic galactic mass would support Peebles and Dicke's [28] view that galaxies result from random coalescence of less massive clouds (unless it should turn out that smaller galaxies can be regarded as a separate later generation). It is obviously important to search determinedly for the intergalactic globular clusters expected on the basis of this theory.

Oort's [39] much-quoted lower limit of $3 \cdot 10^{-31}$ g cm^{-3} (*), for the smoothed-out density of galaxies plainly leaves room for a large amount of material in some form which is less readily detectable. However the observed night sky brightness sets some restrictions on the luminosity of this « missing matter ». By analysing how the sky brightness depends on galactic latitude, ROACH and SMITH [41] have set a remarkably stringent limit to the extragalactic background at optical wavelengths. According to PEEBLES and PARTRIDGE [42], the amount of matter in faint galaxies (or globular clusters) cannot be sufficient to close the universe, without violating this limit, unless the average mass-to-light ratio exceeds ~ 80 solar units.

Clusters of galaxies would presumably arise from smaller amplitude fluctuations than those (with smaller scales) that form galaxies. Being unaffected by photon damping before t_{rec}, they provide fairly direct information on the initial spectrum of irregularities. Although there is no doubt whatsoever about the existence of clusters of galaxies, the situation regarding irregularities with still larger scales remains very controversial. If there are « superclusters » on scales ~ 30 Mpc (ABELL [43]), it is tempting to relate them to the critical mass M_3.

Clustering on even larger scales may be investigated by studying the distribution of quasi-stellar objects with redshift and direction: if they are « cosmological » the typical separation of those QSO$_s$ at $z \simeq 1$ with high enough lu-

(*) This figure was estimated on the basis of a Hubble constant of $H \simeq \simeq 75$ km s^{-1} Mpc^{-1} and scales as H_0^2, so that for $H_0 \simeq 100$ km s^{-1} Mpc^{-1} we have $5 \cdot 10^{31}$ g cm^{-3}. This means that it yields a lower limit of 0.015 for σ_0 which is independent of the actual value of H_0. For a sceptical discussion of the value of this number even as a lower limit, see DE VAUCOULEURS [40].

minosity to have apparent magnitude $m < 19$ is ~ 50 Mpc. When a larger number of QSO_s has been discovered, they should serve as probes for density irregularities on scales $(100 \div 1000)$ Mpc (*). An alternative method of detecting large-scale irregularities would be to search for angular variations in the microwave background temperature attributable to gravitational effects (this is discussed under *c*) below).

Fuller information on these very large scale irregularities could provide us with an estimate of the density parameter σ_0. Any gravitationally bound system must have a density which exceeds the mean density of an Einstein-de Sitter universe. Thus, the existence of systems which seem gravitationally bound, but whose density does not appear greatly enhanced above the background, would be an indication that σ_0 was not much less than $\frac{1}{2}$. Conversely, if apparently freely expanding (and so unbound) superclusters were observed with a density substantially exceeding the mean, this would argue in favour of a low-density universe. When $\sigma_0 \ll \frac{1}{2}$, the Hubble law is obeyed accurately even when there are large irregularities in the density.

b) There are reasons for expecting protogalaxies to be much more luminous than ordinary galaxies. If they collapsed by a factor $\gtrsim 2$ from their maximum radius (as our Galaxy seems to have done), $\sim 10^{61}$ erg of binding energy must have been released in a free-fall time of only $\sim 10^8$ y. This energy was probably dissipated via shock waves in the uncondensed gas, and emitted mainly as Lyman-line radiation. Also, the first stellar populations to form would be expected to contain a high proportion of *O* and *B* stars. Some of the radiation from these stars would escape, and the rest would energize any gas that remains uncondensed. (When these stars evolve to the supernova stage, the explosions might constitute a further heat input into the gas.)

The high luminosities ($\sim 10^{46}$ erg s^{-1}) predicted by the above arguments (**) led FIELD [44] to identify QSO_s with galaxies in the process of formation. If, however, most galaxies formed at redshifts of, say, ~ 10, the associated energy output, which would be emitted at ultraviolet wavelengths, would be shifted into the visible or near infra-red. PARTRIDGE and PEEBLES [45] find that these

---

(*) The density of active quasars per *co-ordinate* volume (*i.e.* over and above the $(1+z)^3$ expansion factor) is a steep function of epoch—according to SCHMIDT (*Astrophys. Journ.*, **151**, 393 (1968)) it varies as $(1+z)^6$ out to $z \simeq 1.5$ at least. The astrophysical interpretation of this evolution is not known, but it is conceivable that the occurrence of the quasar phenomenon is sensitive to the ambient density in the surroundings. A $(1+z)^6$ law would then imply that the density of QSO went as the *cube* of the gas density. A two-fold increase in the mean density might then amplify into an eight-fold increase in the QSO density, in which case the QSO would be exceedingly sensitive indicators of large-scale irregularities.

(**) If helium had been synthesized during the early stages of galaxies, rather than being mainly primeval, even higher luminosities might be expected.

objects may make an appreciable contribution to the infrared background; the precise spectrum would depend on the epoch at which the galaxies formed. Individual protogalaxies may even be detectable—with angular diameters up to $\sim 20''$, because they may be far into the redshift range where angular sizes *increase* with z. WEYMAN [46], in a slightly different analysis, finds that the integrated background in the visible band would exceed Roach and Smith's limit unless the birth of galaxies was concentrated at redshifts $z \gtrsim 11$.

c) It is already possible to measure the isotropy of the microwave background on small angular scales with a precision of $\lesssim 0.1\%$ (CONKLIN and BRACEWELL [47], PENZIAS, SCHRAML and WILSON [48]). No anisotropies have been definitely detected, but it should be feasible to achieve even better precision. Various authors have considered whether this technique may enable us to detect embryonic galaxies or clusters right back at the epoch of recombination, and also to detect very large scale density inhomogeneities.

If a fraction f of the material in the universe remained in the form of an intergalactic gas which was re-ionized at a redshift, the optical depth, due to Compton scattering back to z_i would be

$$\tau_T(z_i) = 4.3\cdot 10^{-2}\sigma_0^{-1} f\{(3\sigma_0 + \sigma_0 z_i - 1)(1 + 2\sigma_0 z_i)^{\frac{1}{2}} - (3\sigma_0 - 1)\} \tag{5.1}$$

for $H_0 = 100$ km s^{-1} Mpc^{-1}. Therefore, unless, roughly speaking $(1+z_i) \gtrsim 5\sigma^{-\frac{1}{3}}$ (in which case the universe could be optically thick to Compton scattering out to z_i) most of the black-body radiation would definitely have propagated freely since t_{rec}. The effect of density perturbations on the radiation was first considered by SACHS and WOLFE [49]. They considered small amplitude perturbations to which a linear treatment could be applied, and assumed that the universe changed *suddenly* from being opaque to being transparent, so that there was a well-defined « surface of last-scattering ». Their discussion was restricted to an Einstein-de Sitter ($\sigma_0 = \frac{1}{2}$) model, but illustrates clearly the nature of the main effects. Consider fluctuations with a single wavelength l_0, and with *present* amplitude $(\delta\varrho/\varrho)_0 \ll 1$. The associated temperature fluctuations can be attributed to gravitational and Doppler effects occurring both *a*) locally and *b*) on the « last-scattering surface »:

a) The Doppler effect arising from a typical observer's peculiar velocity would yield a 24 h period anisotropy with amplitude

$$\left|\frac{\Delta T}{T}\right| \simeq \frac{l_0}{ct_0}\left(\frac{\delta\varrho}{\varrho}\right)_0 . \tag{5.2}$$

There would be a gravitational potential of

$$\sim c^2 \left(\frac{l_0}{ct_0}\right)^2 \left(\frac{\delta\varrho}{\varrho}\right)_0 \tag{5.3}$$

associated with the fluctuation. This would cause the observed radiation temperature to differ from the mean temperature that would be measured at the presesent epoch by an ensemble of observers randomly located through space. The magnitude of the effect is l_0/ct_0 smaller than (5.2) but in any case it has no angular dependence.

b) If we take the last scattering to have occurred at a redshift of ~ 1000, then, from (2.3)

$$\left(\frac{\delta\varrho}{\varrho}\right)_{\text{l.s.}} \simeq 10^{-3}\left(\frac{\delta\varrho}{\varrho}\right)_0 , \tag{5.4}$$

and the peculiar velocities would be $\sim (1000)^{\frac{1}{2}} \simeq 30$ times smaller than they are at present, so that the value of $|\Delta T/T|$ associated with the motion of the « last-scattering surface » would be ~ 30 times smaller than (5.2). However the gravitational potential fluctuation would be *of the same order* as (5.3), since change in l compensates exactly for the factor 10^{-3} in (5.4). The angular scale of these effects would be of the order of the angular diameter of the fluctuations on the « last-scattering surface »; in the Einstein-de Sitter model this is in fact $\sim l_0/ct_0$.

For any density irregularities with $l_0 < ct_0$ (*i.e.* with a scale smaller than the presently observable universe) the largest effect is (5.2), which gives a 24 h anisotropy in the background temperature. Of the two effects *b*) attributable to the « last scattering surface », the gravitational $\gtrless$ the Doppler according as $l_0 \gtrless 100$ Mpc—that is, according as the irregularities are $\gtrless M_H$ at t_{rec}. To quote a numerical example from SACHS and WOLFE, for $l_0 \simeq 1000$ Mpc and

$$\left(\frac{\delta\varrho}{\varrho}\right)_0 \simeq 0.1\,, \qquad \left|\frac{\Delta T}{T}\right| \simeq 0.01\ (^*)\,.$$

These results already set significant limits on the permissible amplitude of very large-scale inhomogeneities. It might seem from this analysis that the temperature fluctuations due to, say, protoclusters may be detectable (since though their scale is smaller, $(\delta\varrho/\varrho)_{\text{l.s.}}$ would be larger than in the cases Sachs and Wolfe were mainly concerned with). However, when dealing with smaller scales it is no longer valid to make the approximation that all the photons

---

(*) These last results can easily be adapted to a hyperbolic universe with $\sigma_0 \ll \frac{1}{2}$. For given values of l_0 and $(\delta\varrho/\varrho)_0$ the Doppler and gravitational contributions to $|\Delta T/T|$ discussed in *b*) are $\propto \sigma_0^{\frac{1}{2}}$ and independent of σ_0, respectively. (The mass associated with a perturbed region is $\propto \sigma_0$, but because of the slower growth in low density models (5.3), the value of $\delta\varrho/\varrho$ at t_{rec} would have been $\sim 10^{-3}\sigma_0^{-1}(\delta\varrho/\varrho)_0$.) The angular scale of the temperature fluctuation, for a given l_0, is $\propto \sigma_0$.

underwent their last scattering at a well-defined instant. In fact the gradual character of the decoupling must be taken into account for all scales $\leqslant 10^{15} M_\odot$ (*i.e.* $\leqslant 10^{-3}$ of M_H at t_{rec}). The expected magnitude of $|\Delta T/T|$ is then *much smaller* than Sachs and Wolfe's calculations would indicate—the photons arriving from a particular direction would have undergone their last scattering in several uncorrelated regions along the line of sight, and the temperature fluctuations due to gravitational and Doppler effects would be smeared out. ZEL'DOVICH and SUNYAEV [50] have given a full discussion of this point. The density fluctuations associated with protoclusters—and *a fortiori* protagalaxies—would be too small to be detected (*). If the background photons have been scattered since the intergalactic gas reheated, the universe would have gradually become transparent over a time of the order of the expansion time-scale, and temperature fluctuations due to any irregularities except those with scales $\gtrsim M_H$ at that epoch would have been smoothed out. The isotropy of the background would not then tell us anything significant about clustering on present scales $\ll 1000$ Mpc.

Within the framework of a linear analysis, radiation is unaffected by the gravitational field of transparent perturbations through which it passes. A discussion of *nonlinear* effects due to spherical irregularities in a « swiss cheese » cosmology (REES and SCIAMA [52]; REES [53]) showed that the potential well associated with the irregularity would increase the time taken by photons to pass through a given distance (leading to a *positive* contribution to ΔT) and that the deepening of the well as R increases leads to a *negative* contribution to ΔT; the net effect can have either sign. It was suggested that such effects might be observed in the directions of apparent « quasar clusters » (STRITTMATTER, FAULKNER and WALMESLEY [54]). Although the problem is confused by contributions to $|\Delta T/T|$ which may be attributable to radio sources (LONGAIR and SUNYAEV [55]), sufficient theoretical work has been done to demonstrate that large scale irregularities may be amenable to systematic study by detailed analysis of the microwave background.

6. – Theory.

It may be helpful to mention some of the more obvious deficiencies in present theories, and some directions in which current investigations might usefully be extended while remaining in the context of « hot big bang » Friedmann models.

(*) The microwave background can, in principle, yield information on the initial amplitude of small scale adiabatic perturbations which damped out before t because the energy dissipated would distort the spectrum from a true black-body (ZEL'DOVICH and SUNYAEV [51]).

6·1. – Studies of radiative damping in the «fireball» have hitherto dealt almost exclusively with a linearized situation in which different wavelengths do not interact, and no shocks develop. An extension of this work to the nonlinear case is especially urgent because the fluctuations which one must postulate in order to explain galaxies would not by any means be infinitesimal in the fireball phase. Consider, as an illustration, the formation of galactic masses of, say, $10^{12} M_\odot$ from adiabatic fluctuations in an Einstein-de Sitter model. We have seen that $\delta\varrho/\varrho$ must be ~ 0.01 at t_{rec}. Galactic masses would suffer a certain amount of damping just before and during the recombination, but let us neglect this for the moment. Following such a perturbation further back into the past, (3.11) tells us that (for $\sigma_0 = \frac{1}{2}$) $\delta\varrho/\varrho$ would be ~ 0.02 at t_{eq}, and (3.10) then tells us that the amplitude would have been 2% right back at the time when a galactic mass first came within the particle horizon. In general, a wave motion can only be validly described by linear theory until $\sim(\delta\varrho/\varrho)^{-1}$ oscillations have been completed: at this stage the wave «breaks» —the wave energy is fed into higher harmonics of the original frequency, shock discontinuities develop, and rapid dissipation ensues. Now the embryonic galaxy we are considering would have time to make $10 \div 20$ oscillations before t (whereas this order-of-magnitude argument would permit up to ~ 50), so that a linear treatment would appear to be only marginally self-consistent (*).

We might expect smaller scale irregularities to have larger $\delta\varrho/\varrho$, but even if they had the *same* amplitude the nonlinear effects would be even more serious because a larger number of oscillations can occur before t_{rec}. ZE'LDOVICH and SUNIAEV [51] have shown that energy injected into the fiireball at times *later* than $t \simeq 100$ y cannot necessarily be thermalized—although Compton scattering would rapidly transform the spectrum into a grey body (WEYMANN [56]) the production of new photons cannot proceed fast enough to establish a genuine black-body spectrum at the temperature appropriate to the enhanced energy density (except at the long-wavelength end of the spectrum). The accuracy with which the microwave background obeys a black-body law then places restriction on the amount of energy dissipated for $t \gtrsim 100$ y. One can use this argument to infer the important result that the adiabatic fluctuations with scales $(10^6 \div 10^{12})\,M_\odot$ which would have dissipated before t_{rec} must have had amplitudes

$$\left(\frac{\delta\varrho}{\varrho}\right) \lesssim\ll 7\cdot 10^{-2}\sigma_0^{\frac{7}{8}}\,. \tag{5.5}$$

Any energy dissipated when $t \ll 100$ y would be completely thermalised,

(*) The situation is no better for $\sigma_0 \ll \frac{1}{2}$. Larger amplitudes are required at t_{rec} than in the $\sigma_0 = \frac{1}{2}$ model, and this aggravation is only partially compensated by other effects.

but it is perhaps worth mentioning that the conventional estimates of helium production in the fireball (PEEBLES [57]; WAGONER, FOWLER and HOYLE [58]) would be modified, because at the relevant epoch the universe would be evolving along an adiabat with lower entropy per baryon than would be estimated if dissipation were left out of account. Large amplitude adiabatic perturbations of $\lesssim 1 M_\odot$, which would lie within the horizon (and therefore oscillate) at the epoch of helium formation but would be dissipated soon afterwards, constitute an additional energy density which would accelerate (*) the expansion. These effects would tend to increase the helium abundance above the $(23 \div 27)\%$ usually calculated.

6·2. – There seems no reason why *rotational* perturbations should not exist, as well as the non-rotational density irregularities to which most authors have restricted their attention. These may decay owing to radiative viscosity in the same manner as adiabatic perturbations, but in the absence of such damping, the angular velocity Ω would vary as

$$\left\{\begin{array}{ll} \Omega \propto R^{-1} & (t \lesssim t_{eq}) \,, \\ \Omega \propto R^{-2} & (t \gtrsim t_{eq}) \,, \end{array}\right. \tag{5.6}$$

(HAWKING [59]). For scales $\ll M_H$, (5.6) can be derived from the elementary argument that the angular momentum within a co-moving volume, which is proportional to $(\varrho_m + \varrho_\gamma) R^5 \Omega$, must be conserved. OZERNOI and CHERNIN [60] have discussed some implications of primordial turbulence, or « photon eddies ». Provided that the rotational velocities are $\ll c_s$ (given by (3.7)) the associated density perturbations can be neglected. After decoupling, however, the sound speed in the matter suddenly drops by a factor $\sim 10^4$, and the turbulence may become supersonic. It would then lead to large amplitude density irregularities, and rapid dissipation via shocks. OZERNOI and CHERNIN propose that structures on the scale of large clusters of galaxies may form in this way. A thorough treatment of viscous damping of vorticity before and during the recombination would be very valuable.

6·3. – The well-known problem of accounting for the build-up of galactic magnetic fields within the time available had led to the widespread supposition that magnetic fields may be primeval. As well as removing the origin of the field beyond the realm of conventional physics, this view raises the possibility that magnetic forces may exert dynamical effects in the early universe.

(*) The relevant time-scale is further contracted by the *special* relativistic time dilatation experienced by a typical small parcel of matter which is undergoing rapid oscillations.

A homogeneous unidirectional field would cause the universe to expand anisotropically, and THORNE [61], JACOBS [62] and others have considered detailed models incorporating magnetic fields. One may alternatively envisage a randomly oriented magnetic field. On scales large compared with the autocorrelation length, the magnetic field can be regarded as exerting an isotropic pressure

$$P_{\text{mag}} \simeq \frac{\langle H^2 \rangle}{8\pi} . \tag{5.7}$$

Provided that the primordial turbulence is incapable of tangling and amplifying the field, $P_{\text{mag}} \propto R^{-4}$ during the expansion. It thus preserves a constant ratio to the radiative pressure. The two would be equal if the present smoothed-out field strength were $\sim 3 \cdot 10^{-6}$ G. This cannot be completely excluded, though it would be difficult to reconcile with—to give two illustrations—the limits on the Faraday rotation measure of extragalactic radio sources and the weakness of the radio background.

A universal field strength of $\sim 10^{-9}$ G would be a more acceptable estimate. The associated pressure would be only $\sim 10^{-7}$ of the radiation pressure, and so would be dynamically negligible throughout the fireball phase (*). However even such a weak random field as this *would* be important after t_{rec}, because it would still be coupled to the predominantly neutral gas (the fraction $\sim 10^{-5}$ of electrons which fail to recombine sufficing to provide a high conductivity) and would exert a *higher* pressure than the thermal gas. It would quickly build up density irregularities, the density tending to be enhanced near neutral points where the field direction reverses. Galactic magnetic fields could have arisen from this primeval « seed field »

6·4. – It would plainly be more satisfactory if a « seed field » could be generated by dynamo processes in the fireball, rather than being relegated to the initial conditions. HARRISON [64] is the first to have discussed a possible mechanism quantitatively. Suppose that there is primordial turbulence in the fireball, as discussed in (**6·2**) above, and that a typical fluid element possesses vorticity. If radiation and matter behave as a single fluid, each eddy spins down according to (5.6). However the coupling must transfer angular momentum from the photon gas to the matter, because in the absence of any coupling these two components would spin down according to $\Omega \propto R^{-1}$ and $\Omega \propto R^{-2}$ respectively. The Thomson cross-section being millions of times larger for electrons than for ions, electrons are tightly coupled to the radiation, but the dif-

(*) H would become arbitrarily large at very early times, but quantum considerations do not appear to set any definite upper limit to the permitted field strength (CHIU and CANUTO [63]).

ferential rotation between the photon-electron gas and the ion gas creates an electric current which tends to generate a magnetic field. Assuming that the primordial turbulence has the *minimum* amplitude needed to accout for the present existence of bound systems, HARRISON calculates that a field $H \simeq 10^{-16}$ G could have been built up. This, however, is too weak to have been amplified by differential rotation into a characteristic galactic field of $(10^{-5} \div 10^{-6})$ G. On the other hand, in a somewhat different scheme where disc galaxies develop from « spinning cores » for which $\delta\varrho/\varrho$ is always of order unity, Harrison's mechanism leads to a seed field of adequate strength.

6'5. – Finally, we still have only a rudimentary understanding of the processes whereby contracting gas clouds may evolve into the first generation of galaxies or clusters. If the collapse occurs at a red shift $z \leqslant 10$, the gas temperature would initially be $\leqslant 1$ °K, but the clouds would be heated by adiabatic compression, and perhaps via shock dissipation. When the temperature rises to ~ 1000 °K (which requires contraction by a factor ~ 30 in radius if adiabatic heating alone occurs, but a smaller contraction factor if there is additional dissipation) molecular hydrogen may form, and prevent the temperature from rising further. (PEEBLES and DICKE [28]; SASLAW and ZIPOY [27]). Otherwise the temperature would continue to rise until, at $\sim 10^4$ °K, atomic hydrogen provided effective cooling. The likelihood of fragmentation into smaller masses is sensitive to the detailed behaviour of the temperature during the collapse. PEEBLES and DICKE discussed the problem in some detail for proto-globular clusters. For galactic masses, the Jeans mass will always be much smaller than the whole cloud.

Any attempt to account for the detailed morphology of galaxies would doubtless involve many complications. Indeed, there is still no convincing explanation for the basic distinction between ellipticals and spirals. The angular momentum is obviously a relevant parameter. This may have been acquired through tidal interactions between neighbouring protogalaxies at the time when their separation was comparable with their dimensions (PEEBLES [65]). On the other hand—and this does not seems to be generally realized—we cannot completely exclude the alternative viewpoint that the angular momentum is primeval. To illustrate this, consider a low-density universe with present mean density $\sim 3 \cdot 10^{-7}$ atoms cm^{-3}. The density at t_{rec} would then be ~ 300 atoms cm^{-3}. Comparing this with the density within a disc galaxy like our own—1 atom cm^{-3}— we see that such galaxies need only have expanded by a factor $\sim (300)^{\frac{1}{3}} \simeq 7$ since decoupling. If they even then had their present angular momentum, the associated rotational velocities (now ~ 250 km s^{-1}) would have to be ~ 1750 km s^{-1}. But in a low-density cosmology c_s is $\sim 1.7 \cdot 10^5$ km s^{-1} right until t_{rec} (3.7), so that just before recombination this angular momentum could exist even if the turbulent velocities were only $\sim 1\%$ of the sound speed. Furthermore as one extrapolates further back, the angular velocity of a turbulent element is proportional to R^{-1}, so that *the rotational velocity at the boundary of the protagalaxy would remain constant* (instead of varying like R^{-1}, as in a dust model). The angular momenta of galaxies can therefore be stored in the fireball, without the associated velocities ever being more than $\sim 1\%$ of c_s. (Since the dynamical importance of rotation is measured by Ω^2/ϱ, it could be progressively *less* important as one extrapolates back towards $t = 0$.) When we recall that the adiabatic perturbations needed to form galaxies would have had $\delta\varrho/\varrho \simeq 0.02$ at t_{rec}, and would give rise to random velocities $\sim 2\%$ of c_s, we realize that the primeval origin of angular momentum is within the spirit of attempts to explain the present structural features of the universe in terms of « small » perturbations of Friedmann models.

If there were a large-scale pre-existing magnetic field, its orientation relative to the rotation axis of the protogalaxy may also control what type of galaxy results (PIDDINGTON [66]).

HOYLE and NARLIKAR [67] proposed an attractive hypothesis to explain the formation and properties of elliptical galaxies. They considered an Einstein-de Sitter model (*), which was perturbed by the insertion of a « point mass » of $\sim 10^8\, M_\odot$. This mass would gradually accrete the surrounding material. If this material condensed into stars before any dissipation had occurred, the star density at a distace r from the central mass would vary as $r^{-\frac{9}{4}}$, which is compatible with the observed surface brightness of ellipticals. There would be no well-defined edge to the galaxy, but the star density would fall off until it merged with the background. (This feature of the model depends on the assumption of an Einstein-de Sitter background universe, in which an *arbitrarily small* perturbation suffices to capture a particle eventually. If $\sigma_0 < \frac{1}{2}$, however, the extra force must exceed a definite nonzero value in order that a particle should be captured, and the galaxy would have a sharp edge, at which the density would correspond to the mean density of the universe when $t \simeq t^*$). It is interesting that the giant elliptical M87 has recently been observed to extend out to a radius of at least $1.5 \cdot 10^5$ pc (ARP and BERTOLA [68], DE VAUCOULEURS [69].)

The spectrum and evolution of primordial fluctuations is probably only relevant to the first generation of galaxies which condense from the background. There is strong evidence that some much younger systems exist (G. R. BURBIDGE, E. M. BURBIDGE and HOYLE [70]), and these may have formed by a different mechanism. Any convincing evidence regarding the existence and properties of intergalactic gas would obviously be important in

---

(*) In fact HOYLE and NARLIKAR were concerned with a modified form of steady-state cosmology, but their theory of elliptical galaxies does not depend on this.

this connection. Thermal instabilities, which cannot have played any role in the formation of the first galaxies, may be effective after the gas has been reionized. It is also possible that existing galaxies may churn up the surrounding gas and initiate the condensation of others in their vicinity (SCIAMA [71]). E. M. BURBIDGE and G. R. BURBIDGE [72] have drawn attention to the curious system NGC 2444/5, where one galaxy could conceivably be forming in the « accretion wake » of another. We plainly cannot expect to account for the extraordinarily varied characteristics of galaxies by any simple unified model, and even a cursory inspection of Arp's [73] « Atlas of Peculiar Galaxies » should be enough to make one suspicious of any scheme that purports to do so.

7. – Discussion.

In view of the abundant aspects of the subject that remain unexplored, we cannot discount the possibility that this may represent a valid and fruitful approach to the problem of the origin of structure in the universe. The major snag seems to be that « initial » fluctuations are required with amplitudes which, on the one hand, must be small enough to be regarded as perturbations of a Friedmann model, but which, on the other hand, are by no means infinitesimally small. The precise initial spectrum is not crucial because of the selective character of the radiative damping and pressure stabilization, except that the amplitude must fall off at the longest wavelengths. The difficulty stems from the power-law, rather than exponential, growth of gravitational instabilities in Friedmann models (except in Lamaître-type models: see the Appendix). One would be somewhat happier if random «statistical» fluctuations could suffice, but this is only possible if there is some efficient nongravitational instability which can provide enhanced growth in the early stages (*) (It is true that such processes could only operate on scales $\lesssim M_H$, which means that at early times the resulting condensations would have very low mass. However $\sqrt{N}/N$ fluctuations in the *number density* of such condensations might yield larger-scale fluctuations with big enough amplitudes to condense gravitationally.)

It is important to stress that there is, at present, no reason whatsoever to postulate one type of initial fluctuation rather than another: unless the expanding universe can be physically related to a preceding collapse phase, it is purely a question of initial conditions. Because of the properties of horizons

(*) The reason why gravitational instabilities grow only as a power of t is that their « e-folding » time precisely equals then expansion time-scale. A nongravitational instability would amplify exponentially, and thus be much more effective than the gravitational instability, even if its growth where only *slightly* faster. (See SASLAW [74], LAYZER [75]).

in Friedmann models, two given particles would not establish causal or thermal contact until a certain time after the expansion started. There is therefore no thermodynamic reason why thermal-type statistical fluctuations should be preferred over any other initial conditions: the actual irregularities may be larger, or even smaller.

Indeed, it could well be argued (McCREA [76]) that it is the universe's present *uniformity* which is puzzling. The assumption that the universe was initially homogeneous—apart from fluctuations which are just large enough to explain its present inhomogeneity—may have little in its favour beyond mathematical simplicity. It involves the remarkable supposition that causally unrelated parts of the universe should have started to expand in an identical fashion. A more realistic—though less tractable—programme would perhaps involve postulating completely chaotic initial conditions, and attempting to understand how dissipative processes might account for the presently observed large-scale uniformity of the universe. The Friedmann models may not constitute a good approximation to the early universe in cases when large inhomogeneities are present, and studies of more general models may provide some additional insight. Of particular interest in this connection is Misner's [20, 77] « mixmaster » universe, which expands anisotropically, but in such a way that the axes of fastest and slowest expansion interchange repeatedly. The remarkable feature of this model is that there are no horizons—all parts are in communication with each other at arbitrarily early times—and so the conceptual problem of explaining the present large-scale uniformity does not arise. This leads one to conjecture that the Friedmann solutions may be peculiar—or « degenerate »—special cases of a wider class of models which generally do not possess particle horizons.

MISNER [20] showed that the anisotropy of a « mixmaster » universe could be essentially destroyed by neutrino viscosity at a temperature $T \simeq 2\cdot 10^{10}$ °K (*). If this type of result could be extended to more general initial conditions, it would mean that our discussion of radiative damping could still be applied, even if the very early stages of the expansion ($t \lesssim 1$ y, say) deviate greatly from a Friedmann model. The goal of all these theoretical investigations should be to establish, for as wide a class of irregularities as possible, the relation between the initial fluctuations and the inhomogeneities emerging from the « fireball » after t_{rec} (which are in turn related to the eventual bound systems). One would hope that any features of the predicted structures which are insensitive to the detailed initial conditions should have some counterpart in the

(*) According to MISNER, the residual anisotropy would now be almost undetectably small, and would not affect the time-scale at the epoch of helium formation. For similar reasons, exotic assumptions regarding physical conditions in the regime when $T \gg 10^{10}$ °K ($t \ll 1$ s) are unlikely to entail any observable consequences.

actual universe. One could, alternatively, invert the « transfer function », and *infer* the characteristics of the early universe from the observed clustering on various scales.

* * *

I have benefited from discussion of these topics with many colleagues, and would especially like to thank J. BARDEEN J. BERGERON, S. W. HAWKING, B. JONES, P. J. E. PEEBLES, D. W. SCIAMA and J. I. SILK.

APPENDIX

The Lemaître model.

The gravitational instability of cosmological models with a Λ-term was first considered by LEMAÎTRE [78]. The discovery of a surprisingly large number of quasars with $z \simeq 2$ has led to a revival of interest in the possibility of Lemaître-type models with a quasi-static period, or « coasting phase », at this redshift (SHKLOVSKI [79]). If the density parameter is $\sigma_0 (< \frac{1}{2})$, suitable choice of Λ allows the universe to have had a long coasting phase at a redshift z_{int} given by

$$z_{int}^2 (3 + z_{int}) \simeq \sigma_0^{-1} . \tag{A.1}$$

In conjunction with Oort's (1958) lower limit $\sigma_0 \gtrsim 0.015$, this implies that $z_{int} \lesssim 3.2$ (CRILLY [80]; BRECHER and SILK [81]). In fact (A.1) is not exact because of the effect of the primeval radiation, but at the permitted values of z_{int} the correction amounts to less than 1 per cent. (A.1) confirms that it is indeed consistent to have $z_{int} \simeq 2$, as is suggested by the quasar data.

Density perturbations would grow exponentially during the coasting phase, and it might appear that, if this phase lasted long enough, condensations could develop from arbitrarily small initial amplitudes. However BRECHER and SILK have shown that there are limits to the duration of the coasting period; and thus to the amplification factor that can be achieved. This is because a model which approximates to the Einstein static universe is unstable to changes in pressure—a reduction of pressure causes expansion, and an increase tends to initiate collapse. Collapsing protogalaxies would radiate away a fraction of their binding energy, and this would increase the pressure. Therefore in order that the universe should not have collapsed, it cannot have been *too* close to the Einstein static model, but must always have been expanding at a strictly positive rate. Unless certain rather artificial requirements are met, the coasting period cannot have lasted more than ~ 5 Hubble times. Although this would not be long enough for galaxies to form from fluctuations which were « statistical » at the epoch of nuclear densities, $\delta\varrho/\varrho$ need be only $\sim 10^{-9}$ at t_{rec}, rather than $\sim 10^{-2}$ as in conventional models. However even this requires adjustment of the values of Λ and σ_0 to within one part in $\sim 10^7$.

After the expansion recommences, the universe approximates to an accelerating de Sitter model ($q = -1$). and there is no possibility of further condensation. Irregularities with a wide range of initial amplitudes would all condense out during the « coasting phase. This contrasts with the situation in cosmologies with $\Lambda = 0$, where the larger amplitude fluctuations (which one would expect to be associated with smaller scales) would condense out at larger redshifts and would tend to form denser bound systems.

REFERENCES

[1] A. R. SANDAGE: *Observatory*, **88**, 91 (The Halley Lecture) (1968).

[2] M. RYLE: *Ann. Rev. Astron. and Astrophys.*, **6**, 249 (1968).

[3] V. A. AMBARTZUMIAN: *Structure and Evolution of the Universe* (Brussels, 1958), p. 241.

[4] E. LIFSHITZ: *Žurn. Ėksp. Teor. Fis.*, **10**, 116 (1946).

[5] Y. B. ZEL'DOVICH: *Adv. in Astron. and Astrophys.*, **3**, 241 (1965).

[6] E. R. HARRISON: *Rev. Mod. Phys.*, **39**, 862 (1967).

[7] G. B. FIELD: *Stars and Stellar Systems*, vol. **9**, edited by A. SANDAGE and M. SANDAGE (in press) (1969).

[8] A. EINSTEIN and E. G. STRAUSS: *Rev. Mod. Phys.*, **17**, 120 (1945).

[9] J. H. JEANS: *Astronomy and Cosmogony* (C.U.P.) (1928).

[10] P. J. E. PEEBLES: *Nature*, **220**, 237 (1968).

[11] G. B. FIELD: *Riv. Nuovo Cimento*, special issue, 87 (1969) (*Proceedings of Inaugural Meeting of E.P.S., Florence, 1969*, in press).

[12] C. W. MISNER: *Nature*, **214**, 40 (1967).

[13] G. B. FIELD and L. C. SHEPLEY: *Astrophys. and Space Sci.*, **1**, 309 (1968).

[14] J. I. SILK: *Astrophys. Journ.*, **151**, 459 (1968).

[15] P. J. E. PEEBLES: *Proc. III Texas Symposium* (in press) (1967).

[16] R. W. MICHIE: unpublished work (1967).

[17] C. W. MISNER and D. H. SHARP: *Phys. Lett.*, **15**, 279 (1965).

[18] P. J. E. PEEBLES: *Astrophys. Journ.*, **153**, 1 (1968).

[19] Y. B. ZEL'DOVICH, V. G. KURT and R. A. SUNYAEV: *Sov. Phys. JETP*, **55**, 278 (1968).

[20] C. W. MISNER: *Astropnys. Journ.*, **451**, 431 (1968).

[21] J. M. STEWART: *Astrophys. Lett.*, **2**, 133 (1968); *Mont. Not. Roy. Astron. Soc.*, **145**, 347 (1969).

[22] R. F. CARSWELL: *Mon. Not. Roy. Astr. Soc.*, **144**, 279 (1969).

[23] J. M. STEWART and J. L. ANDERSON: in preparation (1969).

[24] G. GAMOW: *Rev. Mod. Phys.*, **21**, 367 (1949).

[25] E. R. HARRISON: *Phys. Rev. Lett.*, **18**, 1011 (1967).

[26] P. J. E. PEEBLES: *Astrophys. Journ.*, **142**, 1317 (1965).

[27] W. C. SASLAW and D. M. ZIPOY: *Nature*, **216**, 976 (1967).

[28] P. J. E. PEEBLES and R. H. DICKE: *Astrophys. Journ.*, **154**, 891 (1968).

[29] O. J. EGGEN, D. LYNDEN-BELL and A. R. SANDAGE: *Astrophys. Journ.*, **136**, 748 (1962).

[30] P. J. E. PEEBLES: *Astrophys. Journ.*, **157**, 1075 (1969).

[31] E. M. BURBIDGE and A. R. SANDAGE: *Astrophys. Journ.*, **127**, 527 (1958).

[32] A. G. DOROSHKEVICH, Y. B. ZEL'DOVICH and I. D. NOVIKOV: *Sov. Astron.*, **41**, 233 (1967).

[33] F. HOYLE and W. A. FOWLER: *Nature*, **197**, 533 (1963).

[34] G. B. Field: *Astrophys. Journ.*, **142**, 531 (1965).
[35] S. Kato, H. Nariai and K. Tomita: *P.A.S.J.*, **19**, 130 (1967)
[36] J. M. Bardeen: in preparation (1969).
[37] H. Arp: *Astrophys. Journ.*, **142**, 402 (1965).
[38] F. Zwicky: *Morphological Astronomy* (Berlin, 1958), p. 226.
[39] J. H. Oort: *Structure and Evolution of the Universe* (Brussels, 1958), p. 163.
[40] G. de Vaucouleurs: *Science*, **167**, 1203 (1970).
[41] F. E. Roach and L. L. Smith: *Geophys. J.R.A.S.*, **15**, 227 (1969).
[42] P. J. E. Peebles and R. B. Partridge: *Astrophys. Journ.*, **148**, 713 (1967).
[43] G. O. Abell: *Ann. Rev. Astron. and Astrophys.*, **3**, 1 (1965).
[44] G. B. Field: *Astrophys. Journ.*, **140**, 1434 (1964).
[45] R. B. Partridge and P. J. E. Peebles: *Astrophys. Journ.*, **147**, 377, 868 (1967).
[46] R. J. Weymann, Steward Observatory preprint (unpublished) (1966).
[47] E. E. Conklin and R. B. Bracewell: *Nature*, **216**, 779 (1967).
[48] A. A. Penzias, J. Schraml and R. W. Wilson: *Astrophys. Journ.*, **157**, L 49 (1969).
[49] R. K. Sachs and A. M. Wolfe: *Astrophys. Journ.*, **147**, 73 (1967).
[50] Y. B. Zel'dovich and R. A. Sunyaev: in press (1970).
[51] Y. B. Zel'dovich and R. A. Sunyaev: *Astrophys. and Space Sci.*, **4**, 301 (1969).
[52] M. J. Rees and D. W. Sciama: *Nature*, **217**, 511 (1968).
[53] M. J. Rees: in preparation (1969).
[54] P. A. Strittmatter, J. Faulkner and M. Walmesley: *Nature*, **243**, 1441 (1966).
[55] M. S. Longair and R. A. Sunyaev: *Nature*, **223**, 719 (1969).
[56] R. J. Weymann: *Phys. Fluids*, **8**, 2112 (1965); *Astrophys. Journ.*, **145**, 560 (1966).
[57] P. J. E. Peebles: *Astrophys. Journ.*, **146**, 542 (1966).
[58] R. V. Wagoner, W. A. Fowler and F. Hoyle: *Astrophys Journ.*, **148**, 3 (1967).
[59] S. W. Hawking: *Astrophys. Journ.*, **145**, 544 (1967).
[60] L. M. Ozernoi and A. D. Chernin: *Sov. Astron.*, **12**, 901 (1969).
[61] K. S. Thorne: *Astrophys. Journ.*, **148**, 51 (1967).
[62] K. C. Jacobs: *Astrophys. Journ.*, **155**, 359 (1969).
[63] H. Y. Chiu and V. Canuto: *Astrophys. Journ.*, **153**, L 157 (1968).
[64] E. R. Harrison: *Astrophys. Lett.*, **3**, 133 (1969); *Mont. Not. Roy. Astr. Soc.* **147**, 279 (1970).
[65] P. J. E. Peebles: *Astrophys. Journ.*, **155**, 393 (1969).
[66] J. H. Piddington: *Mont. Not. Roy. Astron. Soc.*, **136**, 165 (1967).
[67] F. Hoyle and J. V. Narlikar: *Proc. Roy. Soc.*, **290** A, 177 (1966).
[68] H. Arp and F. Bertola: *Astrophys. Lett.*, **4**, 23 (1969).
[69] G. de Vaucouleurs: *Astrophys. Lett.*, **4**, 17 (1969).
[70] G. R. Burbidge, E. M. Burbidge and F. Hoyle: *Astrophys. Journ.*, **138**, 873 (1968).
[71] D. W. Sciama: *Mon. Not. Roy. Astron. Soc.*, **115**, 3 (1955).
[72] E. M. Burbidge and G. R. Burbidge: *Astrophys. Journ.*, **130**, 12 (1959).
[73] H. Arp: *Astrophys. Journ.*, *Suppl.*, **44**, 1 (1966).
[74] W. C. Saslaw: *Mon. Not. Roy. Astron. Soc.*, **136**, 39 (1967).
[75] D. Layzer: *Proc. Brandeis Summer School* (in press) (1968).
[76] W. H. McCrea: *Science*, **160**, 1295 (1968).
[77] C. W. Misner: *Phys. Rev. Lett.*, **22**, 1071 (1969).
[78] A. Lemaître: *Mont. Not. Roy. Astron. Soc.*, **90**, 490 (1931).
[79] J. Shklovski: *Astrophys. Journ.*, **150**, L 1 (1967).
[80] A. J. Crilly: *Mont. Not. Roy. Astr. Soc.*, **141**$_c$ 435 (1969).
[81] K. Brecher and J. I. Silk: *Astrophys. Journ.* **158**, 91 (1969).

Detection of Gravitational Waves.

B. BERTOTTI

European Space Research Institute - Frascati

1. – The recent confirmation [1] that bursts of gravitational waves have been detected in coincidence by two receivers constructed by WEBER at College Park, Md., and Chicago brings to the fore a new, fascinating field of research with important cosmological implications. I would like to rewiev in this talk the problem of the detection of gravitational waves, with special attention on its cosmological aspects.

2. – Gravitational waves differ essentially from other kinds of waves because of the principle of equivalence, according to which any gravitational force is indistinguishable locally from an apparent force, due to the lack of inertial character of the frame of reference; hence only measurements of the *difference* of gravitational force acting on two test bodies is meaningful. Mathematically, this corresponds to the circumstance that the object of the measurement is the *curvature tensor* $R_{\mu\nu\varrho\sigma}$.

For the same reason, the notion of energy flux, which is crucial in classical wave theory, does not have an invariant meaning: one can indeed construct a set of 16 quantities whose divergence vanishes (the energy-momentum pseudotensor $t_{\varrho\sigma}$), but in empty space this object does not contain derivatives of the metric tensor higher than the first and vanishes locally in an inertial frame.

In the weak-field approximation we can write expressions of the kind

$$g = \eta + \int \mathrm{d}^3 k\, h(\boldsymbol{k}) \exp\left[-ickt + i\,\boldsymbol{k}\cdot\boldsymbol{r}\right], \tag{1}$$

$$R = \int \mathrm{d}^3 k\, k^2 h(\boldsymbol{k}) \exp\left[-ickt + i\,\boldsymbol{k}\cdot\boldsymbol{r}\right] \tag{2}$$

for the metric and the curvature tensor, respectively. We neglect consideration of all indices and accordingly content ourselves with rough order-of-magnitude results. Writing in the random phase approximation the stochastic average

$$\langle h(\boldsymbol{k})\, h^*(\boldsymbol{k}')\rangle = |h(\boldsymbol{k})|^2 \delta(\boldsymbol{k} - \boldsymbol{k}') \,, \tag{3}$$

we have for the spectrum of the curvature tensor

$$|R(\boldsymbol{k})|^2 = k^4 |h(\boldsymbol{k})|^2 \,. \tag{4}$$

This has the property

$$\int \mathrm{d}^3 k |R(\boldsymbol{k})|^2 = \langle R^2 \rangle \,, \tag{5}$$

which shows how the mean square Riemann tensor is decomposed in its Fourier spectral components.

Similarly the pseudotensor

$$t = \frac{c^4}{G} \left(\frac{\partial g}{\partial x}\right)^2 \tag{6}$$

has a spectrum

$$t(\boldsymbol{k}) = \frac{c^4}{G}\, k^2 |h(\boldsymbol{k})|^2 . \tag{7}$$

We shall use systematically only the frequency spectrum which, in the case of isotropic radiation, is given by

$$P(\omega) = \frac{k^2\, t(\boldsymbol{k})}{c} = \frac{c^3}{G}\, k^4 |h(\boldsymbol{k})|^2 \,, \qquad \omega = ck \,. \tag{7'}$$

We see that in the simplest case of a homogeneous distribution of waves the curvature spectrum and the pseudotensor spectrum are simply related to each other and we may as well describe the gravitational waves by the latter concept, which is connected to the mass loss of a source of gravitational waves by *global* conservation theorems.

Many and unforeseen complications arise when one wants to build a satis factory and precise theory of the gravitational waves; the analogies with Maxwell's equations are useful only up to a certain point and, in particular, in the treatment of weak, linearized perturbation of a flat space-time. We wish to recall only that gravitational waves have only two transversal degrees of freedom; for example, for a wave which propagates along the x-direction, a simple representation of the metric tensor is given by

$$h_{22}(k, 0, 0) = h_{33}(k, 0, 0)\,, \qquad h_{23}(k, 0, 0) = h_{32}(k, 0, 0)\,, \tag{8}$$

all the other components being nil. At the present stage of the game we do not need anything more refined than this linear theory.

3. – Aside from Weber's recent measurements, we possess only two rather poor, *negative* pieces of information concerning the spectrum of gravitational waves. First of all this energy content must be smaller than the amount required by the overall curvature of space-time (assuming that the cosmological constant vanishes). One can describe a universe filled with gravitational radiation by the ansatz

$$g = g^{(0)} + h\,, \tag{9}$$

where $g^{(0)}_{ij}$ is the Robertson-Walker metric, obtained by averaging the field equations over regions large compared with the wavelength. The nonlinear terms give then a nonvanishing contribution, which one can interpret as a « right-hand side » and add to the energy-momentum tensor for matter. It is clear that, if cH^{-1} is the radius of the universe, we must have order-of-magnitude-wise (assuming a homogeneous situation)

$$\frac{G}{c^4}\int d\omega\, P(\omega) = \int d^3k\; k^2 |h(\boldsymbol{k})|^2 \leqslant \frac{H^2}{c^2}\,. \tag{10}$$

Secondly, an upper limit to $h(\boldsymbol{k})$ in our neighbourhood at the frequencies corresponding to the normal modes of the Earth can be obtained in the following way [2, 3]. The Earth can be considered as a huge test body, whose proper oscillations can be excited by an impinging gravitational wave; the principal mode has a period of about one hour. Such modes have always a small degree of excitation; an upper limit to $P(\omega)$ for $\omega^5 = 2\cdot 10^{-3}\ \text{s}^{-1}$ (the frequency of the principal mode) can be obtained ascribing it entirely to gravitational waves. From the statistical measurement of microseisms one can thus deduce

$$P(2\cdot 10^{-3}) \leqslant 2\cdot 10^{-3}\ \text{erg cm}^{-3}\,\text{s}\,. \tag{11}$$

A more precise evaluation of the numerical coefficients lowers this estimate appreciably (see [2]).

4. – When using material test bodies (we shall discuss later the case of photons) the curvature tensor can be measured by the change δl in their distance l or the change in relative velocity δv. In the former case one uses the equation of geodesic deviation to arrive at a formula of the type

$$\frac{\delta l}{l} = \frac{c^2 R}{\omega^2}, \tag{12}$$

where ω is the frequency of the gravitational wave. The second case corresponds to a measurement of the Doppler shift z between two points separated by a distance l. This problem has been worked out in [4] and leads to a Fourier component $z(\omega)$

$$\frac{\delta v}{v} = z(\omega) = \frac{clR}{\omega}, \tag{13}$$

under the condtion that l is much smaller than the wavelength c/ω of the gravitational perturbation. Comparing (12) and (13) one sees at once that the first method gives, for a given R, a greater fractional change in the quantity to be measured.

The use of free test bodies, however, is bry far inferior to a resonant system at the frequency ω we want to detect. The response is amplified by the Q of the system, so that instead of (12) we have

$$\frac{\delta l(\omega)}{l} = \frac{c^2 R(\omega)}{\omega^2} Q \tag{12'}$$

within a band ω/Q wide.

Weber's apparatus [5, 6] consists essentially in a piezoelectric transducer which transforms into electrical energy the oscillation energy of an aluminum cylinder about a meter long, tuned at 1660 Hz. The first problems to be overcome experimentally are the acoustic isolation and the seismic effects. The noise is reduced to a level corresponding to a displacement in the measured band width of the faces of the cylinder of 10^{-14} cm!—It is important to note that the noise is not random, in the sense that the frequency with which strong fluctuations occur varies very much from day to day. The experimental novelty consists in the fact that sometimes strong fluctuations are seen to occur at the same time on two detectors at College Park, Md., and Chicago, Ill., 1000 km apart.

The time resolution of Weber's detection system is about 0.4 s, enough to exclude a seismic effect by the following argument. If such a coincidence were due to an earthquake, the position of the source in the interior of the Earth must be such that the travel times of the disturbance to the two detectors differ by less than 0.4 s; this defines a region approximately bounded by two hyperboloids with Chicago and College Park as foci; its volume is about one thousandth of the total earth volume. But then all the other disturbances coming from the rest would produce many noncoincident fluctuations, which are not observed.

During an 81-day period from December 30, 1968 to March, 21, 1969, 17 coincidences were detected [1], each corresponding to an energy flux of more than 10^4 erg cm^{-2} s^{-1}; in particular, on February 16, 1969, three « events » were seen in half an hour.

The antenna beam width is 360° along the meridian and 70° along the parallel; in spite of the rather poor definition a correlation with the *sidereal time* was found [7].

Let us summarize Weber's findings.

Frequency of gravitational wave	$\omega = 10^4$ rad s^{-1}
Frequency band	$\Delta\omega = 0.1$ rad s^{-1}
Energy flux over the band	$cP(\omega)\,\Delta\omega = 10^4$ erg cm^{-2} s^{-1}
Energy flux from 0 to ω, assuming $P = \text{const}$	$c\omega P(\omega) = 10^9$ erg cm^{-2} s^{-1}
Burst frequency	$\nu = 3\cdot 10^{-6}$ s^{-1}
Mean energy density	$\nu\omega P(\omega) = 10^{-7}$ erg cm^{-3}
Duration of burst	$\tau \leqslant 0.4$ s

We see that $P(10^4) = 3\cdot 10^{-6}$ erg cm^{-3} s, to be compared with the much larger upper bound (11).

It is too early to try to assess the nature and the origin of these events; but note that a source at 1 kpc must release about 10^{52} erg per burst to give the required intensity. This is if the order of magnitude of the energy release in a supernova explosion; if the bursts are due to such a collapse, we should have in the whole galaxy a collapse frequency of about $10^3\nu \sim 3\cdot 10^{-3}$ s^{-1} which seems to be by far too great.

The bursts seem to be more frequent when the center of the Galaxy is within the beam [7]; our poor knowledge of its structure makes it a wonderful candidate for the theoreticians' speculations.

5. – The convenience of using very extended probes to increase the differential gravitational force to be measured has suggested to some people [8] the consideration of the effect of a gravitational wave on the intensity of a very distant source. One would argue qualitatively, the gravitational field has the same effect of an index of refraction $n = 1 + h$, so that the fractional change in intensity is (for a ray propagating along z)

$$\frac{\Delta I}{I} = -\int_0^L \mathrm{d}z \int_0^z \mathrm{d}\xi \left[\frac{\partial^2 h(x, y, \xi)}{\partial x^2} + \frac{\partial^2 h(x, y, \xi)}{\partial y^2}\right], \tag{14}$$

according to the usual laws of geometrical optics [9]. A fluctuating index of refraction corresponding to a gravitational wave would thus produce a scintillation in the source, superimposed upon the one due to atmospheric turbulence. Observation from a satellite or a correlation experiment between two telescopes at large distances (the correlation length for atmospheric turbulence is a few meters) would provide a measurement of an *extraterrestrial component of star scintillation*, which may be due to gravitational waves. It is easy to see from eq. (14) that the root mean square deviation in the intensity due to a random wave field is proportional to the power $\frac{3}{2}$ of the distance of the source and would provide a very sensitive test for the presence of gravitational fluctuations.

It is instructive to see that this promising program in reality does not work because eq. (14) is wrong and *there is no first-order effect* [10, 11].

This can be understood writing down the optical scalar equations, which control the propagation along the light ray of the divergence θ, the rotation Ω and the complex shear σ of the light beam. Such equations are derived with a complicated algebra from the equation of geodesic deviation of neighbouring null rays and read [12, 13]

$$\left\{\begin{aligned} &\frac{\mathrm{d}\theta}{\mathrm{d}r} + \theta^2 - \Omega^2 + |\sigma|^2 = \frac{1}{2}R\,, \\ &\frac{\mathrm{d}\Omega}{\mathrm{d}r} + 2\Omega\theta = 0\,, \\ &\frac{\mathrm{d}\sigma}{\mathrm{d}r} + 2\theta\sigma = C\,. \end{aligned}\right. \tag{15}$$

Here r is the affine parameter along the null ray; R and C are respectively some components of Ricci tensor and Weyl conformal curvature tensor.

The appropriate solution for a point source in a weak gravitational field is of the form [14] $\theta = 1/r + \delta\theta$, $\Omega = 0$ and fulfills

$$\left\{\begin{aligned} &\frac{1}{r^2}\frac{\mathrm{d}}{\mathrm{d}\lambda}(r^2\,\delta\theta) + |\sigma|^2 = \frac{1}{2}R\,, \\ &\frac{\mathrm{d}}{\mathrm{d}r}(r^2\sigma) = r^2 C\,. \end{aligned}\right. \tag{16}$$

In this approximation r is the ordinary spatial distance from the source.

We see that the lowest-order change induced in θ (and hence in the intensity) is *due to R and vanishes in empty space.* We could say, a beam of light traversing an empty, weakly curved space, suffers a change in shape, but not in cross-section.

6. – Let us compute the order of magnitude of the *second-order effect* produced by C. According to eq. (2) for a ray propagating along the z-direction

$$C(r) = \int \mathrm{d}^3k\, k^2\, h(\boldsymbol{k}) \exp\left[-i\frac{\omega' r}{c}\right], \tag{17}$$

where $\omega' = (1 - k_z/k)\,\omega$ is the Doppler-shifted frequency of the gravitational wave. Neglecting terms of order $c/\omega r$ we obtain from (16)

$$\sigma(r) = c\int \mathrm{d}^3k\,k^2 h(\boldsymbol{k})\,\frac{\exp[-i(\omega' r/c)]}{-i\omega'}$$

and, taking statistical averages and using (3) and (7′),

$$\langle|\sigma(r)|^2\rangle = \frac{G}{c^4}\int \mathrm{d}\omega\,P(\omega)\,. \tag{18}$$

The quadratic mean shear is constant along the path; we then get from (16_1)

$$\delta(\theta) = -\frac{r}{3}\frac{G}{c^4}\int \mathrm{d}\omega\,P(\omega)\,. \tag{19}$$

The fractional change in intensity is found to be proportional to the *square* of the distance:

$$\frac{\delta I}{I} = -2\int_0^r \mathrm{d}r'\,\delta\theta(r') = \frac{1}{3}\,r^2\,\frac{G}{c^4}\int \mathrm{d}\omega\,P(\omega) \leqslant \frac{1}{3}\,r^2\,\frac{H^2}{c^2}\,. \tag{20}$$

The last inequality is obtained from the theoretical condition (10) and shows that even in the extreme case of a universe closed by radiation the effect is negligible, except for the most distant galaxies.

7. – This calculation, however, is valid only in the geometrical-optics approximation and is insufficient for high-frequency fluctuations. Assume that the scintillation is caused by random distributions of gravitational perturbation of size $c/\omega = 2\pi/k$; over a screen at a distance r interference fringes will appear, with dimensions of order $r(\lambda\omega/c)$ (λ being the wavelength of light). The geometrical-optics approximation breaks down when this length is of the same order of magnitude as the geometrical shadow, which is c/ω for a source at infinity. Thus when

$$\frac{c}{\omega} \leqslant \sqrt{r\lambda} \tag{20}$$

physical optics must be used (see Figure.)

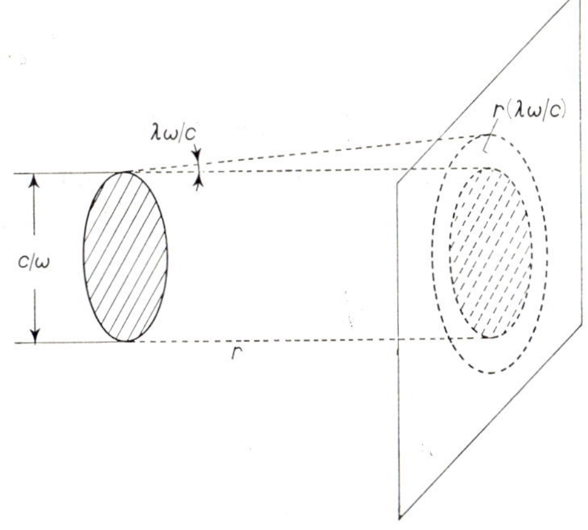

Fig. 1.

We want to show that even in this case there is *no first-order effect when the gradients of the metric are neglected.* We write for the vector potential of the electromagnetic field

$$A_\mu = (a^0_\mu + a^1_\mu) \exp [il_\varrho x^\varrho] , \tag{21}$$

where a^0_μ is a constant vector and l_ϱ is the wave vector of the light. In *empty space* A_μ is a harmonic vector (in the covariant sense); hence a^1_μ satisfies an equation of the type

$$\frac{\partial^2 a^1_\mu}{\partial x_\nu \partial x^\nu} + 2il^\nu \frac{\partial a^1_\mu}{\partial x^\nu} = L_\mu(h_{\varrho\sigma}) . \tag{22}$$

The indices are raised and lowered with the flat metric $\eta_{\mu\nu}$. We have neglected—in agreement with the intention of looking for linearized effects—squares of h and products of h and a^1; L_μ denotes a linear, second-order differential operator. When $\lambda \ll c/\omega$ it reduces to a *real*, algebraic expression:

$$\frac{\partial^2 a^1_\mu}{\partial x_\nu \partial x^\nu} + 2il^\nu \frac{\partial a^1_\mu}{\partial x^\nu} = l_\varrho l_\sigma h^{\varrho\sigma} a^0_\mu . \tag{23}$$

Taking the Fourier transform of both sides we have

$$(- k^\nu k_\nu + 2il^\nu k_\nu)\, a^1_\mu = l_\varrho l_\sigma h^{\varrho\sigma} a^0_\mu .$$

Since $h_{\varrho\sigma}$ represents a gravitational wave

$$0 = k^\nu k_\nu l_\varrho l_\sigma h^{\varrho\sigma} a^0_\mu = k^\nu k_\nu (- k^\varrho k_\varrho + 2il^\varrho k_\varrho)\, a^1_\mu .$$

We cannot take $- k^\varrho k_\varrho + 2il^\varrho k_\varrho = 0$, which corresponds to a trivial, slightly different plane wave; hence

$$k^\varrho k_\varrho a^1_\mu = 0$$

and, going back to space-time co-ordinates, eq. (22) reads

$$2il^\nu \frac{\partial a^1_\mu}{\partial x^\nu} = l_\varrho l_\sigma h^{\varrho\sigma} a^0_\mu . \tag{24}$$

Consequently a^1_μ is *purely imaginary* and corresponds to a small change in the *phase* of the wave, leaving its amplitude unchanged.

A scintillation in intensity is produced only by the next, much smaller term in L_μ, which is purely imaginary and of the form $ia^0 l(\partial h/\partial x)$; the ratio of this term to the right-hand side of (24) is of order $\lambda\omega/c$.

In spite of this pessimistic result it may be worth-while to investigate the problem in more detail.

REFERENCES

[1] J. Weber: *Phys. Rev. Lett.*, **22**, 1320 (1969).
[2] R. L. Forward, D. Zipoy, J. Weber, S. Smith and H. Benioff: *Nature*, **189**, 473 (1961).
[3] J. Weber: *Phys. Rev. Lett.*, **18**, 498 (1967).
[4] B. Bertotti: *Journ. Math. Phys.*, **7**, 1349 (1966).
[5] J. Weber: *Phys. Rev.*, **117**, 306 (1960).
[6] J. Weber: *General Relativity and Gravitational Waves* (New York, 1961).
[7] J. Weber: *Haifa Relativity Conference, July 15-17, 1969.*
[8] F. Winterberg: *Nuovo Cimento*, **53** B, 264 (1968).
[9] Y. I. Tatarski: *Wave Propagation in a Turbulent Medium* (New York, 1961), p. 97.
[10] D. M. Zipoy and B. Bertotti: *Nuovo Cimento*, **56** B, 195 (1968).
[11] D. M. Zipoy: *Phys. Rev.*, **142**, 825 (1966).
[12] R. K. Sachs: *Proc. Roy. Soc.*, **264** A, 309 (1961).
[13] P. Jordan, J. Ehlers and R. K. Sachs: *Akad. Wiss. Mainz*, No. 1 (1961).
[14] B. Bertotti: *Proc. Roy. Soc.*, **294** A, 195 (1966).

The Stability of Relativistic Star Clusters (*).

J. R. IPSER
California Institute of Technology - Pasadena, Cal.

Recent developments in astronomy and astrophysics—*e.g.*, the discovery and study of the quasi-stellar sources, of explosions in galactic nuclei, of strong extrasolar X-ray sources, and of pulsating radio sources—have led to a re-awakening of interest in the possible roles in nature of relativistic systems with strong gravitational fields. Thus far, this interest has concentrated largely on the roles which relativistic stars might possibly play in various astrophysical situations. As a result, much effort has been expended on studies of the structures and stabilities of relativistic stars, and of the gravitational radiation emitted by relativistic stars. (These studies are reviewed by THORNE in his lectures in this volume.)

One can ask if, in addition to relativistic stars, there might exist in nature other systems with strong gravitational fields, and for which relativistic effects would be important. HOYLE and FOWLER [1] offered an answer to this question. They suggested that each quasi-stellar source might lie at the center of a massive relativistic star cluster, and might derive its red-shift from the gravitational field of the cluster.

A rough measure of the importance of relativity for a star cluster is the parameter

$$\alpha = \frac{2GM}{c^2 R} \approx 0.01 \left(\frac{M/10^{11}\, m_\odot}{R/1\ \text{pc}}\right).$$

Here M is the mass of the cluster, and R is some mean radius of the star distribution. If $\alpha \geqslant 0.01$ relativistic effects may be important. Notice that this value of α corresponds to 10^9 stars in 0.01 pc, 10^{11} stars in 1.0 pc, etc. That such star densities are enormous is evident from the estimate that our galaxy has about 10^{11} stars in 10^4 pc ($\alpha \approx 10^{-6}$).

(*) Supported in part by the National Science Foundation [GP-9433, GP-9114] and the Office of Naval Research [Nonr-220(47)].

Consequently, it appears that, of all known astronomical systems, relativistic star clusters could possibly be associated with only the quasi-stellar sources and the nuclei of certain galaxies (Seyfert galaxies, N galaxies, and compact galaxies).

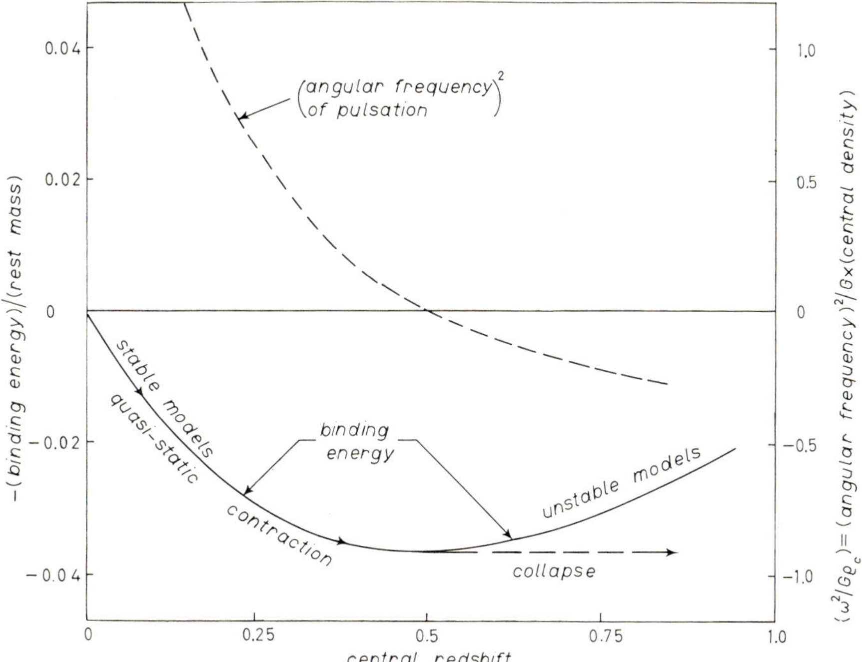

Fig. 1. – Idealized history of the evolution of a relativistic star cluster with a truncated Maxwellian velocity distribution. Plotted horizontally is z_c, the red-shift of a photon emitted from the center of the cluster and received at infinity. Plotted vertically are the fractional binding energy of the cluster, and the squared frequency, ω^2, of its fundamental radial mode of oscillation. The cluster initially contracts quasi-statically, toward states of tighter and tighter binding. When the fractional binding energy reaches a maximum, gravitational collapse sets in.

If relativistic star clusters do exist in nature, a relativistic effect, besides the gravitational red-shift, which should be important for them is the onset of gravitational collapse. Just as a gas sphere (star) becomes unstable against gravitational collapse when it contracts beyond a certain critical point, so also a star cluster should become unstable when it evolves past a certain critical density.

The point of onset of collapse has been calculated recently [2] for a variety of spherical clusters with both isotropic and anisotropic velocity distributions. As time proceeds, collisions between stars and the evaporation of stars cause a cluster to evolve through states of larger and larger density. When the density becomes so large that the red-shift of a photon emitted from the center of the cluster and received at infinity is $z_c \sim 0.5$, relativistic collapse sets in. The stars spiral in toward the center of the cluster, leaving behind a black hole with, perhaps, some stars orbiting it. All spherical clusters that have been studied are unstable against collapse if they have $z_c \gtrsim 0.5$, and are stable if they have $z_c \lesssim 0.5$.

These results suggest that it will be difficult, and perhaps impossible, to construct stable clusters with red-shifts as large as 2.4. Such clusters are needed in the Hoyle-Fowler star-cluster model for the quasars. However, there remains the possibility that cluster collapse is involved in the explanation of violent events in quasars and in the nuclei of certain galaxies.

REFERENCES

[1] F. Hoyle and W. A. Fowler: *Nature*, **213**, 373 (1967).

[2] E. D. Fackerell: paper in preparation (1969); J. R. Ipser: *Astrophys. Journ.*, **156**, 509 (1969); **158**, 17 (1969); J. R. Ipser and K. S. Thorne: *Astrophys. Journ.*, **154**, 251 (1968).

General Relativistic Neutron and Hyperon Star Models.

H. Heintzmann

Institut für Theoretische Physik der Universität - Köln.

In this book Thorne has reviewed the main features of what are believed to be realistic neutron star models. Perhaps the two main results concerning general relativistic cold neutron stars are that there is an upper limit for *a*) the number of baryons and *b*) the central density of a stable neutron star. It should however be pointed out that in the absence of a complete theory of nuclear interactions we must be sceptical about any equation of state used to describe stellar models for high densities. In this short note we show of how much the general features of neutron star models change if we are willing to accept any equation of state which does not contradict present-day physics. We performed our analysis of neutron star models and of their stability along the lines given in Chapters 5 and 6 of Thorne's 1966 lectures. We used a new exact solution of Einstein's field equations (the derivation of which will be given elsewhere) which is flexible enough to represent many equations of state, the use of which can be justified by different assumptions about the underlying nuclear forces.

In the hierarchy of known interactions, gravitation is by far the weakest. Is it therefore possible that weak interactions which are stronger than gravitational forces, but so weak that they have escaped experimental detection, can stop gravitational collapse of stars with many solar masses? We get a rough estimate of the influence of weak interactions upon the equation of state in the following way: their range cannot be macroscopic because of the Cavendish and because of the Eötvös experiments and their strength must be smaller than that of molecular forces. For our estimate we take the Zel'dovich (1961) equation of state

$$\varepsilon = mc^2 n + 2\pi g^2 \frac{n^2}{\mu^2},$$

$$P = 2\pi g^2 \frac{n^2}{\mu^2}$$

(where ε is the total mass-energy density, n the baryon number density, g the coupling constant and $1/\mu$ the range of interaction) and put $\mu = 10^8\ \text{cm}^{-1}$ and $g^2 = 10^{-23}\ \text{g cm}^3\ \text{s}^{-2}$ which corresponds roughly to quanta of 1/400 rest mass of the electron and to a force $K = 10^{-4} K_{\text{el}}$, where K_{el} is the Coulomb force between two electrons. With these weak forces we would have then $P \sim \varepsilon$ for a mass-density of the order of $10^{13}\ \text{g cm}^{-3}$, well below nuclear density. From Table I, part *a*) we see that with these forces the maximal stable mass

TABLE I. – *We list the following quantities*: m gravitational radius, N number of baryons, Δ mass defect due to gravitational binding, σ measure for the red-shift z on the surface of the star: $z = (1 - \sqrt{1-\sigma})/\sqrt{1-\sigma}$, ϱ_c density at the centre of the star, R_b eigen-radius of the star, $\langle (\mathrm{d}P/\mathrm{d}\varepsilon)^{\frac{1}{2}} \rangle$ measure for the stiffness of the equation of state, T period of the lowest mode of oscillation. Values marked with an asterisk have been derived by CAMERON and are given in Thorne's lecture.

	$m/m_\odot$	$N/N_\odot$	Δ	σ	$\varrho_c \cdot 10^{-14}$	R_b (km)	$\left\langle \left(\frac{\mathrm{d}P}{\mathrm{d}\varepsilon}\right)^{\frac{1}{2}} \right\rangle$	T (s)
	7.9	9.1	0.133	0.4	0.26	63	0.61	$1.6 \cdot 10^{-3}$
a)	11.3	13.5	0.167	0.5	0.26	74	0.82	$3.4 \cdot 10^{-3}$
	14.7	18.6	0.212	0.6	0.26	83	0.98	$2.2 \cdot 10^{-3}$
	2.5	2.9	0.133	0.4	2.6	20	0.61	$5.0 \cdot 10^{-4}$
	2.0	2.4	0.167	0.5	8.6	13	0.82	$3.0 \cdot 10^{-4}$
b)	2.56	3.7	0.212	0.6	8.6	14.5	0.98	$3.8 \cdot 10^{-4}$
	3.3	4.4	0.264	0.72	8.6	16	1.4	$4.4 \cdot 10^{-4}$
	5.4	8.3	0.35	0.86	8.6	27	3.2	10^{-3}
	1.95*	2.32	0.16	0.58	30.0	9.88	N.A.	$5.0 \cdot 10^{-4}$
c)	2.0	2.54	0.21	0.6	14.0	11.4	0.98	$3.0 \cdot 10^{-4}$
	1.70*	2.07	0.18	N.A.	48.0	N.A.	N.A.	$3.0 \cdot 10^{-4}$
	1.70	2.04	0.17	0.5	13.0	10.8	0.82	$2.5 \cdot 10^{-4}$
	0.63	0.8	0.21	0.6	140.0	3.1	0.98	10^{-4}
	0.75	1.0	0.26	0.72	160.0	3.6	1.4	10^{-4}
d)	0.7	1.1	0.35	0.86	500.0	2.1	3.2	$1.4 \cdot 10^{-4}$
	0.495	0.78	0.35	0.86	1000.0	1.47	3.2	10^{-4}
	0.97	1.57	0.39	0.663	∞	4.33	9.9	$10^{-}{}_{p}$

would be of the order of twenty solar masses—more than twenty times larger than without weak interactions. Thus we see that weak interactions are able to affect drastically our ideas about neutron stars. As regards strong interactions we may ask what happens if we are willing to abandon special relativity at small distances and if we allow for equations of state of the hard-core type with $P \sim \varepsilon^a$ and $a = \frac{3}{2}$ or 2. The adiabatic index γ is then no longer smaller than 2. The effect of such equations of state of neutron star models is given in part *b*) of Table I. The highest stable mass is seen to be 4 times larger than with ordinary nuclear forces where the equation of state obeys $P \leqslant \varepsilon$. The other question we are able to answer with our exact solution is whether or not there exist acceptable equations of state which give stability of stellar models whose central mass-energy density is higher than 10^{16} g cm^{-3}. Part *d*) shows that this is indeed the case even if the equation of state is restricted by $P \leqslant \varepsilon$. It is therefore possible that hyperon stars exist with mean densities hundred times larger than nuclear matter. We conclude this short note with the remark that all stellar models considered give pulsation rates faster than the fastest pulsars discovered to date.

* * *

The author is very much indebted to Profs. P. MITTELSTAEDT, E. SALPETER and K. S. THORNE for many illuminating discussions and to W. HILLEBRANDT for performing the numerical calculations and the stability analysis.

Elementary Particles in De Sitter Space.

G. Börner

Max-Planck-Institut für Physik und Asirophysik - München

In the attempts to develop a quantum field theory of elementary particles in general Riemannian manifolds [1] the Minkowski space-time picture is closely followed. The field equations are obtained by covariant generalization of the Minkowski invariant ones, and quantization is introduced via specific solutions of these equations. A clear physical understanding of what one is doing—intepretation of quantum numbers, etc.—is, however difficult to achieve.

An attractive feature of the de Sitter space is that with its ten-parameter symmetry group $SO_{1,4}$ the classification of physical states according to physically meaningful quantum numbers presents no problem. Despite its simplicity this model also will reveal some of the changes one may expect to occur in quantum field theory when departing from flat space.

The eigenvalues of the two Casimir operators I_1 and I_2 of $SO_{1,4}$ are connected with the mass $\varkappa^2$ and spin s of a field φ in de Sitter space by the equations [2]

$$I_1 \varphi = \varkappa^2 \varphi \,, \tag{1}$$

$$I_2 \varphi = s(s+1)(\varkappa^2 + c(s))\, \varphi \,. \tag{2}$$

For scalar fields $s = 0$, and in the co-ordinate system (λ, y) with $g_{\mu\nu} = \lambda^{-2}$ diag $(R^2, -1, -1, -1)$ (R: radius of curvature [2]) I_1 takes the form

$$I_1 = -\frac{\lambda^2}{R^2}\frac{\partial}{\partial\lambda}\frac{\partial}{\partial\lambda} + \frac{2\lambda}{R^2}\frac{\partial}{\partial\lambda} + \lambda^2 \frac{\partial}{\partial y^\alpha}\frac{\partial}{\partial y^\alpha} \,. \tag{3}$$

So (1) is just the covariantly generalized Klein-Gordon-equation of Minkowski space: $\Box\varphi$ is covariantly generalized to

$$\varphi_{;\mu}{}^{;\mu} \equiv g^{-\frac{1}{2}}\, \partial_\mu (g^{\frac{1}{2}} g^{\mu\nu} \partial_\nu)\, \varphi \,, \tag{4}$$

which is equal to $-I_1\varphi$ in $(\lambda, \boldsymbol{y})$ co-ordinates. Since the fields must give rise to specific representations of $SO_{1,4}$ the spectrum of $\varkappa^2$ is confined to [2]

$$(5) \qquad 0 < (\varkappa^2 R^2 - \tfrac{9}{4}) \equiv \varrho^2 < \infty \quad \text{and} \quad \varkappa^2 R^2 - \tfrac{9}{4} = -(n + \tfrac{1}{2})^2, \qquad n = 0, 1, 2, \ldots$$

We see that it is not correct simply to generalize the K-G-equation and take over the mass spectrum from Minkowski space. Just looking at (1) as a differential equation, does not exclude any complex value of $\varkappa^2$. The group-theoretical formalism has to be employed to give the correct spectrum. The solutions to (1) giving rise to a representation are completely determined. For the discrete part of the spectrum there is only one solution for each value of $\varkappa^2$

$$(6) \qquad \varphi_n(w, \boldsymbol{y}) = N_n w^2 \exp[i\boldsymbol{k}\cdot\boldsymbol{y}]\, j_n(w)$$

(N_n: normalization factor, $w = |\boldsymbol{k}|\,\lambda R$, j_n: spherical Bessel function)

while for the continuous part we have two solutions for each value of $\varkappa^2$.

$$(7) \qquad \varphi_{\varrho^\pm}(w, \boldsymbol{y}) = N_\varrho[\theta(w) \pm \theta(-w)] w^2 \exp[i\boldsymbol{k}\cdot\boldsymbol{y}] \cdot \left\{ \begin{matrix} \text{Im} \\ \text{Re} \end{matrix} \right. \exp[i(\pi/4)]\, h^{(1)}_{-\frac{1}{2}+i\varrho}(|w|)\,,$$

$$\theta(w) = \begin{cases} +1\,, & w > 0\,, \\ 0\,, & w < 0\,. \end{cases}$$

To introduce quantization we have to construct the commutation function for a fixed value of $\varkappa^2$ from the system of solutions (6) or (7). It turns out that a causal $\big(\equiv$ zero for $s = (1/\lambda\lambda')\,(R^2(\lambda - \lambda')^2 - (\boldsymbol{y} - \boldsymbol{y}')^2) < 0$, *i.e.* outside the light cone$\big)$ commutation function can only be obtained from the solutions (7)

$$(8) \qquad \Delta(\lambda, \boldsymbol{y}; \lambda', \boldsymbol{y}') = i\int \mathrm{d}^3\boldsymbol{k}\big(\varphi^*_{\varrho+}(w', \boldsymbol{y}')\varphi_{\varrho-}(w, \boldsymbol{y}) - \varphi^*_{\varrho-}(w', \boldsymbol{y}')\varphi_{\varrho+}(w, \boldsymbol{y})\big)\, \text{ctgh}\, \pi\varrho\,.$$

The solutions of the discrete spectrum do not allow this construction of a causal commutation function.

It is clear that only those masses $\varkappa^2$ are physically admissible, which permit a causal propagation of the field operators. We thus have arrived at the spectrum condition $\varkappa^2 > \frac{9}{4}R^2$ for physical states. It is interesting to note that the solution with $\varkappa^2 = 2/R^2$, contained in (5) $(n = 0)$, invariant under the conformal group $SO_{2,4}$, has been excluded from the physical state space. This solution possesses only the acausal two-point function $D \sim 1/s$. In the de Sitter space there does not exist the usual conformal invariant particle of spin zero, which shows a causal propagation in an appropriate conformal compactification of Minkowski space [3].

Similar considerations can be applied to the spin-1 particles in de Sitter space [4]. One arrives at a spectrum condition $\varkappa^2 > \frac{1}{4}R^2$ for physical states. So the photons do not have a zero rest mass, they do not propagate along null geodesics.

REFERENCES

[1] A. LICHNÉROWICZ: *Les Houches 1963, Relativity, Groups, Topology* (New York, 1964); B. S. DE WITT: *Les Houches 1963, Relativity, Groups, Topology* (New York, 1964).

[2] G. BÖRNER and H. P. DÜRR: *Nuovo Cimento*, **64** A, 669 (1969). This paper contains an extensive bibliography.

[3] L. CASTELL: *Nucl. Phys.*, **5** B, 601 (1968).

[4] G. BÖRNER and L. CASTELL: *Photons in De Sitter space* (in preparation).

Cosmological Density Fluctuations During Hadron Stage.

W. Kundt

I. Institut für Theoretische Physik der Universität Hamburg - Hamburg

1. – Introduction.

One of the main tasks in modern cosmology is to understand how all the objects seen (or not seen) in the sky could form. Such (gravitationally bound) objects are stars, quasars, globular clusters, galaxies, galaxy clusters. None of these objects is believed to have existed since the big-bang, *i.e.* since the primeval high-density and high-temperature phase of the universe. Neutrino and photon viscosity would have evened out a great deal of density inhomogeneities on scales comparable to the effective range of these particles, which extends up to and beyond the galactic scale. And galaxies and galactic clusters, if as old as the universe, would have collapsed a long time ago due to gravitational instability. Though as yet one cannot rule out the possibility that the universe was highly inhomogeneous at its birth, and that the present structure in the sky is due to the (unknown) initial data, a much more attractive explanation would be that all the structure formed from thermal random fluctuations of a homogeneous initial state.

The latter explanation has been lucidly surveyed by Rees in his lecture. Let me briefly repeat the main steps of the argument: An isothermal gravitating gas is unstable against collapse if its mass is bigger than the Jeans mass. This follows from the fact that the attractive gravitational force of density concentrations grows linearly with the radius (for constant density contrast), and eventually overcomes the repulsive heat motion. In our universe, the expansion acts as a damping force in the growth equation of irregularities, so that density fluctuations grow according to a power law as long as radiation drags are irrelevant. The latter are important on cosmologically small scales if we live in a universe with a particle horizon, *i.e.* if a torch lit during the big-bang cannot illuminate the whole of the universe. This is the case for the isotropic Friedmann models but possibly not for all homogeneous world models as pointed out by Misner [1].

The history of our universe shows several important epochs. They are the 1) hadron stage, 2) lepton stage, 3) photon stage, and 4) baryon stage called after the particle species that govern the energy balance. Their dividing temperatures are roughly 10^{12}, 10^{10}, 10^{5} in degrees, corresponding roughly to ages 10^{-5}, 1, 10^{11} in s.

If there is a particle horizon, matter cannot condense before it enters it. This happens for galaxies at about a year, *i.e.* during the photon stage. After coming into the horizon, density fluctuations have to sit and wait until the Jeans length decreases below their scale; which happens for galaxies during recombination, at about 10^{5} y.

There is the important question of how big were the density fluctuations when matter became unstable against condensation. Before the baryon stage, density fluctuations above the Jeans scale grow in direct proportion to time; see PEEBLES [2, 3], or REES. The density fluctuations of an (nongravitating) adiabatically expanding Planck gas are time independent. Consequently, the further back in time one starts with thermal fluctuations, the larger are the cosmic fluctuations when the baryon stage sets in.

How far back in time can one go, or does one have to go? Some physicists speak of oscillating universes (with an upper limit to the energy density). I doubt this possibility, for reasons given in my survey article [4]. How far back in time does the hadron stage extend? One may doubt the validity of high-density equations of state when the thermal de Broglie wavelength of a particle reaches the size of the 3-space curvature radius. A much more stringent limit is set by the requirement that the de Broglie wavelength should be smaller than the Hubble radius ct (so that a particle be a causal unit). An even stronger requirement for applicability of equilibrium thermodynamics would be that the (Hubble) time t be larger than the average time of strong interactions (10^{-23} s).

The main concern of this article are the cosmic density fluctuations during the hadron phase. Here PEEBLES has offered an estimate [3] which is based on the model of relativistic free particles with negligibly small chemical potential. The model depends on two unknown functions of temperature, namely the sums of the spin statistical weights g_j for each species of fermions and bosons respectively, (antiparticles counted separately), whose rest energy is smaller than kT. These statistical weights g_j are assumed to be slowly increasing functions of temperature, extrapolating the known mass spectrum of elementary particles. As a result, PEEBLES obtains much stronger fluctuations than needed when applying his model back to the time when $ct = \lambda_{dB}$, but weaker fluctuations than needed when only applying it back to 10^{-23} s.

In the meantime, HAGEDORN has proposed a different model for a high-temperature hadron gas [5, 6]. Contrary to PEEBLES he assumes—or to some extent derives from simpler assumptions—that the statistical weights grow exponentially; which is in accord with the presently known mass spectrum. This leads to the existence of a highest temperature $T_0 = 1.8 \cdot 10^{12}$ degrees, and (as a consequence) to an equation of state for which the pressure becomes negligible against the energy density as $T \to T_0$. Moreover, Hagedorn's model implies density fluctuations which grow in proportion to $(T_0 - T)^{-1}$, yielding much higher values than the more orthodox model of PEEBLES.

In the next Section we give a short review of Hagedorn's high-energy thermodynamics, and in the last Section we discuss its cosmological implications.

2. – Hagedorn's theory.

HAGEDORN observes that in very high-energy laboratory collisions, the transverse momenta of particles emerging from a collision obey (approximately) a Boltzmann distribution (in the center-of-mass frame) corresponding to a temperature T_0 which is almost independent of the primary energies. He therefore suggests that in the interaction region during the collision time, particles are created with thermal energies corresponding to temperatures $\lesssim T_0$. An increase in primary energies results in an increase in the number, and masses of the newly created particles but not in their kinetic energies. Secondly, HAGEDORN observes that excited states of hadrons appear in no way different from new particles unless they have very large angular momentum. He therefore assumes that asymptotically for large energies, the (rest) mass spectrum of elementary particles approaches the energy spectrum up to a polynomial factor. More completely, he makes the following assumptions for his hadron gas: (we set $c = 1 = \hbar = k$)

1) interaction energy $\ll$ total energy, *i.e.* nearly free particles;

2) all hadrons admitted, hadrons of the same kind indistinguishable;

3) variable numbers of particles;

4) number of elementary particles approaches asymptotically the number of excited states (up to a polynomial factor);

5) the interaction takes place in a natural volume V given by the range of strong interactions; *i.e.* $V = (4\pi/3)\, m_\pi^{-3}$: (with $m_\pi :=$ pion mass);

6) the hadronic mass spectrum starts with π, K, $\mathcal{N}$, ...:

Conservation laws do not contradict assumption (3) because pair creation is possible.

The partition sum Z describing thermal equilibrium reads

$$Z = \sum_{N_{ab} \geqslant 0} \exp\left[-\sum_{a,b} N_{ab} E_{ab}/T\right], \tag{7}$$

where according to assumption (3) there is no restriction on the total number $\sum_{a,b} N_{ab}$ of occupied states, and where according to (1), (3)

$$E_{ab} = \sqrt{\boldsymbol{p}_a^2 + m_b^2} \tag{8}$$

is the energy of a particle of species b with 3-momentum $\boldsymbol{p}_a$ and rest mass m_b. Taking account of (2), and distinguishing between bosons and fermions one gets by summing geometric series

$$Z = \prod_{a,b} (1 - \exp[-E_{ab}/T])^{-1} \prod_{a,f} (1 + \exp[-E_{af}/T]) , \tag{9}$$

where b stands now for bosons, and f for fermions. When we form the logarithm of Z, products become sums, and the latter are approximated by integrals:

$$\begin{cases} \sum_a \to \dfrac{V}{2\pi^2}\displaystyle\int_0^\infty \mathrm{d}p\, p^2 , \\ \sum_b \to \displaystyle\int_0^\infty \mathrm{d}m\, \varrho_b(m) , \end{cases} \tag{10}$$

where $\varrho_b(m)$ describes the boson mass spectrum, and $\varrho_f(m)$ analogously the fermion mass spectrum. When finally we expand the logarithms of the parentheses into their Taylor series we obtain

$$Z = \exp\left[\frac{V}{2\cdot\pi^2}\sum_{n=1}^\infty \frac{1}{n}\int_0^\infty \mathrm{d}m\, \varrho(m;n)\int_0^\infty \mathrm{d}p\, p^2 \exp[-n\sqrt{\boldsymbol{p}^2 + m^2}/T]\right], \tag{11}$$

$$\text{with } \varrho(m;n) := \left\{\begin{array}{ll} \varrho_b(m) + \varrho_f(m) =: \varrho(m) & \text{for } n \text{ odd} \\ \varrho_b(m) - \varrho_f(m) & \text{for } n \text{ even} \end{array}\right\}.$$

On the other hand, Z is by definition of the form

$$Z = \int_0^\infty \mathrm{d}E\, \sigma(E) \exp[-E/T] . \tag{12}$$

Formulae (11), (12) contain the two unknown functions $\varrho(m;n)$, $\sigma(E)$ which are related through assumption (4). After long calculations, HAGEDORN comes to the conclusion that (4) implies [5] exponential growth:

$$
\begin{cases} \varrho(m) \simeq a m^{\alpha} \exp\left[-m/T_0\right], \\ \sigma(m) \simeq c m^{\gamma} \exp\left[-m/T_1\right]. \end{cases} \tag{13}
$$

This implies $Z(T)\to\infty$ for $T\to T_0$, whence $T_1 = T_0$, and again by comparison of (11) and (12), (possibly leaving out an alternative):

$$
Z \simeq c\Gamma(\gamma+1)\left(\frac{T_0^2}{T_0-T}\right)^{\gamma+1} \qquad \text{for } T\to T_0\,, \tag{14}
$$

with

$$
\alpha = -\frac{5}{2}\,, \qquad \gamma = aV\left(\frac{T_0}{2\pi}\right)^{\frac{3}{2}} - 1\,. \tag{15}
$$

Formula (14) describes Hagedorn's model; it contains the three unknown parameters T_0, γ, and c. In order to determine two of them, HAGEDORN suggests a second model which is obtained from the above one by replacing assumption (2) by (2′): all particles are distinguishable. This assumption is justified by observing that the above model yields average occupation numbers $\langle N_{ab}\rangle$ which are small compared to one. Going through the equivalent analysis, and using assumption (6) one finds (Hagedorn is sloppy about this point)

$$
T_0 \approx \sqrt[3]{\frac{\pi}{V}} \approx m_\pi = 1.4\cdot 10^8\ \text{eV}\,, \tag{16}
$$

$$
\gamma = 0\,, \qquad c = (3T_0)^{-1}\,. \tag{17}
$$

As a result one gets from (14)

$$
\ln Z \approx \ln\frac{T_0}{3(T_0-T)}\,, \qquad \text{for } T\to T_0\,. \tag{18}
$$

Whence by standard formulae:

$$
\langle E\rangle = T^2\partial_T \ln Z = \frac{T_0^2}{T_0-T}\,, \tag{19}
$$

$$
\mu := \frac{\langle E\rangle}{V} = \frac{AT_0}{T_0-T}\,, \qquad A := \frac{T_0}{V}\,, \tag{20}
$$

$$
\frac{\langle E^2\rangle - \langle E\rangle^2}{\langle E\rangle^2} = \frac{T^2}{\langle E\rangle^2}\,\partial_T\langle E\rangle = 1\,, \tag{21}
$$

$$
p = T\partial_V \ln Z = A\ln\frac{\mu}{A} \tag{22}
$$

for the average particle energy, the energy density, the mean squared energy fluctuation, and the pressure p. The validity of eq. (22)—obtained by comparison with eq. (11)—may be questioned because all above derivations were performed for fixed interaction volume V; we shall do without it.

It should be stressed that HAGEDORN has succeeded in fitting his theory to both the observed mass spectrum (which contained $\int \mathrm{d}m\,\varrho(m) = 1432$ particles by January 1967; $\varrho(m) := \varrho_b(m) + \varrho_f(m)$), and to the observed scattering cross-sections above 1 GeV, ranging over ten orders of magnitude [6]. The best experimental values of the parameters are:

$$T_0 = 1.6\cdot 10^8\ \mathrm{eV} = 1.8\cdot 10^{12}\ \text{degrees}\,, \qquad a = 2.6\cdot 10^4\ \mathrm{MeV}^{\frac{3}{2}}\,, \qquad \ln(cT_0) = -1.7\,.$$

3. – Application to cosmology.

As explained in the introduction, PEEBLES has concluded that the thermal de Broglie wavelength was equal to the Hubble radius at the time $10^{-41}\sqrt{g}$ s, corresponding to a temperature of $(10^{31}/\sqrt{g})$ degrees; where g stands short for an average spin statistical weight of estimated order 10^6. At that time, thermal density fluctuations on a galactic scale were of the order of $\delta\mu/\mu = 10^{-38}$, whereas galaxy formation only needs $\delta\mu/\mu = 10^{-51}\sqrt{g}$ (according to present theory).

As a tentative suggestion we apply Hagedorn's theory to this problem. Or, more carefully, we only assume his formula (20) relating the energy density μ to the temperature. As a consequence, $\ln Z$ grows infinite for $T \to T_0$. From this we conclude that $p/\mu \to 0$ for $T \to T_0$. For, consider the curves of constant $\ln Z$ in the $(V\text{-}T)$-plane; they satisfy

$$0 = \mathrm{d}\ln Z = \mathrm{d}T\partial_T \ln Z + \mathrm{d}V\partial_V \ln Z\,,$$

whence

$$-\frac{\mathrm{d}\ln V}{\mathrm{d}\ln T} = \frac{T\partial_T \ln Z}{V\partial_V \ln Z} = \frac{\mu}{p}\,. \tag{23}$$

But the curves $V(T)$ of constant $\ln Z$ cannot intersect the straight line $T = T_0$, and all terms in the last equation grow infinite for $T \to T_0$.

Consequently, for T nearly equal to T_0 we deal with the approximate equation of state $p = 0$; as opposed to $p = \mu/3$ for a radiation gas. And according to eq. (21), density fluctuations in an arbitrary volume V are given by

$$\left(\frac{\delta\mu}{\mu}\right)^2 = \frac{V_S}{V}\,, \tag{24}$$

where we have written V_s for the strong interaction volume called V in Sect. **2**. Formula (24) assumes that fluctuations in different cells of volume V_s are statistically independent.

In Fig. 1 we compare our results with the tentative ones of PEEBLES.

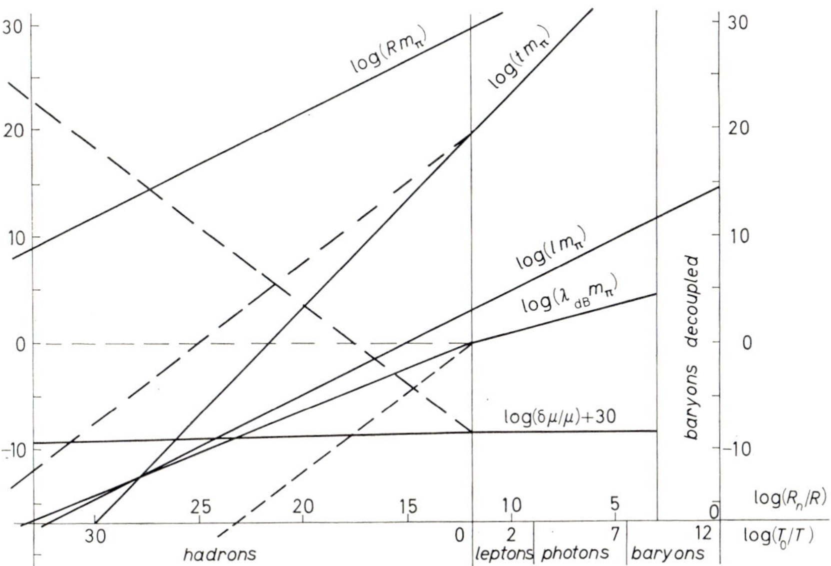

Fig. 1. – Double logarithmic plot of curvature radius R, world time t, mean distance between nucleons l, and de Broglie wavelength λ_{dB} of nucleons—in units of the range of strong interactions m_π^{-1}—as functions of the world radius R, and temperature T. R_n is the present world radius, $T_0 = 1.8 \cdot 10^{12}$ degrees. Solid lines correspond to the orthodox model discussed by PEEBLES, broken lines correspond to Hagedorn's model. During the hadron phase, time and temperature are different functions of R in the two theories. Moreover, $\delta\mu/\mu$ is the thermal equilibrium density contrast.

Assumed is an Einstein-de Sitter model with present radiation temperature 3 degrees, and present matter density $2 \cdot 10^{-29}$ g/cm³ corresponding to 10^{-5} protons/cm³. The total energy density μ as a function of $x := T/T_0$ is given by

$$\frac{\mu(x)}{\text{erg cm}^{-3}} = \left\{ \begin{array}{lll} g(x) B x^4 \,, & B = 10^{35} \,, & \text{for } T \geqslant 10^5 \text{ degrees} \\ A/(1-x) \,, & A = 2 \cdot 10^{34} \,, & \text{for } T \geqslant 10^{12} \text{ degrees} \end{array} \right\}, \tag{25}$$

where B is obtained from Planck's law, and A was determined above. The first line is valid for all temperatures $\geqslant 10^5$ degrees if PEEBLES is right, and $g(x)$ is a slowly increasing function of x. The associated density fluctuations on a galactic scale are calculated in analogy to (21)

$$\left(\frac{\delta\mu}{\mu}\right)^2 \approx \left\{\begin{array}{ll} g^{-\frac{1}{2}}\dfrac{V_S}{2V_0} \approx g^{-\frac{1}{2}}\cdot 10^{-75} & \text{for } T \geqslant 10^{-5} \text{ degrees} \\ \dfrac{V_S}{V} & \text{for } T \geqslant 10^{12} \text{ degrees} \end{array}\right\}. \tag{26}$$

From Fig. 1 one gathers that already at $t = 10^{-23}$ s, (looking backward in time), density fluctuations seem to exist in abundance for eventual galaxy formation. A largely improved presentation will appear shortly in *General Relativity and Gravitation.*

* * *

I am indebted to Prof. K. S. THORNE for clarifying discussions.

REFERENCES

[1] C. W. MISNER: *Phys. Rev. Lett.*, **22**, 1071 (1969).
[2] P. J. E. PEEBLES: in *Proceedings of the IV Texas Conf. on Relativist. Astrophysics, Jan. 1967*, in press.
[3] P. J. E. PEEBLES: *Nature*, **220**, 237 (1968).
[4] W. KUNDT: *Springer Tracts in Mod. Phys.*, **47**, 111 (1968).
[5] R. HAGEDORN: *Suppl. Nuovo Cimento*, **3**, 147 (1965).
[6] R. HAGEDORN: *Nuovo Cimento*, **56** A, 1027 (1968).

Antimatter and Cosmology (*).

G. STEIGMAN

Institute of Theoretical Astronomy - Cambridge

1. – Introduction.

A consequence of combining quantum mechanics and special relativity is the prediction that particles come in pairs. To each particle there corresponds an antiparticle. The partners of these pairs have equal masses, opposite electric charges, opposite quantum numbers. When a particle and its antiparticle meet, they annihilate. Antiprotons and antineutrons can bind together forming antinuclei which will gather positrons around them to form antiatoms. Antiatoms will group together to form antichemical compounds, and so on. Has nature exploited this symmetry in the laws of physics and created a universe consisting of equal numbers of particles and antiparticles? For every star is there an antistar, for every galaxy an antigalaxy?

Antiatoms and the chemical compounds they form have the same properties as atoms and their chemical compounds. They emit spectral lines of the same wavelength, which in the presence of magnetic or electric fields exhibit Zeeman or Stark splitting. Thus, the properties of an antistar are indistinguishable from those of a star, an antigalaxy indistinguishable from a galaxy.

The best hope for detecting the presence of antimatter in the universe is to search for the remnants of annihilation. Two gamma rays of 0.51 MeV each are emitted in electron-positron annihilation at rest. Pions (π^+, π^-, π^0), which ultimately decay into electrons, positrons, neutrinos and gamma rays are the products of nucleon-antinucleon annihilations. The neutrinos are essentially undetectable. The electrons and positrons will remain close to their regions of origin and can only be detected indirectly. The gamma rays can provide the only direct « observation » of antimatter. Upper limits on, or observations of gamma rays of the appropriate energy enable limits to be set on the amount of mixed matter and antimatter. Indeed, recent gamma-ray

(*) A more detailed review of this subject has appeared in Nature, **224**, 477 (1969).

measurements [1] place severe limits on the amount of antimatter in our own galaxy as well as in the metagalaxy. Keeping these limits in mind we shall review the status of various charge symmetric (particle-antiparticle symmetric) cosmologies.

2. – Gamma rays.

Before discussing the various cosmologies let us briefly review the information that the gamma-ray measurements give us.

An electron-positron pair annihilates predominantly into two gamma rays. Since most annihilations occur at rest, the particles having been slowed down first, a narrow line of gammas with energy 0.51 MeV is produced. A nucleon-antinucleon pair annihilates predominantly into several pions [2]. About five pions are produced per annihilation, with approximately equal numbers of π^+, π^-, π^0's. The $\pi^\pm$'s decay in flight to $\mu^\pm+\nu_\mu$ and the $\mu^\pm$'s decay in flight to $e^\pm+\nu_e+\nu_\mu$. In this manner a spectrum of $e^\pm$'s is produced whose average energy per particle in the laboratory frame is ~ 100 MeV. The π^0's decay in flight into two gammas yielding a gamma-ray spectrum which extends from several tens of MeV to several hundred MeV in the laboratory frame. The spectrum peaks at about 70 MeV and the average gamma ray receives about 180 MeV [3]. Unless the annihilation occurs in very dense regions, the gammas and neutrinos will escape. The electrons and positrons may be tied by magnetic fields to such regions and deposit all their energy there. Thus, the energy deposited in a given region (where annihilation is occurring) is connected with the number of gammas leaving the region. For each annihilation $\sim 3\div 4$ gammas are produced. Approximately $(10\div 20)\%$ of the annihilation energy is in the form of relativistic electrons and positrons.

One might guess that the line emission at 0.51 MeV from e^+-e^- annihilation might be easier to detect than the gamma ray spectrum produced in nucleon-antinucleon annihilation. However, because the gamma-ray spectrum rises steeply as the energy decreases, better limits are set by the higher-energy gammas from nucleon-antinucleon annihilation.

Consider a region in which matter and antimatter are mixed. The annihilation rate per unit volume is given by

$$S = n_+ n_- \sigma v \,, \tag{1}$$

$n_\pm$ are the particle, antiparticle number densities, v is the relative velocity and σ is the annihilation cross-section. The nucleon-antinucleon annihilation cross-section has been measured in the energy range: 25 MeV$\div$7 GeV [2]. For the entire range $\sigma v \sim 10^{-15}$ cm^3/s is an excellent approximation. If the annihilating particles are atoms and antiatoms and if $v \ll c$, the annihilation cross-section is considerably larger, being of atomic dimensions [4]. Under these conditions $\sigma v \sim 10^{-10}$ cm³/s. For each annihilation approximately $3 \cdot 10^{-4}$ erg are given to the relativistic electrons and positrons. Thus, for each gamma ray produced by an annihilation, $\sim 10^{-4}$ erg of energy in the form of relativistic electrons and positrons are also produced. Hence, the rate at which energy is deposited per unit volume is

$$\frac{\mathrm{d}\mathscr{E}}{\mathrm{d}t} = \varepsilon S \text{ erg/cm}^3\text{/s}\,, \quad \varepsilon \sim 3 \cdot 10^{-4} \text{ erg}\,. \tag{2}$$

It has been speculated the Q.S.O.'s radio galaxies, Seyfert galaxies and similar objects derive their energy from annihilation [5]. If all other emission (infra-red, radio, etc.) is assumed due to the 100 MeV electrons and positrons, there is a relation between the gamma-ray flux (expressed in photons per cm² per s) and the flux of all other radiation (expressed in ergs per cm² per s).

$$\mathscr{F}_\gamma(\text{photons/cm}^2\text{/s}) \approx 10^4 \mathscr{F}_{\text{other}}(\text{erg/cm}^2\text{/s})\,. \tag{3}$$

On the other hand, if there exists an estimate of the rate at which energy is deposited in these objects $(V(\mathrm{d}\mathscr{E}/\mathrm{d}t))$, then the expected gamma-ray flux can be estimated

$$\mathscr{F}_\gamma \approx \frac{10^4}{4\pi R^2}\left(V\frac{\mathrm{d}\mathscr{E}}{\mathrm{d}t}\right). \tag{4}$$

R is the distance between us and the object. This argument has been applied to 3C273 [6]. The predicted value of the gamma-ray flux exceeds the upper limit to the observed flux by a factor of three. Indeed, for any Q.S.O. which radiates $\sim 10^{48}$ erg/s, the expected gamma-ray flux exceeds $\sim 10^{-5}$ photons/cm²/s. Such fluxes are within present limits of detectability; their absence argues against the annihilation hypothesis. In fact, any source whose measured intensity exceeds 10^{-9} erg/cm²/s, should also be detectable as a gamma-ray source, independent of its distance, provided that annihilation is the energy source.

Annihilations in intergalactic space would produce an isotropic gamma-ray flux

$$\frac{\mathrm{d}\mathscr{F}_\gamma}{\mathrm{d}\Omega} \approx \left(\frac{g_\gamma R_0}{4\pi}\right) S_{\mathrm{IG}} \text{ photons/cm}^2\text{/s/sr} \tag{5}$$

g_γ is the number of gammas produced per annihilation ($g_\gamma \sim 3 \div 4$). R_0 is the Hubble radius ($R_0 \approx 10^{28}$ cm) Recent measurements [1] of an isotropic gamma-ray background at energies $\gtrsim 100$ MeV yield $\mathrm{d}\mathscr{F}_\gamma/\mathrm{d}\Omega = (1.1 \pm 0.2)\cdot 10^{-4}\ \mathrm{cm^{-2}\ s^{-1}\ sr^{-1}}$. This limits the intergalactic annihilation rate per unit volume S_{IG}.

$$S_{IG} \lesssim 4\cdot 10^{-32}\ \text{annihilations/cm}^3\text{/s}\ . \tag{6}$$

If the intergalactic gas is assumed to be ionized $\sigma v \sim 10^{-15}\ \mathrm{cm^3/s}$. In this case $n_+ n_- \lesssim 4\cdot 10^{-17}$. If symmetry is assumed: $n_+ = n_- \lesssim 6\cdot 10^{-9}\ \mathrm{cm^{-3}}$ and the mass necessary to close the universe ($n_0 \approx 10^{-5}\ \mathrm{cm^{-3}}$) is not to be found in the intergalactic gas. Conversely, if the universe is assumed closed ($n_+ + n_- = n_0$), then $n_+ \approx n_0$, $n_- \lesssim 4\cdot 10^{-12}\ \mathrm{cm^{-3}}$ and the universe is not charge symmetric. On the other hand, if the intergalactic gas is neutral, $\sigma v \sim 10^{-10}\ \mathrm{cm^3/s}$, and so $n_+ n_- \lesssim 4\cdot 10^{-22}$. If $n_+ = n_-$, then $n_\pm \lesssim 2\cdot 10^{-11}\ \mathrm{cm^{-3}}$, a number in agreement with the limits on the density of a neutral intergalactic gas set by the lack of Ly_α absorption in the spectra of Q.S.O.'s with large red shifts [7].

Annihilations in our own galaxy would tend to produce a line source in the galactic plane.

$$\frac{\mathrm{d}\mathscr{F}_\gamma}{\mathrm{d}\varphi} \approx \left(\frac{g_\gamma R_G}{2\pi}\right) S_G\ \text{photons/cm}^2\text{/s/rad} \tag{7}$$

R_G is the radius of the Galaxy ($R_G \approx 3\cdot 10^{22}$ cm). Measurements [1] indicate a line source of gamma rays in the galactic plane $\mathrm{d}\mathscr{F}_\gamma/\mathrm{d}\varphi \approx 2\cdot 10^{-4}\ \mathrm{cm^{-2}\ s^{-1}\ rad^{-1}}$. For the neutral interstellar medium the following limit may be set: $n_+ n_- \lesssim 10^{-16}$. For a symmetric interstellar medium, this implies $n_+ = n_- \lesssim 10^{-8}\ \mathrm{cm^{-3}}$ which is unacceptable. Since $n = n_+ + n_- \approx 1\ \mathrm{cm^{-3}}$ it follows that $n_+ \approx n$, $n_- \lesssim 10^{-16}\ \mathrm{cm^{-3}}$. Thus, the interstellar gas is overwhelmingly of one kind.

Finally, there are indications of a small source in the direction of the galactic center. $\mathscr{F}_\gamma \approx 10^{-4}\ \mathrm{cm^{-2}\ s^{-1}}$. A small source due to annihilations occurring in a volume V at a distance R would produce a flux:

$$\mathscr{F}_\gamma \approx \left(\frac{g_\gamma}{4\pi}\right)\frac{SV}{R^2}\ \text{photons/cm}^2\text{/s}\ . \tag{8}$$

Thus, $SV \lesssim 4\cdot 10^{41}$ annihilations/s which corresponds to a mass loss of only $\sim 10^{-8}\ M_\odot$/y. For $n_+ \sigma v \gtrsim 10^{-10}\ \mathrm{s^{-1}}$ (*i.e.* $n_+ \gtrsim 1\ \mathrm{cm^{-3}}$) the amount of annihilating antimatter is only $M_- \lesssim 3\cdot 10^{-6}\ M_\odot$. Although it is quite difficult to exclude this possibility, it is difficult to imagine how antimatter could have survived the early stages of galactic evolution before condensed objects had formed and when the annihilation time scale is small compared with the evolution time scale.

With the results of this section in mind we proceed in what follows to investigate, in the context of charge symmetric models of the universe, steady-state cosmology [8], the Alfvén-Klein cosmology [9], and the big-bang cosmology [10, 11].

3. – Steady-state cosmology.

In a charge symmetric, steady-state cosmology the metagalactic particle density is: $n = n_+ + n_- \approx 10^{-5}\ \text{cm}^{-3}$, $n_+ = n_-$. The creation rate per unit volume (of particle-antiparticle pairs) $q \approx \frac{1}{2}\cdot 10^{-22}$ pairs/cm^3/s is fixed by the expansion of the universe and is practically unaffected by annihilations. Now consider our own Galaxy. The gamma-ray measurements previously discussed limit the number of annihilations which occur per unit volume per second: $S_G \lesssim 2\cdot 10^{-26}$ annihilations/cm³/s. In addition we have seen that our Galaxy is overwhelmingly of one kind of matter ($n_+ \approx n \approx 1\ \text{cm}^{-3}$, $n_- \ll n_+$) so that $\Delta n = n_+ - n_- \approx n_+ \approx 1\ \text{cm}^{-3}$. Now, in our galaxy the total particle density $n = n_+ + n_-$ satisfies

$$\frac{\mathrm{d}n}{\mathrm{d}t} = \left[2q + \frac{\sigma v(\Delta n)^2}{2}\right] - \frac{\sigma v n^2}{2}\,. \tag{9}$$

Note that since baryon number is conserved Δn is a constant. The solution of (9) with the initial condition $n = n_0$ at $t = 0$ is

$$\left\{\begin{aligned} n(t) &= (A/B)^{\frac{1}{2}}[D\exp[t/\tau] - 1][D\exp[t/\tau] + 1]^{-1}\,,\\ A &= 2q + \frac{\sigma v(\Delta n)^2}{2}\,, \qquad B = \frac{\sigma v}{2}\,, \qquad \tau = (4AB)^{-\frac{1}{2}}\,,\\ D &= (A^{\frac{1}{2}} + B^{\frac{1}{2}}n_0)(A^{\frac{1}{2}} - B^{\frac{1}{2}}n_0)^{-1}\,. \end{aligned}\right. \tag{10}$$

For $t \gg \tau$ a steady state sets in whereby the annihilation rate balances the creation rate

$$q = \sigma v n_+ n_- = \tfrac{1}{4}\sigma v[n^2 - (\Delta n)^2] = S_G\,. \tag{11}$$

But this implies that $q \lesssim 2\cdot 10^{-26}$ pairs/cm³/s, which is four orders of magnitude lower than the value needed to keep the metagalactic density constant. To put the result another way, if matter (and antimatter) were created in our Galaxy at the rate required by steady-state theory, the gamma radiation would be four orders of magnitude greater than what is observed.

The above conclusions can be avoided in either of two ways. It can be argued that matter is created where there is matter and antimatter is created where there is antimatter thus avoiding annihilation altogether. This, however, amounts to postulating a new law of physics, for each creation event would locally violate conservation of baryon number. It is irrelevant that globally, baryon number would be conserved. Due to the finite speed of light, a nonlocal conservation law is simply not a conservation law. A second alternative is to hide the creation in dense regions, such as the nuclei of galaxies. The annihilation gammas would then be absorbed and thus never detected. Such a scheme has recently been proposed by HOYLE [12]. It is difficult to see how one could rule out such models.

4. – Alfvén-Klein cosmology.

In the Alfvén-Klein cosmology [9] the initial state of the universe is taken to be a thin cloud of protons, antiprotons, electrons and positrons at extremely low density. The gas contracts gravitationally increasing the density. Annihilations then occur at an increasing rate. The annihilation radiation slows the collapse, brakes it, and finally turns it to the present expansion of the universe. A process similar to electrolysis, employing a primeval magnetic field, is proposed to separate matter from antimatter. Because of the small scale of such processes, they are led to conclude that each galaxy consists of equal amounts of matter and antimatter. Appealing though this theory may be, a number of objections may be raised against it.

a) At the turning point (when the contraction of the universe is turned into expansion), the universe is, in fact, inside its Schwarzschild radius. Therefore, it could never have bounced.

b) Throughout the contraction the mean free path of the radiation is larger than the radius of the universe. How this radiation could have exerted a significant pressure on the infalling matter is difficult to understand.

c) Indeed, there exist quite general theorems [13] which state that a homogeneous isotropic collapsing universe with no Λ term and no negative energy density fields can never bounce.

d) Galaxy and star formation are supposed to occur when the metagalaxy is at or near maximum density. But this is when radiation pressure is supposed to be strong enough to brake the collapse of the universe and turn it into an expansion. It is difficult to conceive how galaxies and stars might form under such violent conditions.

e) There exists no mechanism in this theory for producing the 3 °K radiation.

f) To separate as much as a solar mass of antimatter from matter on a galactic scale, very strong magnetic fields are required $(\gtrsim(1\div100)\text{ G})$. Such large scale, strong fields have not been observed.

g) The observed gamma rays place severe limits on the amount of mixed matter and antimatter in our Galaxy. This difficulty is avoided by assuming that regions of matter and antimatter are kept separated by a leidenfrost phenomenon. This can be countered by an argument due to WOLTJER [14]. It is pointed out that large scale Faraday rotation is observed in the galactic disc. A patchy distribution of matter and antimatter would tend to wipe out any large scale effects.

The indications are that Alfvén-Klein cosmology raises more problems than it solves. In its present form it cannot be an acceptable cosmology.

5. – Big-bang cosmology.

The discovery of the 3 °K, black-body radiation [15] lends dramatic support to the « big-bang » model of the universe [10]. Conventionally, a non-charge symmetric theory, it may be taken as the strongest argument in favor of the view that the universe does not contain equal numbers of particles and antiparticles. This follows from an analysis due independently to ZEL'DOVICH and CHIU [16] which we shall briefly outline.

In the early universe ($t \lesssim 10^{-3}$ s) when the temperature and density were high (*e.g.*: $kT \gtrsim Mc^2$), nucleons, antinucleons and radiation are in equilibrium. At this stage annihilations are balanced by pair creations. As the universe expands adiabatically ($VT^3 = \text{constant}$), the temperature drops ($kT < Mc^2$) and pair creations can no longer balance annihilations and most nucleon-antinucleon pairs annihilate. Since the density is also decreasing the annihilation rate becomes negligible before all pairs annihilate and as a result a certain number of nucleons and antinucleons are frozen out. The preceding argument of course applies to all particle-antiparticle pairs although the temperature at which the freezing out occurs will be different. For a charge symmetric universe this freezing out of the nucleon-antinucleon pairs occurs within $\sim(10^{-3}\div10^{-2})$ s of the singularity, when there are $\sim 10^{18}$ photons for each particle.

$$\left(\frac{n}{n_\gamma}\right)_0 \approx 10^{-18}\,. \tag{12}$$

As the universe expands the number of photons N_γ in a comoving volume V remains constant (*i.e.*: $n_\gamma = N_\gamma/V \sim (kT/\hbar c)^3$). When the annihilation rate becomes negligible the number of particles in a comoving volume V also remains constant. It follows that

$$\left(\frac{N}{N_\gamma}\right)_{\text{now}} = \left(\frac{n}{n_\gamma}\right)_{\text{now}} = \left(\frac{n}{n_\gamma}\right)_0 \approx 10^{-18}\,. \tag{13}$$

But, at present, $n > 10^{-7}\ \text{cm}^{-3}$ and $n_\gamma \sim 10^3\ \text{cm}^{-3}$ so that, in fact: $(n/n_\gamma)_{\text{now}} \gtrsim 10^{-10}$. Had the universe been charge symmetric it would contain $\sim (10^8 \div 10^{10})$ times more radiation than is observed. Unless this radiation catastrophe can be avoided, big-bang cosmology and charge symmetry are incompatible.

HARRISON [10] has suggested a way out of this dilemma. Suppose in the early state of the universe there were spatial fluctuations in the baryon number $\Delta N = N_+ - N_-$. Since baryon number is conserved, annihilations amplify any initial inhomogeneity $(\Delta N_{\text{then}} = \Delta N_{\text{now}} = \pm N_{\text{now}} = \pm (N_+ + N_-)_{\text{now}})$. Thus, as the universe expands, separate regions of matter and antimatter are produced. If the initial fluctuations occur when matter, antimatter and radiation are still in equilibrium, then: $N_{\text{then}} \approx (N_\gamma)_{\text{then}} = (N_\gamma)_{\text{now}}$ so that

$$\left(\frac{\Delta N}{N}\right)_{\text{then}} \approx \left(\frac{N}{N_\gamma}\right)_{\text{now}} = \left(\frac{n}{n_\gamma}\right)_{\text{now}} \approx 10^{-9}\,. \tag{14}$$

And so, a small spatial baryon number fluctuation $\Delta N/N \sim 10^{-9}$ is amplified by annihilation to $(\Delta N/N)_{\text{now}} = \pm 1$. Thus, initial spatial inhomogeneities in a charge symmetric universe prevent a radiation catastrophe and, in addition facilitate Galaxy formation by providing separate regions of matter and antimatter. In conventional big-bang cosmology it is quite difficult to build galaxies; extremely large density fluctuations are required $(\delta\varrho/\varrho \sim 10^{-3})$ [17]. Because annihilations act as such an efficient amplification mechanism, much smaller baryon number fluctuations do the same job.

Notice, however, that even such small fluctuations could not be statistical. For, if $\Delta N \sim N^{\frac{1}{2}} \sim 10^{-9} N$, then $N \sim 10^{18}$ particles which is less than 10^{-6} g! Put another way, suppose we ask for the largest possible statistical fluctuation:

$$\Delta N_{\text{then}} \sim N^{\frac{1}{2}}_{\text{then}} \sim (N_\gamma)^{\frac{1}{2}}_{\text{then}} \sim (N_\gamma)^{\frac{1}{2}}_{\text{now}} \approx 10^{44}\,. \tag{15}$$

The initial fluctuation would be $(\Delta N/N)_{\text{then}} \sim 10^{-44}$! Fluctuations $\Delta N/N \sim 10^{-9}$ must be nonstatistical in origin.

6. – Conclusion.

The most practical, direct method for establishing the presence of antimatter in the universe is to wait until it meets up with matter and annihilates, then search for the annihilation products. Indeed, as we have seen, gamma-ray observations enable limits to be set on the amounts of mixed matter and antimatter.

In steady-state cosmology the laws of physics as determined by terrestrial experiments force us to admit creation of particle-antiparticle pairs. The limits set by the gamma-ray measurements preclude this creation from occurring uniformly in space. Creation must occur in regions of high density so that the subsequent gamma rays will be absorbed and hence escape detection. Otherwise, steady-state cosmology and charge symmetry are incompatible.

Since a big-bang cosmology has a « beginning » it can be argued that the universe is asymmetric because the initial conditions were asymmetric. Had the initial conditions been symmetric, a radiation catastrophe would have occurred. An ensemble of initial universes can be imagined. The symmetric ones evolved into universes which are filled with too much radiation to be identified with our universe which must have evolved from one of the asymmetric initial universes.

If, for esthetic or philosophic reasons, one still insists on a symmetric universe; an efficient, nonstatistical, particle-antiparticle separation mechanism must be found. The alternative is a universe with $10^8 \div 10^{10}$ times more radiation than ours.

REFERENCES

[1] G. W. Clark, G. P. Garmire and W. L. Kraushaar: *Astrophys. Journ. Lett.*, **153**, L203 (1968); G. W. Clark, G. P. Garmire and W. L. Kraushaar, *I.A.U. Symposium*, No. 37 (to be published), L. Gratton editor) C. E. Fichtel, D. A. Kniffen and H. B. Ogelman: *I.A.U. Symposium*, No. 37, (to be published). L. Gratton editor).

[2] G. G. Fazio: *Ann. Rev. Astron. Astrophys.*, **5**, 481 (1967).

[3] G. M. Frye and L. H. Smith: *Phys. Rev. Lett.*, **17**, 733 (1966); F. W. Stecker: *Smithsonian Astrophysical Observatory Special Report* 261 (1967).

[4] G. Steigman: Ph. D. Thesis, New York University (1968).

[5] H. Alfvén and A. Elvius: *Science*, **164**, 911 (1969); G. R. Burbidge and F. Hoyle: *Nuovo Cimento*, **4**, 558 (1956); E. Teller: *Perspectives in Modern Physics*, edited by R. E. Marshak (New York, 1966), p. 449.

[6] G. M. Frye and C. P. Wang: *Can. Journ. Phys.*, **46**, S488 (1968).

[7] G. Burbidge and M. Burbidge: *Quasi-Stellar Objects* (1967), p. 145.

[8] F. Hoyle and J. V. Narlikar: *Proc. Roy. Soc.*, **273**, 1 (1963).

[9] H. Alfvén and O. Klein: *Ark. f. Fys.*, **23**, 187 (1962); H. Alfvén: *Rev. Mod. Phys.*, **37**, 652 (1965).

[10] G. Gamow: *Phys. Rev.*, **70**, 572 (1946).

[11] E. R. Harrison: *Phys. Rev. Lett.*, **18**, 1011 (1967); *Rev. Mod. Phys.*, **39**, 862 (1967); *Phys. Rev.*, **167**, 1170 (1968); *Phys. Today*, **21**, No. 6, 3 (1968).

[12] F. Hoyle: *Nature* **224**, 477 (1969)

[13] S. HAWKING and G. F. R. ELLIS: *Astrophys. Journ.*, **151**, 25 (1968).

[14] L. WOLTJER: *I.A.U. Symposium*, No. 31, 480 (1967).

[15] A. A. PENZIAS and R. W. WILSON: *Astrophys. Journ.*, **142**, 420 (1965); P. G. ROLL. and D. T. WILKINSON: *Phys. Rev. Lett.*, **16**, 405 (1966); G. B. FIELD and J. L. HITCHCOCK: *Phys. Rev. Lett.*, **16**, 817 (1966); P. THADDEUS and J. F. CLAUSER: *Phys. Rev. Lett.*, **16**, 819 (1966).

[16] YA. B. ZEL'DOVICH: *Adv. Astron. Astrophys.*, **3**, 241 (1965); H. Y. CHIU: *Phys. Rev. Lett.*, **17**, 712 (1966).

[17] G. B. FIELD: *Astrophys. Journ.*, **142**, 531 (1965).

Creation of Particles by Gravitational Fields.

H. URBANTKE

University of Vienna - Vienna

In this seminar contact will be made with the seminars given by SEXL, KUNDT, STEIGMAN, BOERNER.

Questions: Origin of particles populating the Universe.
Origin of the peculiar particle-antiparticle asymmetry observed at least in our neighborhood.

We can look at these questions in view of the following cosmological models:

1) Big-bang: Origin and asymmetry; no creation required at present.

2) Klein-Alfvén (see STEIGMAN's seminar): Symmetry.

3) Steady state: Particle creation always required, asymmetrical.

1. – Mechanisms of creation.

As far as one can see, quantum field theory is the only reasonable framework for the calculation of creation processes; classical calculations are misleading. Possibilities:

a) Existence of a negative-energy field in nature, so that the vacuum becomes unstable and particles are created together with quanta of the negative-energy field: « spontaneous creation ». Example: Hoyle's *C*-field, but treated as a quantum field with a modified type of coupling (this was indicated in the seminar by SEXL; a fuller discussion can be found in ref. [1]). If one does not go to nonlocal field theories, one gets in this way symmetric creation, although the elementary processes need not be symmetric.—But to have symmetric creation, one need not surpass essentially ordinary GR.

b) Creation by external fields: « continuous creation ». Here we concentrate upon the gravitational field as the external field. This is basically similar to pair production by electromagnetic fields, and the framework of GR is not left.

2. – Particle creation by gravitational fields.

We start from the action principle $\delta W = 0$,

$$W = \int \left(\frac{1}{\varkappa} R + \Lambda\right) \sqrt{-g}\, \mathrm{d}^4 x\,; \qquad \varkappa = 8\pi G, \qquad \hbar = c = 1\,,$$

where Λ is the matter Lagrangian depending on some matter field variables Φ. Fields equations:

$$\frac{\delta W}{\delta g_{ab}} = 0 \dots R^{ab} - \frac{1}{2} g^{ab} R + \varkappa \mathcal{T}^{ab} \qquad \frac{\delta W}{\delta \Phi} = 0 \dots, \quad \text{field equations for matter.}$$

Since it is in general impossible to solve this system, we do something like the Born-Oppenheimer approximation in molecular physics. Namely, we separate some amount of matter already present, described by an energy-momentum tensor $\mathcal{T}_{ab}$, which generates a certain gravitational field g_{ab}, according to the first equation. Now we study the effect of this *g*-field on some quantized matter field and look whether particles of this field are created and at what rate. Formally: field equation

$$\left(\partial_a \frac{\partial}{\partial \Phi_{,a}} - \frac{\partial}{\partial \Phi}\right) \Lambda \sqrt{-g} = 0\,,$$

where $\Lambda = \Lambda(g_{ab}, \Phi)$ with given g_{ab}.

This amounts to studying quantized fields in a given curved space. There exists some literature on this, *e.g.* [2]-[12]. In refs. [2], [3], [4], [6], [7] the special case of de Sitter space is treated; [8], [9] discuss some general properties; [5], [10], [11] deal with creation. The quantization procedure seems to be consistent, but an invariant physical interpretation is not always clear. Explicitly invariant results have been obtained so far only in the case where the gravitational field is treated in the linear approximation: we put $g_{ab} = \eta_{ab} - 2f\Psi_{ab}$, where $f = \sqrt{\varkappa}$, and expand the Lagrangian

$$\Lambda(g_{ab}, \Phi) = L(\Phi) + f T^{ab} \Psi_{ab} = L(\Phi) + L_{\text{int}}\,,$$

where L, T^{ab} are the corresponding flat space Lagrangian and energy-momentum tensor (note that the external field approximation introduces such a distinction between $\mathcal{T}^{ab}$ and T^{ab}!). Now since L_{int} is quadratic in Φ, particle creation is in pairs, and we can ask for the probability of a pair to be created:

$$W_2 = \sum_2 |(2, \text{out}\,|0)|^2 = \sum_2 |(2|S|0)|^2 ,$$

where we have to use our first-order S-matrix $S = 1 + if\int d^4x :L_{\text{int}}:$. This gives

$$W_2 = \sum_2 (0| -if\int ... |2)(2| if\int ... |0) = f^2 \int d^4x\, d^4y\, \Psi_{ab}(x) K^{abcd}(x, y) \Psi_{cd}(y) ,$$

where $K_{abcd}(x, y) = (0|:T^{ab}(x)::T^{cd}(y):|0)$ (we were able to extend the intermediate 2-state to a sum over a complete set of states). Poincaré invariance and conservation laws tell us that K^{abcd} depends only on $x - y$ and must be of the form

$$K^{abcd}(x) = D^{ab} D^{cd} K(x) + D^{a(c} D^{d)b} I(x) ,$$

where $D^{ab} = \eta^{ab} - \partial^a \partial^b$. $I(x)$, $K(x)$ are two invariant functions determined by the contractions

$$\Box^2 K = \frac{1}{15}(2K_{a\ c}^{\ a\ c} - K_{ab}^{\ \ ab}) , \qquad \Box^2 I = \frac{1}{15}(3K_{ab}^{\ \ ab} - K_{a\ c}^{\ a\ c}) .$$

Vacuum expectation values of this type have spectral representations which we write in our case as

$$\frac{-i}{16\pi^2} \int_{4m^2}^{\infty} ds \sqrt{1 - \frac{4m^2}{s}}\, \varrho(s) \Delta^+(x, \sqrt{s}) ;$$

m = mass of particles created. We call the spectral function $\varrho(s) = f_2(s)$ in the case of $\Box^2 I$, and $\varrho(s) = 4f_1(s) + f_2(s)$ for $\Box^2 K$. The normalizations are taken the same as in ref. [14]. These functions are the only thing that depends on the special type of particles considered.

Inserting into the formula for W_2, taking Fourier transforms and using the equations for the gravitational field in their linearized version in Fourier components,

$$f\mathcal{T}_{ab}(k)/k^2 = 2P_{a[b} P_{c]d} \Psi^{cd} ,$$

contracted:

$$f\mathcal{T}(k)/k^2 = 2P_{ab} \Psi^{ab} \qquad (P^{ab} = \eta^{ab} - k^a k^b / k^2) ,$$

allows one to express W_2 finally as

$$W_2 = \frac{\varkappa^2}{8\pi} \int\limits_{k^2>4m^2, k_0>0} \frac{d^4k}{(2\pi)^4} \sqrt{1 - \frac{4m^2}{k^2}} (k^2)^{-2} \{f_1(k^2)\mathcal{T}(k)\mathcal{T}^*(k) + f_2(k^2)\mathcal{T}^{ab}(k)\mathcal{T}^*_{ab}(k)\} .$$

Note that only the source $\mathcal{T}_{ab}$ of the gravitational field enters, so that the result is certainly invariant. We can draw the following conclusions:

1) Only the high-frequency parts $k_0 \gtrsim 2m$ are effective (adiabatic theorem). For protons, $k_0 \simeq 10^{23}$ Hz.

2) Situations where such high-frequency parts have sufficiently large amplitudes do not exist at present. In particular, the expansion of the universe as a whole cannot—via this mechanism—be responsible for a continuous creation as required by steady state.

3) The only exceptional situation where this might be so occurs during the first 10^{-23} s of a big bang. Electron production would take place about 1000 times longer, and afterwards only radiation is created.

These last statements are to be taken only as qualitative in nature since in this case we are leaving the weak field approximation as well as the external field approximation. One will be able to make them more precise only if one could treat the gravitational field as dynamic, which most probably implies that it has to be quantized, too.

Qualitatively, one gets the following picture of the early universe: The particles emerge from an originally empty universe, being created by their own gravitational field. The question wether the particles or the creating gravitational field existed earlier is meaningless, however, because of the time scale which is $\simeq 10^{-23}$ s. Particles and universe originate together in one self-consistent explosion (« gravitational bootstrap »). Essentially all matter was created at that time, later on only radiation with wavelengths of the order of the radius of the universe arose from this mechanism. Questions of energy conservation might be answered by pointing out that the universe has zero total energy—as far as this dubious concept can be formulated at all—, the rest mass of the particles being approximately compensated by the gravitational binding energy. No further field like the C-field is needed.

There remains the question of an early separation of particles and antiparticles, so that the rapid expansion then is able to practically isolate regions with particles and those with antiparticles. One would try to look for a short-range interaction which could effect this. Indeed, a preliminary answer has been given by OMNÈS [13], who showed that the existence of the $(\frac{3}{2}, \frac{3}{2})$-resonance in the pion-nucleon system leads to a separation between neutrons and antineutrons in the thermodynamics of elementary particle reactions. In this way one obtains a kind of marriage between big bang and the Klein-Alfvén theory.

REFERENCES

[1] R. U. SEXL and H. URBANTKE: *Acta Phys. Austr.*, **26**, 339 (1968).

[2] M. GUTZWILLER: *Helv. Phys. Acta*, **29**, 313 (1956).

[3] H. URBANTKE: Thesis, University of Vienna (1965 (unpublished).

[4] O. NACHTMANN: *Commun. Math. Phys.*, **3**, 1 (1967).

[5] O. NACHTMANN: *Dynamische Stabilität im de Sitter Raum*, preprint, Vienna, 1967; *Zeits. f. Phys.*, **208**, 113 (1968).

[6] W. E. THIRRING· *Proceedings of the VI Internationale Hochschulwochen für Kernphysik*, edited by P. URBAN (Schladming, 1967).

[7] G. BÖRNER: This volume p. 332.

[8] A. LICHNÉROWICZ: in *Relativity, Groups and Topology*, edited by DE WITT and DE WITT (New York, 1964).

[9] H. URBANTKE: *Nuovo Cimento*, **63** B, 203 (1969).

[10] T. IMAMURA: *Phys. Rev.*, **118**, 1430 (1960).

[11] F. L. SCARF: CNRS, 1962.

[12] L. PARKER: *Phys. Rev. Lett.*, **21**, 562 (1968).

[13] R. OMNÈS: several preprints; also: *Phys. Rev. Lett.*, **23**, 38 (1969).

[14] W. E. THIRRING: *Principles of Quantum Electrodynamics* (New York, 1958).

PROCEEDINGS OF THE INTERNATIONAL SCHOOL OF PHYSICS «ENRICO FERMI»

Course XIV
Ergodic Theories
edited by P. CALDIROLA

Course XV
Nuclear Spectroscopy
edited by G. RACAH

Course XVI
Physicomathematical Aspects of Biology
edited by N. RASHEVSKY

Course XVII
Topics of Radiofrequency Spectroscopy
edited by A. GOZZINI

Course XVIII
Physics of Solids (Radiation Damage in Solids)
edited by D. S. BILLINGTON

Course XIX
Cosmic Rays, Solar Particles and Space Research
edited by B. PETERS

Course XX
Evidence for Gravitational Theories
edited by C. MØLLER

Course XXI
Liquid Helium
edited by G. CARERI

Course XXII
Semiconductors
edited by R. A. SMITH

Course XXIII
Nuclear Physics
edited by V. F. WEISSKOPF

Course XXIV
Space Exploration and the Solar System
edited by B. ROSSI

Course XXV
Advanced Plasma Theory
edited by M. N. ROSENBLUTH

Course XXVI
Selected Topics on Elementary Particle Physics
edited by M. CONVERSI

Course XXVII
Dispersion and Absorption of Sound by Molecular Processes
edited by D. SETTE

Course XXVIII
Star Evolution
edited by L. GRATTON

Course XXIX
Dispersion Relations and Their Connection with Causality
edited by E. P. WIGNER

Course XXX
Radiation Dosimetry
edited by F. W. SPIERS and G. W. REED

Course XXXI
Quantum Electronics and Coherent Light
edited by C. H. TOWNES and P. A. MILES

Course XXXII
Weak Interactions and High-Energy Neutrino Physics
edited by T. D. LEE

Course XXXIII
Strong Interactions
edited by L. W. ALVAREZ

Course XXXIV
The Optical Properties of Solids
edited by J. TAUC

Information about Courses I-XIII may be obtained from the Italian Physical Society.

Course XXXV
High-Energy Astrophysics
edited by L. GRATTON

Course XXXVI
Many-Body Description of Nuclear Structure and Reactions
edited by C. BLOCH

Course XXXVII
Theory of Magnetism in Transition Metals
edited by W. MARSHALL

Course XXXVIII
Interaction of High-Energy Particles with Nuclei
edited by T. E. O. ERICSON

Course XXXIX
Plasma Astrophysics
edited by P. A. STURROCK

Course XL
Nuclear Structure and Nuclear Reactions
edited by M. JEAN

Course XLI
Selected Topics in Particle Physics
edited by J. STEINBERGER

Course XLII
Quantum Optics
edited by R. J. GLAUBER

Course XLIII
Processing of Optical Data by Organisms and by Machines
edited by W. REICHARDT

Course XLIV
Molecular Beams and Reaction Kinetics
edited by CH. SCHLIER

Course XLV
Local Quantum Theory
edited by R. JOST

Course XLVI
Physics with Storage Rings
edited by B. TOUSCHEK

Tipografia Compositori - Bologna - Italy